ENCYCLOPEDIA OF GENOMICS

Vol. 4

MOLECULAR GENOMICS

By

Dr. Shiv Kant Prasad

School of Studies of Zoology & Biotechnology

Vikram University

Ujjain

(India)

D P H

DISCOVERY PUBLISHING HOUSE PVT. LTD.

NEW DELHI-110 002

Published by:

Tilak Wasan

DISCOVERY PUBLISHING HOUSE PVT. LTD.

4383/4A, Ansari Road, Darya Ganj

New Delhi-110 002 (India)

Phone: +91-11-23279245, 43596064-65

Fax : +91-11-23253475

E-mail: parul.wasan@gmail.com

discoverypublishinghouse@gmail.com

web : www.discoverypublishinggroup.com

***First Edition:* 2012**

ISBN: 978-93-5056-032-7 (Set)

978-93-5056-033-4 (Vol. 1)

978-93-5056-034-1 (Vol. 2)

978-93-5056-035-8 (Vol. 3)

978-93-5056-036-5 (Vol. 4)

978-93-5056-037-2 (Vol. 5)

Encyclopedia of Genomics

Printed at:

Shree Balaji Art Press

Delhi

Preface

Progress in any scientific discipline is dependent on the availability of techniques and methods that extend the range and sophistication of experiments which may be performed. Over the last 30 years or so this has been demonstrated in a spectacular way by the emergence of genetic engineering. This field has grown rapidly to the point where, in many laboratories around the world, it is now routine practice to isolate a specific DNA fragment from the genome of an organism, determine its base sequence, and assess its function. The technology is also now used in many other applications, including pharmaceutical analysis of samples, paternity disputes, medical diagnosis, genomic and the biotechnology industry.

The present title **"Encyclopedia of Genomics"** is a fast growing area of research for undergraduate and post-graduate students in life sciences, chemistry, engineering and medicine. The overall objective of this publication is to provide a professional level reference work with comprehensive coverage of the molecular basis of genome and the application of that knowledge in genetics, evolution, medicine, and agriculture. It deals with the life processes at a molecular level genetic disease diagnosis and genetic therapy; the theory and techniques for understanding manipulating, and synthesizing biological molecules and their aggregates; and the application of biological processes to make or modify products to improve plants or animals or to develop microorganisms for specific uses.

The development of techniques for genomic science is based on an understanding of the molecular aspect of genomic processes occurring in nature. These techniques have, in turn, enabled expansion of our knowledge and understanding of how genes are organized and expressed in living cells and has provided major impetus to the development of biotechnological enterprises.

The present title has been written to provide readers with a thorough understanding of the techniques which underpin the developments in the field of genomic science. By gaining this understanding, readers will be able to apply the techniques of genomic manipulation to tackle specific problems or to achieve desired objectives in a wide variety of fields.

There can be no claim to originality except in the manner or treatment and much of the information has been obtained from the books and scientific journals available in the different libraries.

The author express his thanks to his friends and colleagues whose continue inspirations have initiated him to bring out this book.

The author expresses her gratitude to Mr. Wasan and staff of M/s Discovery Publishing House Pvt. Ltd. for their whole hearted co-operation in the publication of this book.

S.K. Prasad

Contents

Chapter 1

Introduction

Small-molecule drugs are a cost-effective way to treat and prevent disease. A study by the Slone Institute published in 2002 estimated that over 50% of the adult population in the United States used at least one pharmaceutical drug during the preceding week. The positive impact of small-molecule drugs on health care has been well documented.

The discovery of novel drugs has traditionally been a combination of clever science, brute force, and good *fortune*. With the advent of high-throughput screening technology, combinatorial chemistry, and the completion of the human genome sequence in the late 1990s, the hope was that *technology* could address the brute-force aspect and the genome sequence would provide insights into the *underlying* science, and good fortune would continue.

Although there are some exceptions, productivity in the industry overall has gone down. Some of this is owing to higher regulatory standards and more difficult therapeutic areas, but a significant portion is the result of the lack of well-validated targets to apply the technology to. The industry portfolio of *pharmaceutical* targets of approx 500 in 1996 has not been significantly *expanded*.

The availability of the human genome sequence and novel biological tools, such as *siRNA*, *antisense*, *knockouts*, and *transgenics*, suggests that over time, the physiological function of many of the genes in the genome may be *deciphered*. However, the time frame for this may be much greater than most people anticipated.

For comparison, the first bacterial genome sequence was completed in 1995, and although we may be able to now classify the majority of the genes by *biochemical* function, we don't know most of their physiological roles. One *conceivable* way to speed up the molecular dissection of the biology underlying various disease states is to use small-molecule compounds that specifically inhibit individual targets.

There are several key factors required to be successful when using small molecules to explore biology.

- The relative selectivity of the chemical probes that will be used must be known.
- The correlations between the cellular readout(s) used and the pathway or *phenotype* that

is being assessed must be *independently* validated.

- All data must be fully integrated, allowing the user to navigate through biological pathways and supporting *literature*, assay results, and detailed information on compounds.

TERMINOLOGY

As with any emerging field, there are often differences of opinion on *terminology* among researchers, sometimes subtle, sometimes not. I will attempt to capture and describe the major variants and apologize if I inadvertently leave out any major themes, or end up misrepresenting some differences in trying to summarize the field.

Chemical Genetics

Chemical genetics, as described by Schreiber et al. in 1999, refers to the use of small molecules to induce alterations in gene products in *mammalian* systems, in a manner similar to using mutations. This approach became feasible through the combination of high-throughput cellular assays and diverse *libraries* of compounds.

The ability to perform genetic screens in cellular assays vastly increases the throughput-traditionally a key limitation when studying higher organisms. It also allows the separation of effects in somatic cells from those in *development.* It is critical to this chemical genetic approach to have a library of compounds that have a high probability of being relatively selective; otherwise, the ability to *interpret* the results becomes at least as complex as *deciphering* highly polygenetic *phenotypes.*

To address this, *diversity-oriented* synthesis has been proposed to provide arrays of complex small molecules that are easily synthesized. The *natural-product* basis for many of the molecules and their *complexity* are believed to contribute to their cellular potency and selectivity. This chemical genetic approach has been applied to identify novel inhibitors of *alphatubulin* and historic *deactylation.*

As in classical genetics, a chemical genetic approach involves screening with probes that potentially could interact with any target in the genome, while trying to identify specific phenotypes. An alternate approach, termed *reverse chemical genetics, is* analogous to introducing specific gene disruptions. Here, compounds that are known to specifically interact with a given target are used in broad phenotype screens to help identify the physiological role of that target.

Reverse Chemical Genetics

The key to a reverse chemical genetic approach is to have a one-to-one link between the small-molecule compound and the target of interest. This can be achieved by optimization of chemical reagents by thorough *profiling* against other potential targets, or alternatively by altering the target itself, to introduce changes that can be *exploited* for specificity.

This approach has been most broadly applied to members of the protein *kinase* family, where specific changes can be made to the residues surrounding the active site that don't significantly alter the affinities or kinetics for natural *substrates,* but now allow the binding of specific inhibitor *analogs.*

Table 1.1: Approaches to Chemical Genetics.

		Approach	Goal	Step 1	Step 2	Step 3
Genetics	*AcB*	Randomly perturbate the system looking for changes in a specific phenotype	Target/pathway identification	Assemble large diverse chemical library	Assay for specific phenotype	Identify targets of compounds that induce the phenotype
Reverse Genetics	*baC*	Broad search for phenotype associated with perturbations of specific genes	New role for target	Select target	Find specific inhibitor	Broadly assay for phenotypes
Screening	*bAc*	Search for compounds which modulate a target with a known phenotype	Compounds/drugs for validated targets	Select target with presumed physiological role	Assemble large diverse chemical library	Screen for modulators of target
Genomics	*ABC*	Screen all targets for specific modulators, then try to identify phenotypes for each target	Target/pathway identification	Assemble large diverse chemical library	Screen for modulators of any targets	Broadly assay for phenotypes
Profiling	*AC*	Broadly screen for phenotypes associated with compounds, regardless of the target toxicity	Link between compounds and phenotype-efficacy or	Assemble large diverse chemical library	Broadly assay for phenotypes	
Chemo-genorics	*BaC*	Use information on targets to identify specific modulators, then look for phenotypes	Compounds/drugs for novel targets	Select targets	Identify inhibitor	Broadly assay for phenotypes

Replacement of the wild-type copies of a given gene with these engineered mutants allows these compounds to be used to inhibit the function of the gene in cellular assays as well as in adult animals. The in vivo assay, complete with the *complications* of *pharmacokinetics* and *pharmacodynamics*, closely mimics the effect that would be expected from dosing an animal with a selective inhibitor against the wild-type target.

Additionally, introducing mutations into the *adenosine triphosphate* (ATP)-binding sites of targets allows the binding of labeled ATP analogs; these can then be used to trace biochemical pathways at the molecular level by looking directly at *phosphorylated* substrates, thus furthering the link between the target and the observed phenotype. A comparison between standard genetics and chemical genetics is shown in Table 1.1.

Screening

Screening of large sets of compounds, often assembled to be quite diverse, is often one of the first steps in a drug discovery project. The assay used for the initial screen will both help define the likelihood of getting potent hits, as well as form the foundation for the follow-up path. Assays that are more physiological require that active molecules pass through additional filters depending on the assay, such as transversing the *cell wall, serum binding, bioavailability,* and *metabolic stability*.

These factors tend to decrease the hit rate, but result in molecules with better overall properties. The downside to this approach is that further compound optimization may be hindered by a lack of knowledge about the molecular target(s) of the initial hits.

Biochemical-based target screening is likely to yield hits that have a clearer path to optimization, but are at higher overall risk because the link between the target and the desired physiological changes may not be strong and these compounds may be far from having the desired physical properties. Smaller compound sets and focused compound libraries can be used to screen more broadly for physiological phenotypes. In the typical high-content screening experiment, the effects off compounds on cellular assays are captured by multiple parameters.

These range from cellular and subcellular morphology, intracellular translocation events, changes in state, such as *phosphorylation*, and *proteomic* and *genomic profiling*.

The advances in recent years of high-throughput screening, *miniaturization*, and imaging have made it *technically* practical to assemble vast databases detailing the effects of specific compounds on a genome-wide scale. The current challenge in the field is to learn how to *interpret* this information to expand our knowledge off the underlying biological pathways.

Whereas the primary goal of the high-content screening in this chemical genomic approach may be elucidation of the link between targets and the underlying *physiology*, the data collected on specific compounds can be used to drive drug programs.

Chemogenomics

Chemogenomics refers to the generation of specific sets of compounds with drug-like properties, which are specific for a given set of targets. The distinction between this and chemical genomics is that the knowledge extracted from sequences, 3D structures, assay results, and known chemical properties are used in the compound-design phase so that the compounds have a high likelihood of being selective and the biological consequences of inhibiting the target are known.

In a typical chemogenomic approach, there is a very focused set of assays used to drive potency and selectivity, and a broader set of assays used to optimize chemical properties. The different chemical genetic approaches described in the above subheadings are summarized in Table elsewhere in this chapter.

An example of a *chemogenomic* approach is shown in Figure elsewhere in this chapter, where the protein sequences of members of a gene family have been overlayed on the 3D structure of an inhibitor-binding site from a representative of one of the targets in this family. The colored surface suggests areas of the active site where specificity is likely to be obtained, and the *superimposed* ligand suggests locations where changes to the core scaffold should be directed.

The overall similarity in sequence and protein structure allows information obtained from one member of the gene family to be transferred to other members. Using this common frame off reference allows knowledge gleaned from the optimization of one chemical scaffold to be extrapolated to guide the tuning of the potency and specificity of other scaffolds that bind to the same site.

IMPROVING HEALTH

The goal of all of these approaches is to use chemical compounds to interrogate biological systems and ultimately provide insights that lead to improved health. A key limitation in the field is the ability to *extrapolate* from a molecular phenotype, to cellular response, to animal *pharmacology*, to efficacy in the clinic.

In a traditional target or cell-based screen, there are a limited number of readouts that are used to determine the efficacy of a compound on a system-certainly not enough data to safely or accurately predict any effects in the clinic. *Broad* screening of specific compounds using technologies such as gene profiling or proteomics can provide vast amounts of data at the cellular detail level, but it is still difficult to translate this into known *biological phenotypes*.

Whereas a bottom-up approach, where events at the molecular level can be used to rationally explain physiology at a *macroscopic*, is an objective of some in systems biology, insights can be made by more *empirical* approaches. *Correlations* discovered between molecular readouts and more phenotypically relevant measurements, such as *toxicity* and *efficacy*, are proving to be quite powerful. These relationships will become more robust as more data are gathered and integrated into systems allowing cross-discipline statistical analyses.

Although there is still no substitute for well-controlled clinical trials, the use of these technologies should improve the quality of the molecules entering the clinic and hopefully reduce failure rates, allowing treatments for more difficult diseases to be considered.

Chapter 2

Laboratory Methods

With the increasing knowledge about genetic *alterations* associated with various cancers, molecular analysis of these alterations on *formalinfixed, paraffin-embedded* (FFPE) tissue is of increasing importance in *diagnostic* pathology as well as cancer research. Different methods are available for examining routinely processed material for copy number changes, including comparative genomic *hybridization* (CGH), *fluorescence* in situ hybridization (FISH), and loss of *heterozygosity* (LOH) analysis.

However, most of these techniques have technical limitations. With FISH and LOH analysis, only one to several loci can be investigated. Although CGH is able to screen copy number changes through-out the whole genome, it requires a relatively large amount of tissues to obtain sufficient DNA (at least 500 ng for a single experiment).

Multiplex ligation-dependent probe *amplification* (MLPA) is a novel technique to measure the copy number of up to 45 nucleic acid sequences in one reaction [1]. This method relies on sequence-specific probe hybridization to genomic DNA followed by *multiplex-PCR* amplification of the *hybridized* probe and semi-quantitative analysis of the resulting PCR products. The method is easy to perform; requiring as little as 50 ng of DNA.

A further advantage of this method is that it can be performed on partially degraded DNA, which makes this technique very suitable for analysis of routinely processed paraffin-embedded sections. A couple of reports, published recently, utilize this method for the analysis of copy number changes in *melanocytic* tumors.

REQUIREMENTS

Separation of DNA

1. Proteinase K solution: 2 mg/mL proteinase K in 25 mM Tris-HCl, pH 8.2 (stock solution kept at -20°C). The stock is diluted 10-fold in 25 mM Tris-HCl, pH 8.2 at 25°C. Make fresh dilutions.

2. 1 *M* NaSCN: 81 g *sodium thiocyanate* (Sigma S7757)/ L in *demineralized* water. Store at room temperature.
3. TE buffer: 10 mM Tris-HCl, pH 8.2, 1 mM EDTA.
4. Xylol (Sigma).
5. Heating block for slides.
6. Ethanol series comprising successively 75, 96, and 100% ethanol.
7. Inverted microscope.

MLPA Reaction

1. The SALSA human chromosome *aberration* kits, P005, P006 and P007 (MRC Holland, Amsterdam, the Netherlands).
2. Each kit contains the following:

   Probe mix.,

   MLPA buffer (l.5 *M* KCl, 300 *mM/L* Tris-HCl, pH 8.5, 1 mM/L EDTA). Ligase-65.

   Ligase-65 buffer A.

   Ligase-65 buffer B.

   PCR buffer.

   FAM-labeled PCR primers including dNTPs.

   Polymerase.

   Enzyme dilution buffer.
3. Ligase-65 mix: This should be made less than 1 h before use and stored on ice. Mix 3L Ligase-65 buffer A, 3L Ligase-65 buffer B, and 25L water. Add 1 µL Ligase-65 and mix again.
4. Polymerase mix: This should be made less than 1 h before use and stored on ice. Mix 2L SALSA PCR-primers, 2L SALSA enzyme dilution buffer, and 5.5 µL water. Add 0.5 µL SALSA polymerase. Mix well, but do not vortex.
5. Thermal cycler with heated lid.

Gel Electrophoresis

1. ROX-labeled internal size standard (ROX-500 Genescan, Applied Biosystems, Foster City, CA).
2. *Formamide, deionized* (nuclease and protease tested, Sigma).

TECHNIQUES

Separation of DNA from FFPE Tissues

1. Prepare 10-µm thick sections of normal and tumor tissues that are mounted on glass slides.

2. Heat a slide for 15 min at 75°C to melt the *paraffin.* Then place the slide in xylol for 5 min. Repeat this step to dissolve the paraffin as much as possible.
3. Immerse the slide *sequentially* in 100% ethanol for 30s, 96% ethanol for 30s, and 75% ethanol for 30s, followed by a final wash in tap water for 30s.
4. Incubate the slide in 1 M NaSCN at 37°C overnight. 5. Rinse the slide with tap water and finally in TE and let it dry.
6. Manually microdissect tumor areas under an inverted microscope, referring to a corresponding *hematoxylin* and eosin-stained *pathology* slide.
7. Pour a few drops (20- 40 ML) of proteinase K solution onto the tissue and scratch it off into a 1.5-mL tube. Add more *proteinase K* solution to make a final volume of 50-100L.
8. Incubate at 37°C overnight.
9. The next morning, heat the tube for 20min at 80°C to inactivate the proteinase K. 10. Centrifuge for 10min at 12,000g.
11. Transfer the supernatant to a clean tube. Use 2-5 μL for each MLPA reaction.

MLPA Reaction

All the reactions should be carried out in a thermal cycler equipped with a heated lid. Use 200-L PCR tubes for the following reactions.

Denaturation and Hybridisation of the DNA

1. Dilute the DNA sample (20-500 ng DNA) with TE buffer to 5 μL.
2. Heat for 5 min at 98 deg;C and cool to 25°C before opening the thermal cycler.
3. Add 1.5μL of SALSA probe mix and 1.5μL of MLPA buffer. Mix with care.
4. Incubate for 1 min at 95°C followed by 16 h at 60°C.

Ligation Reaction

1. Reduce temperature of the *thermal cycler* to 54°C. While at 54°C, add 32μL of Ligase-65 mix to each sample and mix well.
2. Incubate at 54°C for 10-15 min, then heat to 98°C for 5 min.

PCR Reaction

1. Add 4μL of 10X SALSA PCR buffer to 26μL of water and 10μL MLPA ligation reaction.
2. While tubes are in the thermal cycler at 60°C, add 10μL polymerise mix to each tube and start the PCR reaction.
3. Cycle the reactions with the following program: 95°C for 30 s, 60°C for 30 s, and 72°C for 60s, for 33 cycles. Follow this by final extension at 72°C for 20 min.

Gel Electrophoresis

The amount of MLPA PCR reaction is quantified by *capillary* electrophoresis or sequence gel *electrophoresis*, depending on the instrument and fluorescent label used. The electrophoresis conditions for the ABA 310 and ABI3100 Genetic Analyzer we use are shown.

One Capillary Reaction

1. Mix: 0.75µL of the PCR reaction, 0.75 µL of water, 0.5µL of ROX-labeled internal size standard, and 12µL of Hi-Di *formamide*. Volume is based on the use of a 0.5-ml, genetic analyzer sample tube (for a 48-well tray).
2. Incubate the mixture for 2min at 94°C and cool on ice.
3. Capillaries: 310 *capillaries*, 47 cm × 50µm. *Polymer*: POP-6. Settings: Run temperature 60'C. Injection time 5-10s. Run voltage l5 kV. Run for 30 min. Filter set D (select the correct dye set for using other internal standard).

16 Capillaries Reaction

1. Mix: 1 µL of the PCR reaction, 0.5 µL of ROX-labeled internal size standard, and 8.5 µL Hi-Di formamide. Volume is based on the use of a 96-well plate (for a 96-well tray).
2. Incubate the mixture for 2 min at 94 °C and cool on ice.
3. Capillaries: 3100 capillary array, 36cm. Polymer: POP-4 or POP-6. Settings: Run temperature 60°C. Cap fill volume: 184 steps. Prerun voltage: l5 kV. Prerun time: 180s. Injection voltage: 3.0kV. Injection time 10-30s. Run voltage 15 kV. Run time 1,500s. Filter set D (select the correct dye set for using other internal standard).

Analyzing Data

Quantative and Qulitative Regulates

DNA quantity control fragments of 64, 70, 76, and 82 by are included in each SALSA probe mix. These generate amplification products much smaller than the probe *amplification* products. The data quality control *fragments* present a warning when the amount of sample DNA used was lower than the 20ng human DNA required for reliable MLPA results. The four data quality control fragments gener-ate amplification products when ligation is *omitted* or when no sample DNA is present.

The four 64-82 by amplification products present a much smaller peak size than the 94 by amplification product that is sample DNA and *ligation-dependent* control product, when 50 ng human DNA or more is used in the MLPA reaction. If the 64, 70. 76, and 82 by amplification products have similar or larger peak sizes than the 94bp fragment and the 130-472bp MLPA probe amplification products, either the ligation reaction failed or the amount of sample DNA was less than 20ng. In either case, the results obtained may not be *reliable* and should not be used for further analysis.

Results

1. Export the gene scan data of sizes and peak areas of multiplex PCR products to an Excel or SPSS file. A data flow manual can be downloaded from the MRC-Holland website.
2. Remove nonspecific amplification products and primer-dimer peaks. They are either low peaks or of short length.
3. Normalize all the expected MLPA products by dividing each peak area by the combined

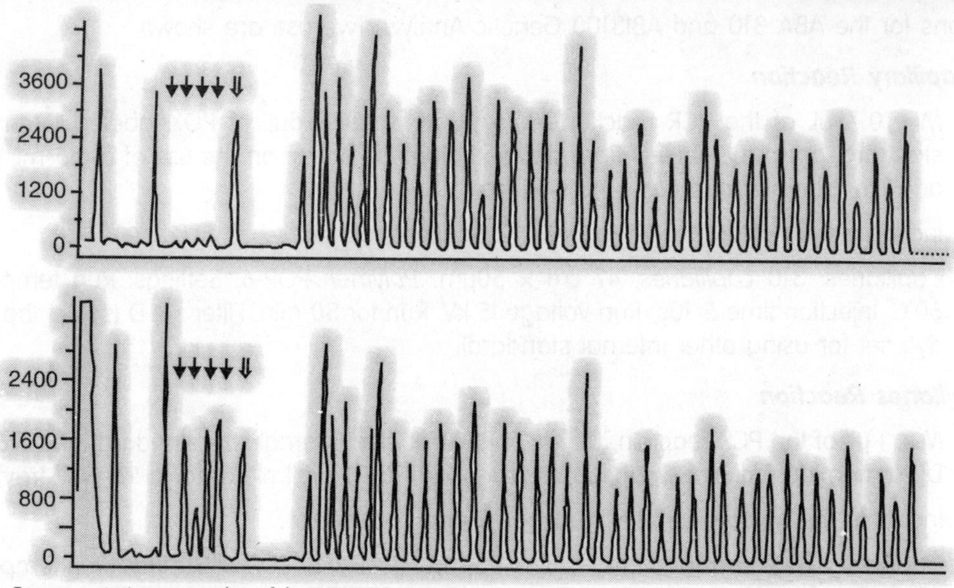

Figure 2.1: Representative examples of the MLPA electrophoresis profiles using a SALSA P006 Chromosome Aberration Kit. (Top) An example of good experiment. The 64-, 70-, 76-, and 82-bp quality control peaks (black arrows) have a much smaller peak size than the 94hp amplification product (a white arrow), which are sample DNA and ligation-dependent control product. (Bottom) An example of bad experiment with insufficient DNA, which should not be used for copy number calcula-tions. The 64, 70, 76, and 82bp amplification products (black arrows) have similar or larger peak sizes than the 94bp fragment (a white arrow) and the 130-472bp MLPA probe amplification products.

peak area of all peaks in that lane (relative peak area).

4. Calculate the relative copy number for each probe, which is expressed as a ratio of the relative peak area for each locus of the sample to that of the relative peak area of the normal *sex-matched* control.
5. Perform similar calculations for peak heights. Normally, there are no significant differences in the outcome of normalized areas and heights.
6. Assess the copy number change for each locus.

POINTS TO REMEMBER

1. A number of protocols and kits are available for DNA extraction from FFPE tis-sues. A critical point in choosing the DNA *extraction* method is to reduce impu-rities, which would inhibit ligation and PCR reactions, as much as possible. The *protocol* described in the text is simple and, as far as we know, the best for an MLPA reaction. Other methods could be used, but the DNA *extraction* protocol should be identical between control and *tumor* samples.
2. Sufficient amounts of DNA can be obtained from a 5 mm × 5 mm × 10µm-thick section. Do not use more starting material than necessary, to reduce the *contamination* of *impurities* that could influence the *polyrnerase* activity.
3. Control sampies of normal male and female DNA should be prepared, respec-tively, from

tissues fixed and processed in the same way as the tumor samples. We use normal lymph node tissues for control samples.

4. This incubation step in the chaotropic NaSCN solution removes impurities from the slides and improves the efficacy of *proteolytic* digestion.
5. This step can be skipped if the section consists of tumor cell populations only. For small tumors or tumors with marked infiltration of *non-neoplastic* cells, laser *microdissection* may be used to obtain pure tumor cell populations.
6 The following program of a *thermal cycler* covers the complete MLAP reaction:

   98°C for 5 min, 25°C pause.

   95°C for 1 min, 60°C pause. 54°C for 15 min.

   98°C for 5 min, 4°C pause. 60°C pause.

   95°C for 30 s, 60°C for 30 s, 72°C for 1 min for 33 cycles.

   72°C for 20 min, 4°C pause.
7. PCR cycles can be increased up to 40.
8. As the MLPA analysis is based on a PCR reaction, the specificity and *reproducibility* of this method needs to be critically evaluated. When the specificity of MLPA was examined by calculating the ratio for each MLPA robe for normal DNA sample, false-positive results of 7.3% were found.

   When ratios were calculated for test samples using normal samples from other MLPA experiments instead of reference samples included in the same experiment, the false-positive results rose to 20%. Therefore, normal DNA samples should be included in each *experiment* to decrease the number of false positives [2].

   Alternatively, the aver-age values of pooled data of normal DNA samples obtained from a number of experiments can be used for calculations. Duplicate or triplicate experiments for one sample decreases the ratio of false-positive results.
9. The copy number ratio, which is considered significant, varies among studies. One study considered a probe ratio below 0.8 or above 1.2 to represent DNA copy number losses and gains, respectively, while another study considered ratios less than 0.7 as loss and higher than 1.3 as gain calculated from both peak areas and peak heights. It is *recommended* that a stricter *criteria* be adopted to reduce false-positive results.

Chapter 3

Transgenesis

Chromosome *rearrangements* cause genomic disorders and cancer in human. Region-specific low-copy repeats (LCRs) can mediate *nonallelic homologous recombination* (NAHR) that results in chromosome rearrangements. Using the Cre-loxP site-specific recombination system, chromosome rearrangements that cause genomic disorders and cancer can be *recapitulated* in the mouse. Technology advancements in mouse genetics, such as *recombineering*, will undoubtedly facilitate *modeling* genetic changes associated with genomic disorders in the mouse.

In the last decade, a novel class of human disorders caused by chromosome rearrangements owing to regional structural characteristics of the genome has been recognized. These conditions are referred to as genomic disorders. A common biological consequence of the chromosome rearrangements in genomic disorders is copy number change of *dosage-sensitive* gene(s) within the rearranged region. The unique genomic architectures associated with genomic disorders include complex region-specific LCRs that are usually 10-500 kb in size and greater than 95% in sequence identity. LCRs can *stimulate* and mediate NAHR, which in turn leads to unstable genomic regions that are prone to chromosome rearrangements.

It is estimated that as much as 5-10% of the human genome might be duplicated, reflecting the recent evolution of LCRs in the primate lineage. In contrast to the *human genome, preliminary* analysis of the mouse genome suggests that only 1.7-2.0% is part of recent large segmental duplications (less than half of what is observed for the human genome).

In general chromosome rearrangements caused by NAHR are usually relatively small in size (i.e., <2-3 Mb) and often difficult to detect cytogenetically. However, these rearrangements are large from a *molecular* viewpoint. Therefore, molecular approaches, such as pulsed-field gel *electrophoresis,* fluorescent *in situ* hybridization, and more recently arraybased comparative genomic hybridization (aCGH), are used for their detection and analysis.

SHARING GENEIC SIMILARITY

The mouse is the most widely used experimental organism to model human disease and to study mammalian development. One important reason is that humans and mice share remarkable genetic similarity. These two species *diverged* from a common ancestor approx 75 million years ago but share a similar genome size (3 × 10^{-9} bp), have a similar number of genes (25,000-30,000), and display a high homology between individual genes.

Furthermore, the two genomes share extensive syntenic regions with orthologous genes present in the same order on the chromosomes. For example, the whole human chromosome 17 is syntenic to the distal part of the mouse chromosome 11. *Similarities* between the mouse and human extend to *anatomical, embryological, physiological,* and other biological traits.

Additional advantages of mice include small size, short life span, and rapid reproductive cycle. Perhaps the biggest advantage of using mice to model human disease is the availability of *tremendous* genetic resources. For example, there are many inbred mouse strains available that have strainspecific characteristics and show *striking* differences in many biological traits.

William Castle and Clarence Cook Little developed the concept of inbred strains of mice where mice are generated by a minimum of 20 consecutive generations of brother-sister mating. As a result, all the loci in the genome of an inbred mouse strain are *homozygous*. The pure genetic background in *inbred mice* is essential for *performing* genetic analyses.

Genetically modified mice became available when transgenic technologies were developed. To generate a transgenic mouse line, a tissue-specific promoter linked to a DNA fragment (usually the cDNA of a gene) is microinjected into the *pronuclei* of fertilized mouse egg to produce founder lines. In general, *transgenic* mouse technology is primarily used to determine the *phenotypic* consequence associated with *overexpression* or ectopic *expression* of a studied gene.

A wealth of knowledge of gene function has been obtained from studying transgenic mice, however, the conventional transgenic mouse technology has its *limitations*. For instance, in addition to being unable to generate loss of function mutations, conventional *transgenesis* does not allow one to control either transgene copy number or integration site. Furthermore, the expression of a transgene is often affected by position effects from surrounding *chromatin*.

The ability to precisely engineer the mouse genome is the *breakthrough* that has brought mice to the forefront of biological research. This breakthrough combined two technology fronts: first, embryonic stem (ES) cells derived from the inner cell mass of normal mouse *embryos* were isolated in cell culture and were later found to retain their *pluripotency* including the ability to colonize the mouse germline.

Second, it became possible to perform homologous recombination in mammalian cells, and later in mouse ES cells, thus, enabling genetic *manipulation* at a specific *locus- targeting*. Since the first true knockout mice were born, gene targeting technology has *revolutionized* mouse genetics and disease-modeling. It is estimated that more than 3000 mouse genes have been targeted in ES cells and knockout mice derived. These *mutant* mice have provided fundamental insights into gene function in mammals.

GENES KNOCKOUT TECHNOLOGY

The conventional gene *knockout* technology can generate *identifiable* genomic deletions up to 19 kb. However, this is not sufficient to model large chromosome rearrangements found in human genomic disorders. Chromosome rearrangements can be induced by either X-ray or chemicals in the mouse. However, the randomness of the endpoints in the *irradiationinduced* chromosome rearrangements has hampered broader *applications* for such genetic approaches

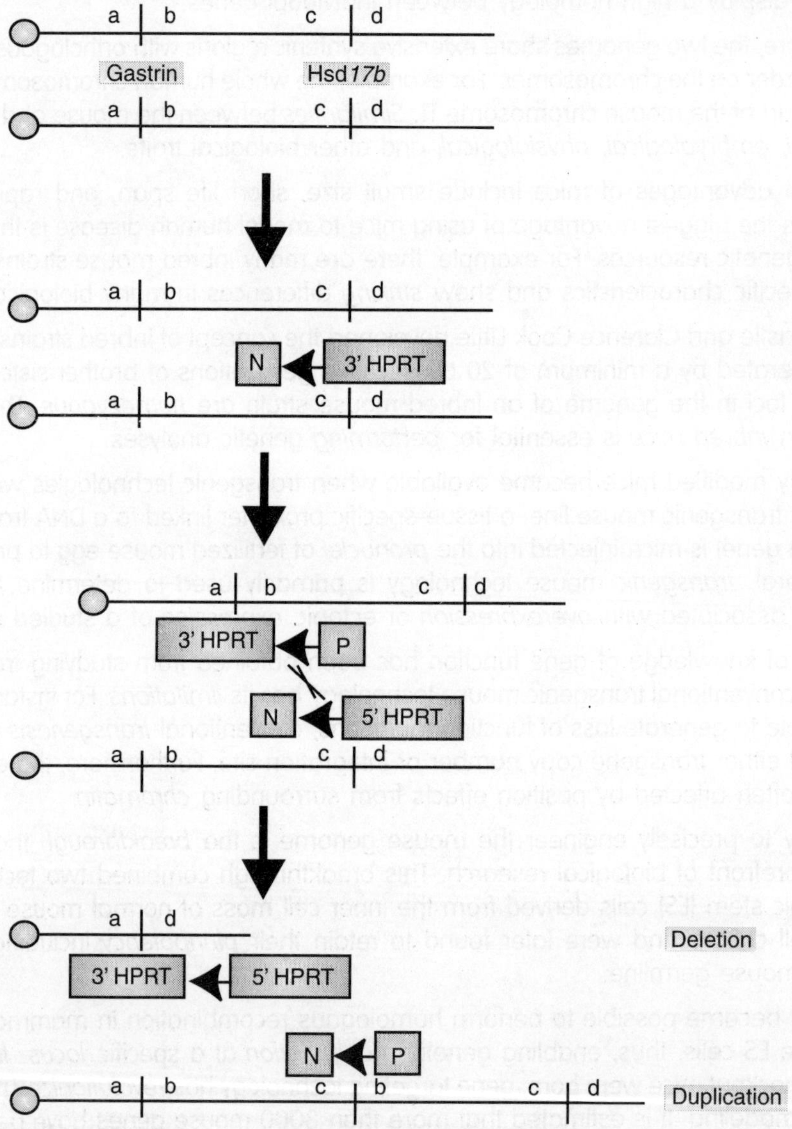

Figure 3.1: Chromosome engineering between Hsd17β and Gastrin *loci on the mouse chromosome 11. (A) Loci on the chromosome 1 1 are represented by a, b, c, and* d. *(B) The two recombination cassettes (Sand 3') are targeted consecutively to the two desired rearrangement endpoints (Hsd1β) and* Gas trin) *in embryonic stem (ES) cells. (C) Recombination between the two* loxP *sites catalyzed by Cre recombinase regenerates a full-length HPRT minigene that enables the deletion to be selected for in HAT medium. (D) When the two cassettes are targeted to the two chromosome 11 homologous, both the deletion and the duplication can be recovered in a single ES cell. Filled arrow, loxP site, P, punomycin resistance cassette; N, neomycin resistant cassette.*

to produce research resources. Chromosome engineering technology was developed to enable the construction of rearrangements of defined sizes in specific genomic locations. It combines gene targeting that defines chromosome rearrangement *endpoints*, and a *site-specific* recombination system from bacteriaphage P1, Cre-loxP, that can mediate recombination between two *loxP* sites.

Cre is a 38-kDa recombinase that alone can catalyse recombination between two 34-bp *loxP* sites in mammalian cells. Prior to its application in chromosome *engineering*, Cre-loxP system had been used to make small genomic deletions in conditional knockout mice. In 1993, while a graduate student in the laboratory of Professor Allan Bradley, together with Ramiro Ramirez-Solis, a *postdoctoral* fellow in the lab, I tested the Cre-loxP system for its ability to induce large chromosomal rearrangements.

It was known that the Cre-loxP system is capable of efficiently inducing smaller deletions of a few thousands basepairs in mammalian cells. We anticipated that the intrachromosomal recombination efficiencies between two *loxP* sites that are millions of basepairs away, or that are on two different chromosomes, could be very low and a positive selection might, therefore, be required to detect and to recover the large recombination products.

In an ideal positive selection system, a selection marker would be activated on Cre-loxP recombination and the recombinants could, thus, be selected out in appropriate medium. We chose the human *hypoxanthine phosphoribosyl transferase* (HPRT) minigene as the positive selection marker for two reasons.

First, an *Hprt* deficient ES cell line, AB2.2, was available in the lab and it had been shown to be capable of highly efficient germline transmission. The second reason is the unique structure of the *HPRT* minigene which contains intron 2 of the human *HPRT* gene that is required for efficient expression in mouse ES cells. *A loxP* site can therefore be conveniently inserted into the middle of this intron *(XbaI* site) without affecting the *HPRT* minigene activity.

We subsequently divided the *HPRT* minigene into two non-functional parts, the 5'. and the 3' cassettes with the intronic *loxP* site embedded in both cassettes. On recombination mediated by *loxP* sites and catalyzed by Cre, a fulllength *HPRT* minigene is reconstituted and the recombinant can be selected out using *hypoxanthine amniopterin thymidine* (HAT) medium.

To generate a defined chromosome rearrangement, the recombination cassettes need to be targeted, on a sequential manner, to two endpoints. We tested the ability of the *HPRT* selection system to generate chromosome rearrangements between the *Hsd17β* and the *Gastrin loci* separated by approx 1 Mb on the mouse chromosome 11.

As illustrated in Figure elsewhere in this chapter, the 5' *HPRT* cassette was targeted to the *Hsd17β locus* in ES cells. The *3' HPRT* cassette was then targeted to the *Gastrin locus* in ES cells with the *5' HPRT* cassette already targeted into the *Hsd17β locus. To* induce the recombination, double-targeted ES cells were transfected with a plasmid that transiently expressed Cre.

We recovered all of the expected recombinants in HAT medium and confirmed the anticipated rearranged genomic structures by Southern hybridization. In order to generate a desired chromosome rearrangement, it is critical to know the orientation of two endpoints relative to each other on the chromosome and to the centromere because of the nature of the selection.

As we described in the initial chromosome engineering experiments, to obtain a deletion without prior knowledge of the endpoint orientation, four types of *double-targeted* ES cells had to be generated and tested. For example, generation of a deletion requires the two recombination cassettes in the correct orientation so that after recombination the *HPRT* minigene is reconstituted on the deletion chromosome in order to survive HAT selection.

The finished mouse genome sequence now enables one to obtain the information regarding the position and *orientation* of the two endpoints for a specific rearrangement. As a result, the recombination substrates can be targeted to the endpoints and the desired rearrangement can be obtained reliably and diagnosed readily.

Cre-loxP-mediated recombination can occur either *in cis* (two *loxP* sites are on the same chromosome) or *in trans* (two *loxP* sites are on different chromosomes). Figure elsewhere in this chapter shows that when two recombination cassettes are targeted to two homologous chromosomes, both the deletion and the duplication can potentially be generated and recovered in a single ES cell.

When these genetically balanced ES cells contribute to the mouse germline, from one *chimaera*, offspring with either the deletion or the duplication chromosomes can be recovered. Following the initial success with 1-Mb chromosomal rearrangements, we found that even larger chromosomal rearrangements could be successfully induced and recovered with the *HPRT* minigene selection system, with sizes ranging from 3 to 22 cM. Most importantly, after multiple rounds of manipulation, ES cells harboring the engineered chromosomal rearrangements still retained the competence for germline transmission.

This finding is often overlooked but represents a key step in chromosome engineering because, after multiple rounds of manipulation, ES cells were originally thought to lose germline competence. Further extension of the *HPRT* minigene selection system has facilitated the development of induced mitotic recombination in mouse ES cells.

The ability to generate designed chromosomal rearrangements opens a new research field and offers new genetic resources. For example, in addition to modeling human disease, chromosome deletions can be used to uncover recessive mutations in genetic screens in vitro and in vivo. Chromosome inversions can be engineered to function as balancers to facilitate N-ethyl-N-nitrosourea mutagenesis screens and to maintain mutant mouse lines.

In addition to Cre-loxP-mediated chromosome engineering, other approaches have been developed to generate chromosome rearrangements. One of these approaches is to irradiate Fl hybrid mouse ES cells. The irradiated ES cells have a negative selection marker, Herpes simplex virus thymidine kinase *(HSV-TK),* targeted by homologous recombination to a defined locus. If deletions caused by irradiation encompass the TK-tagged locus, the cells will survive in 2'-fluoro-2'-deoxy-5-iodo-l-β-D-arabinofuranosyluracil medium. The advantage of this approach is that from a single experiment, multiple deletions of various sizes centered on the *tagged* locus can potentially be recovered.

These deletions are useful for estimating haploinsufficiency tolerance of various chromosome regions in ES cells and may be used directly for in vitro *genetic screens.* Importantly, ES cells carrying these irradiation-induced deletions still retain their ability to contribute to *germline* development in *chimaeras.*

Thus, large genomic deletions from these experiments are useful for estimating *haploinsufficiency* in the mouse prior to making precise chromosomal rearrangements in this *genomic* region using Cre-loxP and for modeling certain genomic disorders. For instance, several deletions on the mouse chromosome 5 syntenic to human 4p 16.3 were produced in one experiment.

The mice heterozygous for these deletions have phenotype similar to Wolf-Hirschhorn syndrome caused by monosomy 4p 16.3. Thus, these deletion mice can directly serve as the mouse models for human disease syndromes. Compared to the TK-anchored deletion strategy, chromosome engineering using Cre-loxP has advantages.

First, defined chromosome rearrangements can only be obtained with the precisely engineered endpoints (preselected loci). Second, other types of rearrangements, such as duplications, inversions, and translocations, can also be generated with chromosome engineering. Third, when a large chromosome rearrangement needs to be induced only in somatic cells, *site-specific* recombination is much more efficient.

MICER and Insertion Vectors

One important practical limitation for chromosome engineering is construction of targeting vectors to introduce the recombination cassettes to the rearrangement endpoints. Previously, genomic DNA libraries in l*-phage were screened with specific probes in order to isolate genomic DNA fragments to serve as the homology regions in gene targeting.

Targeting vectors are then constructed by ligating two homology arms to one of the recombination substrates (5' or 3' *HPRT-loxP* cassettes). However, it could take one person a couple of months to build a targeting vector with this approach. To improve the efficiency of this step, the Bradley laboratory at Baylor College of Medicine constructed two mouse genomic phage libraries in which 10-kb genomic DNA fragments are cloned into two complementary backbones.

The cloning backbone for the first library has the 5' *HPRT-loxP* cassette, *a PGKNeobpA* positive selection cassette, and the *Tyrosinase* coat color gene for easy identification of the mice. On the other hand, the *3' HPRT-loxP* cassette, the *PGKpurobpA* selection cassette, and the *Agouti* coat color gene are cloned into the backbone of the *3' HPRT library*. As a result, clones from these two libraries can be used directly as premade gene targeting vectors for chromosome engineering.

Recently, The Wellcome Trust Sanger Institute has sequenced the two *HPRT-loxP* genomic libraries (Mutagenic Insertion and Chromosome Engineering Resource [MICER]) (http://www.sanger.ac.uk/PostGenomics/mousegenomics/). Individual MICER clones can now be viewed directly on the Ensembl website and ordered through the Sanger Institute (http://www.sanger.ac.uk/cg.i-bin/teams/team38/ CloneRequest/CloneRequest).

To make a defined chromosome rearrangement, MICER clones corresponding to the two endpoints are linearized within the genomic inserts to generate a DNA double strand-break that stimulates efficient *homologous* recombination. Gene targeting in ES cells with these vectors has been shown to be very efficient.

Targeting events in ES cells using MICER clones can be identified using a DNA fragment that is removed from the genomic DNA insert in the linearization process, or using an external probe, in Southern hybridization. Although MICER clones provide a *convenient* way to make targeting vectors for chromosome engineering, they are all *insertion* targeting vectors that can introduce challenges compared to replacement vectors.

Because there is no negative selection marker present on insertion vectors, on homologous recombination, these vectors can potentially form *concatemers* that may greatly complicate analysis of targeted events. Concatemerization of the recombination *substrates* may also *complicate* the ensuing Cre-loxP *recombination* because multiple *loxP* sites in a concatemer might have different orientations.

To obtain clean targeting events, replacement vectors are usually used. Genomic DNA fragments serving as the two homology-arms are isolated so that the recombination cassettes can be placed between them. This can be difficult in many cases because appropriate restriction

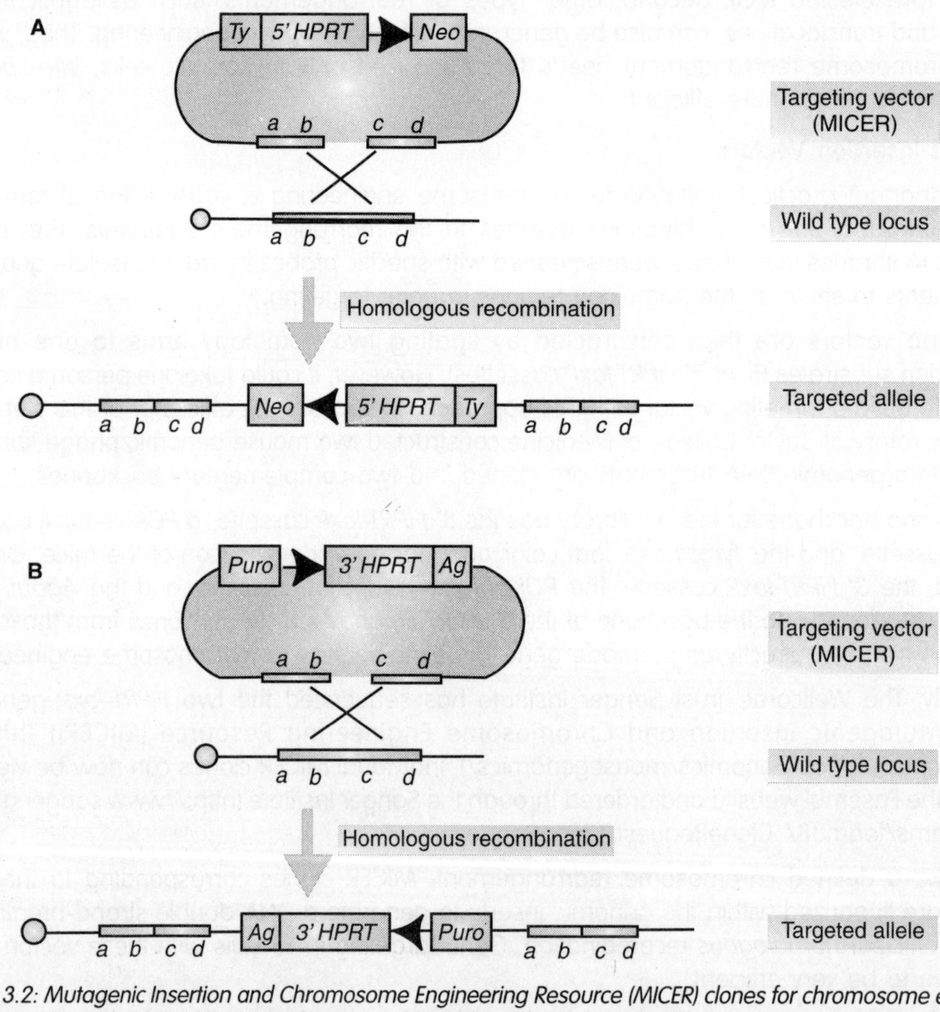

Figure 3.2: Mutagenic Insertion and Chromosome Engineering Resource (MICER) clones for chromosome engineering. (A) Clones from the first MICER library carry the 5'HPRT cassette, Neo and a Tyrosinase minigene (Ty). (B) Clones from the second MICER library have the 3' HPRT cassette, Puromycin resistance cassette (Puro) and a K14-agouti (Ag) cassette. Vectors are linearized in the middle of the genomic fragments (a, b, c, d) to create a DNA double strand break for homologous recombination. Filled arrow, loxP site.

enzyme sites have to be found or engineered in order to ligate all the DNA fragments together.

Manipulating Technology

In the last few years, a highly *efficient* approach to manipulate DNA in *Escherichia coli,* termed recombineering, has become available. This technology utilizes the bacteria phage X-homologous recombination system, called Red. Recombineering allows DNA with homology as short as 50 by to be recombined efficiently in *E. coli* and as a result has facilitated many kinds of genomic manipulations and experiments that previously were difficult to implement.

Three phage proteins are required for the Red functions: Exo, Beta, and Gam. Exo is a 5'>3' exonuclease that acts on linear double-strand DNA (dsDNA) to generate 3' singlestrand DNA (ssDNA) overhangs for recombination. β protein binds to the ssDNA overhangs created by Exo and stimulates complementary strand annealing. The recombination is further enhanced by Gam, which inhibits the RecBCD activity of the host cell. RecBCD is a major nuclease that *degrades exogenous* double strand DNA in *E. coli.*

In 1998, Francis Stewart's laboratory first discovered that homologous recombination function encoded by a prophage (Rac, encodes RecE and RecT) can be used to engineer cloned DNA. Soon after this seminal work, an alternative system was developed in a collaboration between Court et al. and Copeland et al. In this new system, a defective l*-phage with a temperature sensitive repressor, *c*I857, is inserted into the *E. coli* genome.

As a result, the λ-Red functions can be induced by shifting culture temperature from 32 to 42°C for 10-15 minutes. Experience from many laboratories showed that this system appears more efficient than the original RecET system. To construct targeting vectors for chromosome *engineering* using recombineering, the original *HPRT* minigene cassettes are modified with selection markers that are functional in both ES cells and in *E. coli* (Chan et al. unpublished).

For example, the *5' HPRT* cassette is linked to an engineered *Neo* gene that is selectable in ES cells and in *E. coli.* Similarly, the 3' *HPRT* cassette is linked with *a Bsd* cassette that is selectable with Blasticidin in both ES cells and in *E. coli.* The next step is to transform a bacterial artificial chromosome (BAC) containing one rearrangement endpoint into recombineering-competent *E. coli* cells such as DY380.

Alternatively, a mini λ construct can be transformed to the original BAC containing host *E. coli* cells. Once the BAC is established in DY380 cells, the recombination cassette (for example the *5' HPRT-Neo)* flanked by short homology sequences can be targeted to a precise position on the BAC by recombineering. The final targeting vector can subsequently be constructed by retrieving the genomic DNA *fragment* containing the corresponding recombination substrate to a *plasmid backbone.*

This process has proven to be very efficient and reliable. The major advantage of using *recombineering* over premade targeting vector libraries is that more *flexible* designs can be easily introduced to a precise genomic location. The *prerequisite* for using recombineering is that the genomic sequence of a given locus is known. The availability of the mouse genome sequences greatly facilitates the broader use of recombineering in mouse genetics.

Nevertheless, the current mouse genome sequence in the public database is from the C57BL6 strain, whereas most mouse ES cell lines are derived from various 129 strains. It is therefore essential that genomic resources based on the 129 strains of mice should be established and made available to the research community. For instance, at least one BAC library from a 129 strain should be end-sequenced and placed into public databases.

These BACs would best serve the research community if they are already in recombineering competent host cells. Stewart's lab has shown recently that a BAC carrying multiple selection cassettes engineered using recombineering can be used to target multiple linked loci *simultaneously*. This approach can be potentially modified to target the 5' and 3' recombination cassettes to the two rearrangement endpoints simultaneously in one cell transfection experiment.

The success of this strategy can further reduce the laborious serial targeting required for chromosome engineering and increase the likelihood that the *manipulated* cells are still competent for germline *transmission*.

CHROMOSOMAL REARRANGEMENTS

During the last decade, technology advances have already made generating large chromosome rearrangements possible in the mouse. Continuous development of new technologies will *undoubtedly* further facilitate our ability to engineer many kinds of chromosome changes associated with human disorders. Some of the recent successes that utilized chromosome engineering to model genomic disorders and to *genetically* dissect human disease are briefly described.

Digeorge *Syndrome*

del22q11.2 syndrome (also known as DiGeorge syndrome [DGS], *velocardiofacial syndrome* [VCFS], and *conotruncal* anomaly face syndrome) with an *incidence* of 1 in every 4000-5000 births, is the most frequent chromosomal *microdeletion syndrome* known. The common deletion is caused by LCR-mediated recombination. Typical patient phenotype includes cardiovascular defects, *thymic, parathyroid, craniofacial anomalies*, and learning *disabilities*.

Many of the features (e.g., heart outflow tract, *craniofacial. velopharyn- geal, ear, thymic,* and *parathyroid abnormalities*) are attributable to developmental defects of the embryonic pharyngeal apparatus. Although at least 30 genes have been mapped to the common deletion region of *del22q 11.2,* until recently, none of these genes had been shown to carry genetic mutations in the patients who do not have the common deletion.

The mouse syntenic region of the *del22q11.2* genomic region is located on the mouse chromosome 16. Using chromosome engineering, a 1.2-Mb deletion *(Es2-Ufd1l)* was generated and the deletion was subsequently established in the mouse germline. Some of the embryos and adult mice carrying the deletion in heterozygous status manifested cardiovascular defects on a mixed genetic background.

However, these deletion mice did not have other phenotypes associated with the 22q11.2 deletion in the human patients. Nevertheless, the mutant mice did *recapitulate* one important phenotype of the *del22q11.2 syndrome,* confirming that the disease-causing gene(s) is within this deletion interval.

To narrow the critical genomic region, Lindsay et al. generated several smaller deletions and found that one deletion containing the transcription factor gene, *Tbx1, is* responsible for the

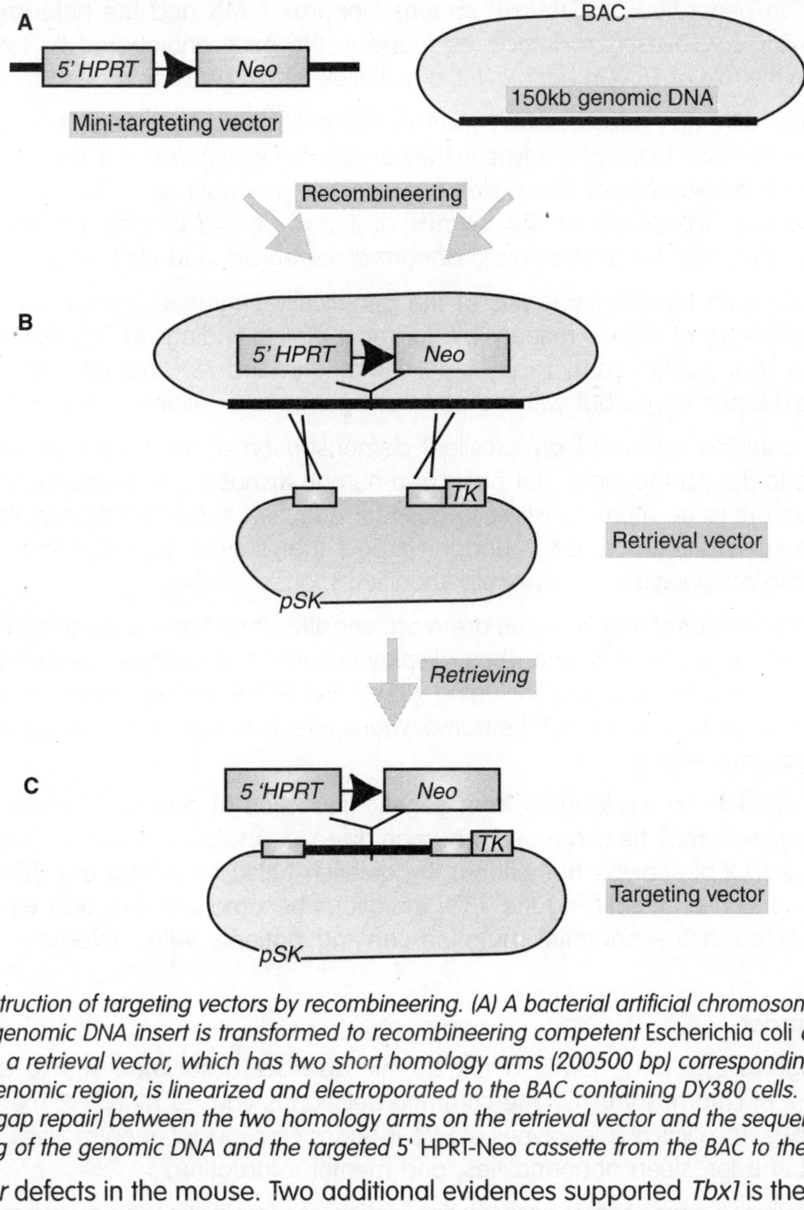

Figure 3.3: Construction of targeting vectors by recombineering. (A) A bacterial artificial chromosome (BAC) with an approx 150-kb genomic DNA insert is transformed to recombineering competent Escherichia coli *cells (DY380). (B) In the next step, a retrieval vector, which has two short homology arms (200500 bp) corresponding to the ends of the interested genomic region, is linearized and electroporated to the BAC containing DY380 cells. (C) Homologous recombination (gap repair) between the two homology arms on the retrieval vector and the sequences on the BAC results in cloning of the genomic DNA and the targeted* 5' HPRT-Neo *cassette from the BAC to the retrieval vector.*

cardiovascular defects in the mouse. Two additional evidences supported *Tbx1* is the causal gene. First, a P1 artificial chromosome containing *Tbx1* rescued the cardiovascular defects in the deltion mice.

Second, *a Tbx1* hypomorphic allele confirmed that *Thx1* is a dosage sensitive gene in the mouse and is required for normal development of the *pharyngeal* arch arteries. In an independent study, a chromosomal deletion *encompassing* the *Thx1* locus was generated by crossing two *loxP* sites, that were in the *del22q11.2* critical region but located on the two *homologous* chromosome 16, to *cis* configuration.

The deletion generated by Cre-loxP spanned approx 1 Mb and the heterozygous deletion mice had similar *cardiovascular* defects as found in the mice engineered by Lindsay et al. To further confirm the role of *TBX1* in *del22q11.2,* a null allele of *Tbx1* was generated by gene targeting.

Interestingly, although mice *heterozygous* for this null allele of *Tbx1* only had cardiac outflow tract anomalies (identical to the phenotype in the deletion *heterozygotes*), the *Tbx1 -/-* mice displayed a wide range of developmental anomalies encompassing almost all of the common DGS/VCFS features, including *hypoplasia* of the thymus and *parathyroid* glands, cardiac outflow tract *abnormalities,* abnormal facial structures, *abnormal vertebrae,* and cleft palate.

Taken these data together, analysis of the genetically engineered mice strongly supported that haploinsufficiency of *TBX1* is responsible for most clinical findings in *del22q11.2* patients. This hypothesis has been confirmed by recent studies that identified *TBX1* mutations in patients with a typical *del22q11.2* phenotype but with no apparent genomic deletions.

The *Tbx1* analyses represent an excellent demonstration of the power of using genetically modified mice to dissect the molecular basis of a human disease. It is interesting to note that the *Tbx1* gene appears to be more sensitive to dosage reduction in human than in the mouse. This may reflect species difference, an important aspect that should be taken into account when interpreting data obtained from genetically modified mouse models.

It has been postulated that humans are more sensitive than the mouse to lower levels of the dosage-sensitive gene products and, thus, display increased phenotype penetrance. In support of this view, in several other cases, including *MSX2* and *PAX9,* heterozygous effects are evident in humans but not in the mouse, but the homozygous effects in mouse are similar in nature to the human heterozygous effects.

Another important lesson learned from genetic dissection of *del22q11.2* in the mouse is that genetic heterogeneity may be common for human disease. Analyses of hundreds of patients with apparent *del22q11.2* phenotype but without the deletion failed to identify any *TBX1* mutations in several previous studies. Identifying the *TBX1* mutations became possible only when the studies were implemented in the potential mutation-carrying patients with extremely careful clinical examination.

Genomic Disorder

Smith-Magenis syndrome (SMS) is a genomic disorder associated with a deletion within subband p 11.2 of chromosome 17. The common deletion is caused by NAHR mediated by LCRs named SMS-REPs. The clinical phenotypes of SMS includes craniofacial abnormalities, brachydactyly, self-injurious behavior, sleep abnormalities, and mental retardation.

Patients with the predicted reciprocal duplication recombination product have less severe clinical phenotype: mild to borderline mental retardation and behavioral difficulties. Using the Cre-loxP chromosome engineering technology, Walz et al. generated mice that carry a deletion on the mouse chromosome 11, which is partially syntenic to the genomic interval frequently deleted in SMS patients.

Heterozygous mice carrying this deletion have craniofacial abnormalities, seizures, marked obesity, and male-specific reduced fertility. On the other hand, mice heterozygous for a duplication of the same genomic region are underweight, do not have seizures, craniofacial abnormalities or fertility defects, but are hyperactive, and have impaired *contextual* fear conditioning. Importantly, mice that have the deletion and the duplication (compound *heterozygous*) do not have any of the major phenotypes consistent with the hypothesis that the traits expression is caused by gene dosage effects and most probably not by position effects from altered chromosomal structures in both rearrangements.

Retinoic acid-induced gene 1 *(RAI1)* has been found recently mutated in rare patients with many features of SMS but without a *recognizable* deletion. Therefore, *RAI1* has been proposed to be the major player in the SMS phenotype. Analysis of *RAI1* knockout mice and mice with increased dosage of *RAI1* should identify the in vivo biological function of this gene and confirm its role in SMS.

In the meantime, it is also necessary to generate nested deletions and duplications within the SMS genomic region to identify genes that might cause minor phenotypes or are genetic modifiers of the phenotypes caused by *RAI1* mutations.

Genomic Rearrangements

Genomic rearrangements also play critical role in cancer. Recurrent genomic aberrations have been found in both human lymphomas and solid tumors. For technical reasons. one type of genomic aberrations, recurrent balanced *chromosomal translocations*, are better characterized in human *leukaemia* and *lymphomas*. Among the most common chromosomal *translocation* breakpoint regions in human *leukaemia* is chromosome 11q23.

These translocations involve the mixed-lineage leukaemia gene, *MLL,* and greater than 30 different *MLL* fusion partners, whereas the identity of these partners determines the etiology of the subsequent leukaemia in humans. Based on the Cre-loxP technology, a chromosomal translocation between *MLL* and eleven nineteen leukaemia *(Enl)* was induced specifically in haematopoietic cells in mice.

These translocation mice developed a rapid onset and high penetrance of leukemogenesis that models many key aspects of the human disease. Other chromosomal aberrations found in human cancer include interstitial *deletions* and *amplifications*. In general, a deletion indicates the presence of a tumor *suppressor* gene, whereas an amplified region contains one or more *oncogenes*.

The chromosomal deletions can be modeled relatively easily using *chromosome engineering*. Because most of the homozygous genomic deletions found in human tumors are large, establishing similar size deletions in the mouse would be difficult owing to possible embryonic lethality. To overcome this technical difficulty, large chromosome rearrangements can be induced by expressing Cre temporally and/or spatially in somatic cells.

It is estimated that Cre can induce up to 10 cM rearrangements in the mouse without significantly reducing the recombination efficiency. Using chromosome engineering, mice with four copies of a chromosome region can be generated. This copy number change is small compared to amplification of a chromosome region in cancer.

However, it has been shown that moderate increase of copy number (three to eight absolute copies) of *P13CA* is associated with activation of P13 kinase activities and ovarian cancer. It is also known that people with trisomy of certain chromosomes have greater cancer risk and poor clinical outcomes in *leukaemia.* Therefore the genomic regions that are found to have moderate copy number increase from array-CGH can potentially be assessed for their relevance to cancer in the mouse with *chromosome* engineering. *Cancer* is a complex genetic disease and multiple genetic events are usually required for cancer development.

Mice harboring the engineered chromosomal rearrangements should be either put on a sensitized genetic background, or be challenged with mutagens to facilitate tumor development in order to assess the relevance of these genetic mutations to cancer development.

CONCLUDING REMARKS

During the last decade, the ability to precisely engineer the mouse genome in ES cells and to establish the mutations in the mouse germline has *revolutionized* mouse genetics and human disease modeling. In particular, chromosome engineering has allowed us to model gross chromosomal changes found in human genomic disorders and in cancer.

Microarray-based CGH has become a powerful method for the genome-wide detection of genomic DNA copy number changes quantitatively and for mapping them directly onto the sequence of the human genome. It is very likely that, with these genomic technologies being more widely applied in clinical *diagnosis* and *analysis,* many other *submicroscopic* genomic rearrangements as well as population-specific polymorphisms mediated by LCRs/ NAHR will be detected.

Similarly, more recurrent chromosomal changes will be found to be involved in cancer *initiation* and *progression.* Furthermore, with systematic analysis of finished human genome sequence, more genomic regions will be predicted to be prone to genomic rearrangements. Modeling these genomic changes in the mouse is exciting and will *undoubtedly* provide us important clues to the *biological* consequences of these genomic mutations.

Chapter 4

Molecular Signalling

We consider a *signal* to be a DNA or RNA sequence pattern that is recognized by a protein or other molecule (or, for RNA, sometimes another region of the same molecule). Binding sites for proteins on DNA are important examples of signals, and in this chapter we focus on these signals.

Particular instances of the sequence pattern may be represented by a sequence of letters in the *pertinent alphabet* (e.g., A, C, G, T for DNA), but for signals recognized by proteins, this is an approximation: proteins actually recognize specific *atoms* or groups of atoms on bases or base pairs.

"Recognize" usually means binding in the thermodynamic sense. This may be strong binding, as in the case of *repressors* binding to operators, or binding may be transient, as with restriction endonucleases. Examples of common signals in DNA are:

- Restriction endonuclease recognition sequences (e.g., GAATTC for EcoRI).
- Binding sites for regulatory proteins: These proteins may function as repressors (e.g., cI protein of *bacteriophage lambda,* which binds to the corresponding operator), may regulate gene expression in response to physiological conditions of the cell (e.g., CRP protein, the cAMP receptor protein of *E. coli),* or may be eukaryotic transcription factors such as the *glucocorticoid* receptor (which binds to the GRE or glucocorticoid receptor element under appropriate conditions).
- Elements within replication origins and termination regions in genomes.
- Promoters: These are sites that determine where transcription is initiated. There are different classes of *promoters,* depending upon the type of RNA polymerase (eukaryotes) or type of specificity factor (*prokaryotes*) that acts at these sites. Within any particular class of promoters, individual examples have been finetuned by evolution to provide the appropriate level of *transcription* for the genes that they control.

This means that promoter sequences display a much greater degree of sequence variability than is observed for other signals (such as restriction endonuclease recognition sequences).

Selection of Binding Sites on DNA

Often the starting point is a cloned or PCR-amplified piece of DNA that is hypothesized to contain a binding site for a particular protein (e.g., a region near the 5' end of a gene, such as the *lac* promoter region). In most cases of interest, the DNA is duplex rather than single-strand. To determine experimentally whether a particular protein binds, it is also necessary to have purified (or partially purified) binding protein (e.g., CRP protein) stored in the freezer.

One way to identify whether a protein-binding site is present or absent on the DNA fragment is an *electrophoretic mobility-shift assay* (or gel-shift assay). In this experiment, duplex DNA in an appropriate buffer for protein binding is mixed with the binding protein and incubated to allow formation of the DNA-protein complex.

The reaction mixture is then loaded onto a polyacrylamide gel, and electrophoresis is performed with naked DNA (no protein supplied) run in an adjacent lane. Binding of the protein to the DNA reduces the mobility of the DNA fragment relative to its mobility when there is no bound protein. The resulting shift in the position of the protein-bound fragment is visible after staining the gel.

The amount of shifted complex depends upon how much protein was supplied and on its binding equilibrium constant. *Footprinting* experiments can locate the positions of protein-binding sites on DNA molecules. In one implementation of this approach, duplex DNA containing the binding site is labeled at one end (on one strand) and is allowed to bind the protein, which covers a small segment of the DNA. The complex is then briefly treated with a nonspecific cleavage reagent such as DNase I, which cuts only in regions of the molecule *not* protected by the protein.

Hydroxyl radicals can also be used for nonspecific cleavage. Each of the many DNA molecules in solution may be cleaved at one or more positions not covered by the protein. By electrophoresis of the treated DNA on a denaturing gel of the type used for analyzing DNA. sequencing reaction products, single-strand fragments whose lengths extend from the labeled end to the various cleaved sites are resolved from each other, generating a "*ladder*" of bands having different lengths.

The cleaved positions are located throughout the DNA *except for that portion of the DNA that was protected by the bound protein.* The region in the ladder where there are few (or fainter) bands is called the *footprint.* The location of the footprint is determined from a control "ladder" having fragments of known size.

Sometimes the investigator may possess proteins known to bind DNA but not instances of the cognate binding site in a DNA of interest. In that case, it is possible to retrieve fragments of DNA containing the binding site or sites by using *chromatin immunoprecipitation* (ChIP). (ChIP experiments are used with other approaches, particularly genomic *microarrays,* to identify genomic locations of sites that bind particular proteins.) For example, if we have purified protein P known to bind to *eukaryotic replication* origins, it is possible to produce in rabbits (or *rats, mice,* or *goats*) specific *antibodies* that recognize protein P.

To identify DNA binding sites for P, we first perform in vivo chemical cross-linking of chromatin proteins, including P, to DNA. The chromatin is then extracted from the cells, and the DNA in the protein-DNA complexes is fragmented into smaller pieces (e.g., 500 by long) by *hydrodynamic* shearing. Complexes of P with the DNA to which they are bound are then precipitated by using the anti-P antibodies (possibly with the aid of a *secondary antibody*).

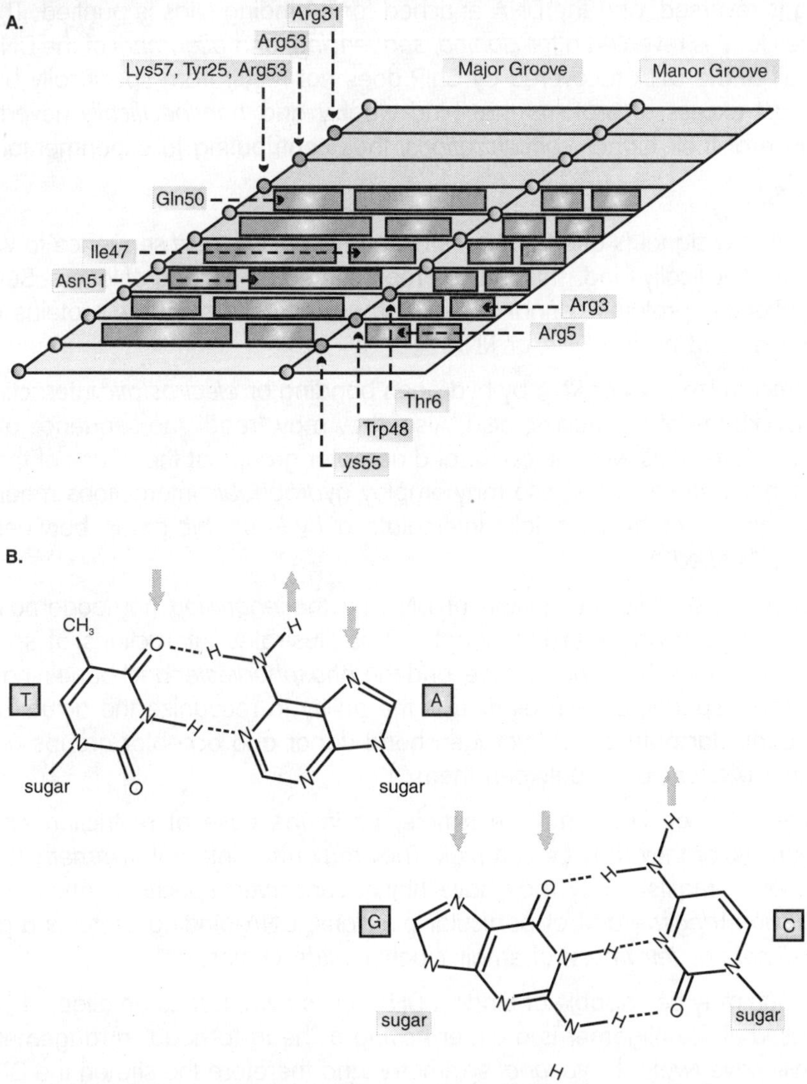

Figure 4.1: How proteins interact with DNA. Panel A: Interactions between specific residues of the engrailed *homeodomain protein with its binding site. The cylindrical surface of the DNA has been cut along one of the phosphodiester backbones and unwrapped for projection onto a plane. Bases A or T are represented by shaded or unshaded rectangles, respectively. Base pairs that do not specifically interact are not shown. Note that the bases indicated in the minor groove are the opposite edges of the same bases appearing in the major groove. Phosphate groups are represented by circles. Shaded circles represent phosphate groups with which* engrailed *protein specifically interacts. Dashed arrows indicate interactions of particular amino acid residues with specific bases or phosphate groups. Note that there are interactions between the protein and the bases in both the major and the minor grooves. Un-derlying data were taken from Kissinger CR et al. (1990)* Cell *63:579-590. Panel B: Locations of hydrogen bond donor groups (arrows pointing away from the struc-ture) and acceptor groups (arrows pointing toward the structure) on the edges of the Watson-Crick base pairs in the major groove. Particular sequences create characteristic patterns of donor and acceptor groups, particularly in the major groove.*

Cross-linking is reversed, and the DNA enriched for P-binding sites is purified. The sequence of the binding sites for P is revealed after *cloning, sequencing,* and *alignment* of the DNA. However, approximately half of the DNA recovered by ChIP does not result from specifically bound P. This is because the vast excess of proteins that bind weakly and *nonspecifically* nevertheless bind somewhat because of their higher *concentrations,* thus contributing to experimental *"noise."*

Protein Binding Site

We indicated that a signal is often a short stretch of a *nucleic acid* sequence to which one or more proteins may specifically bind. This may be represented by a string of letters. But what is the signal actually? After all, proteins cannot "read" Roman letters! Obviously, proteins are sensing chemical groups exposed on the DNA or RNA.

Proteins can recognize DNA or RNA by hydrogen bonding or *electrostatic* interactions with the *phosphodiester* backbone of the nucleic acid. Also, they may "read" the sequence of base pairs by hydrogen bond interactions with the donor and acceptor groups at the edges of the base pairs in the major or minor *grooves.* They also may employ *hydrophobic* interactions mediated by the methyl group of thymine or may partially intercalate a *hydrophobic* group between two base pairs (if the DNA is "*kinked*").

An example of site-specific recognition of DNA by the *engrailed* homeodomain protein is diagrammed in Figure elsewhere in this chapter. This illustrates interactions of specific amino acids with the major groove, the *minor groove,* and the *phosphodiester* backbone (shaded circles). What really counts for specific protein binding is the ability to recognize the actual sequence of base pairs, and particular patterns of hydrogen bond donor and acceptor groups on the edges of base pairs allow *discrimination* between them.

Protein-binding sites on DNA may be simple, as in the case of restriction *endonuclease* recognition sequences, or they may be complex. They may have internal *symmetry* (e.g., inverted repetition of sequence motifs). They may have highly conserved specific sequence patterns or may tolerate *variation* (*degeneracy*) at particular positions. DNA-binding proteins are commonly *oligomers* (e.g., *dimers* or *tetramers*) of smaller polypeptide chains.

Each *polypeptide* may be capable of binding DNA on its own, but as an oligomer, the binding strength is increased. If the oligomer is a dimer having a "head-to-head" arrangement, then the binding protein will have twofold *rotational* symmetry, and therefore the site on the DNA to which it binds will also have twofold rotational symmetry, recognized as inverted repetition of the DNA sequence. An example of what is meant by inverted repetition is the GRE (glucocorticoid receptor element):

5'-NNNNNAGAACANNNTGTTCTNNNNN-3'

3'-NNNNNTCTTGTNNNACAAGANNNNN-5'

The underlined bases appear in an opposite direction relative to the laboratory frame of reference, hence the "*inverted*" adjective. These repeated bases are on opposite strands. The two symmetrically disposed strings of *conserved* bases in a binding site are called *half-sites.* Notice that, in this case, the binding site can be recognized by examining either strand of the DNA for AGAACA.

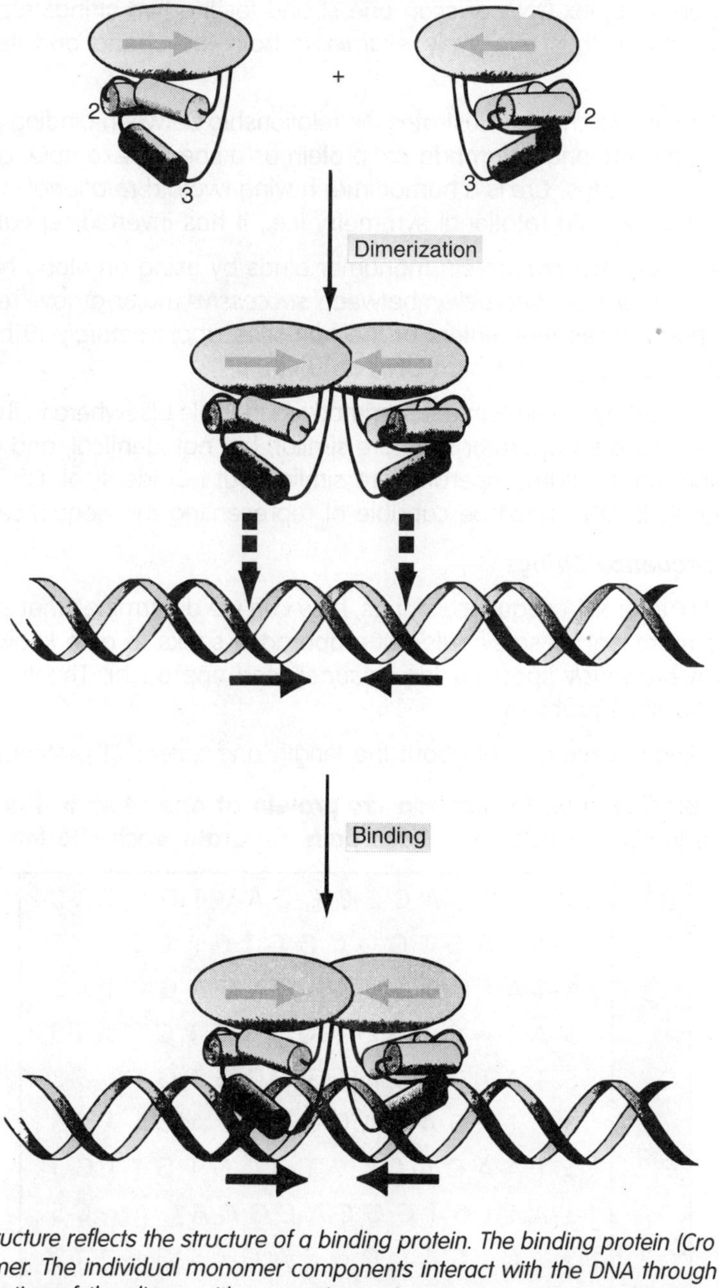

Figure 4.2: How site structure reflects the structure of a binding protein. The binding protein (Cro in this case) binds head-to-head as a dimer. The individual monomer components interact with the DNA through the major groove (bottom panel). In-teraction of the dimer with successive major groove regions on the same face of the helix implies that the centers of the two interaction regions will be spaced 10 to 11 by apart the pitch of B-form DNA. The equivalence of interactions between helices 3 of both monomers and DNA imposes inverted symmetry within the DNA sequence to which Cro binds.

For sites that lack this symmetry, it is necessary either to scan each strand separately to find the binding sites on a duplex DNA or scan one strand for the two strings representing both the site and its complement, thus effectively *examining* both the strand and its *complement* in a single pass.

Figure elsewhere in this chapter illustrates the relationship between binding protein and binding site structure using bacteriophage lambda *cro* protein as a specific example. *cro* protein (Cro) can bind to lambda operator sites. *Cro* is a homodimer having twofold *rotational symmetry*. Therefore, the operator also has twofold rotational symmetry (i.e., it has inverted repeat structure).

Moreover, because each *cro* protein monomer binds by using an alpha helical region that is inserted into the major *groove, interaction* between *successive* major groove regions on the same side of the DNA helix places the centers of the half-sites *approximately* 10 by away from each other. (The average helix pitch of B-form DNA is 10.4bp.)

The sites recognized by *cro* protein are summarized in Table elsewhere in this chapter. Observe that the sequences of the six operator sites are similar, but not identical, and that the sequences of half-sites comprising the same operator are similar, but not identical. Obviously, any realistic description of signals in DNA must be capable of representing this sequence variation.

DNA and RNA Sequence Strings

Given a set of DNA or RNA sequence strings, how can we determine what *signals* they encode, and how can we represent these signals? One approach seeks to gain knowledge of a pattern that has not been *previously* specified (an *unsupervised* approach). This is a daunting, and in strictest terms impossible problem.

However, we know a good deal about the length and extent of protein-binding patterns in

Table 4.1: Binding sites for lambda *cro* protein at operators in the cI region of bacteriophage lambda. Underlined base pairs separate each site into two half-sites.

O_L1	5'-T A T C A C C G C C A G T G G T A-3'
	3'-A T A G T G G C G G T C A C C A T-5'
O_R1	5'-T A T C A C C G C C A G A G G T A-3'
	3'-A T A G T G G C G G T C T C C A T-5'
O_L2	5'-T A T C T C T G G C G G T G T T G-3'
	3'-A T A G A G A C C G C C A C A A C-5'
O_R2	5'-T A A C A C C G T G C G T G T T G-3'
	3'-A T T G T G G C A C G C A C A A C-5'
O_L3	5'-T A T C A C C G C A G A T G G T T-3'
	3'-A T A G T G G C G T C T A C C A A-5'
O_R3	5'-T A T C A C C G C A A G G G A T A - 3'
	3'-A T A G T G G C G T T C C C T A T - 5'

DNA, as illustrated in the previous section. An example of unsupervised pattern discovery is identifying *k*-words that are *over-represented* in a set of functional sequences such as promoters.

In that case, we expected there to be signals in the regions immediately upstream of the 5' end of transcribed gene regions, but we did not specify what they were. Also, we did not require that the signals be present in all members of the set. We were able to identify over-represented k-words such as TAAT and ATAA that are contained within TATAAT (-10 region) without *alignment* or *experimental* specification of the signal.

Unsupervised methods have been used to identify k-words representing both known and previously *unrecognized* regulatory sequences for yeast genes. A statistical approach devised by Bussemaker et al. (2000) can be used to generate a k-word dictionary of regulatory sites or other significant motifs. *Unsupervised* methods are not discussed further here, but see Section elsewhere in this chapter.

Another approach, *supervised* learning, includes prior knowledge of the pattern to be found. For example, multiple instances of a known and aligned DNA signal can be used to define parameters of *probabilistic* models describing that signal. This is the subject of the remainder of this chapter.

Consensus Sequence

There are several ways of representing signals in DNA, ranging from very simple to rather complex. The simplest method of representing a binding site is as a *consensus sequence*—a string of characters corresponding to the most common occurrences of bases at each position. This is *adequate* for sites such as *restriction endonuclease* recognition sequences and *reasonably* good for sites such as GRE *half-sites* (examples aligned below).

Position: 1 2 3 4 5 6

AGAACA

ACAACA

AGAACA

AGAAGA

AGAACA

AGAACT

AGAACA

Consensus: AGAACA

In this particular collection of sites, one of the four bases is much more likely to occur at any position in the site than any of the three others. (Exceptions to the consensus base at any position are underlined.) The consensus *sequence* is a listing of the preferred bases at each position. If there are *positions* at which either of two bases is commonly found, then the consensus might be written to indicate both *possibilities*. So, for the example above, we might write A (C/G) AA (C/G) (A/T) if either of the indicated letters is *tolerated* at positions 2, 5, and 6.

There are cases, however, where there is much more sequence variation at some or all of the positions within the binding site. For such cases, consensus methods are not appropriate and a probabilistic description is required.

A string of iid letters does not contain signals, except by chance. To generate a *string* of iid letters, a base is assigned at each position according to a probability distribution, which might be represented as a column vector (4 ×1 matrix) whose entries correspond to the *probabilities* of A, C, G, or T *determined* by the overall base *composition* of the DNA.

In the example of the GRE half-site above, it should be obvious that there is no single probability distribution for describing the site. Instead, we must specify (in principle) a different distribution at every single position (in other words, we are removing the "id" from "iid").

This suggests another representation of protein-binding sites. We create a matrix of probabilities by "*binding*" together (the cbind function in R) the column vectors that represent the probability distributions (probabilities of each letter for any position) at each position in a site. Therefore, each element in this *matrix* represents the probability of finding a particular base at a given position.

For example, the probability distribution at position 2 for the limited sample of GRE half-sites listed above, with rows corresponding to A, C, G, and T, respectively, is_

$$\begin{bmatrix} 0.00 \\ 0.14 \\ 0.86 \\ 0.00 \end{bmatrix}$$

and the matrix representing all six positions in the half-site is

$$\begin{bmatrix} 1.00 & 0.00 & 1.00 & 1.00 & 0.00 & 0.86 \\ 0.00 & 0.14 & 0.00 & 0.00 & 0.86 & 0.00 \\ 0.00 & 0.86 & 0.00 & 0.00 & 0.14 & 0.00 \\ 0.00 & 0.00 & 0.00 & 0.00 & 0.00 & 0.14 \end{bmatrix}$$

where rows $i = 1, \ldots, 4$ correspond to A, C, G, and T, respectively, and columns $j = 1, \ldots, 6$ correspond to the positions within the site. This representation of the sites is called a *positional weight matrix,* or *PWM.* (See Stormo, 2000a for a review of the history of PWMs.)

Note that the six lambda operator half-sites Table elsewhere in this chapter display more sequence variation at each position than is seen at any position in the GRE half-sites, and thus the positional weight matrix provides a much more precise representation of these half-sites than would a consensus sequence.

There are other cases where *consensus* methods are very imprecise and inappropriate, such as for promoter sequences. In preparation for describing and analyzing a signal in DNA, we gather together *and align* representatives of that signal (see below). A *training set* is a collection of *bona fide* signals (or sites) used to produce the probabilistic model.

A second set of bona fide signals or sites is necessary for testing the mathematical description or model. This is called the *validation set.* Some-times the construction of the model requires a *challenge set* of sequences that are not sites. The *challenge* set is the one to which the real sequences in the training set are contrasted, such as a set of "*nonsites*" to be contrasted with sites.

In some circumstances, we might use a probabilistic model to produce the challenge set (for example, an iid model or a Markov chain model). The process of estimating the probability distributions from the training set is called *learning* in the machine-learning world.

String Representation

Our practical goal is to recognize a signal of length w within a string representing a DNA sequence. To do this, we parse the sequence using windows of width *w* and ask how well the *w* letters in that window correspond to a particular signal. This means that we need to assign some type of score to the sequence in each successive window.

We assume that we have a collection of aligned DNA sequences of the same length w and having no gaps, and that we know that all members of this collection are sites for binding a particular DNA-binding protein. How do we assign a score to any particular sequence to describe its level of correspondence with the training set?

We follow the logical steps presented in Durbin et al. (1998). Given a collection of aligned sites, each of length *w*, let p_{ai} be the probability of any particular letter *a* from {A, C, G, T} at position *i* measured within a site. Identities of letters at each position in the sequence are assumed to be independent of identities of their *neighbors*. (Remember that since we are talking about *signals*, the probability distributions for the letters at each position are *not* identical in the general case.)

The probability of a sequence $\mathbf{A} = a_1\, a_2 \ldots a_w$, given hypothesis B that it is a binding site is

$$P(\mathbf{A} \mid B) = \prod_{i=1}^{w} p_{ai} \tag{4.1}$$

Suppose that the same sequence was chosen from a random collection of bases (hypothesis *R)* with probabilities q_{ai}. (The random model is iid.) Then the probability of the sequence is given by

$$P(\mathbf{A} \mid R) = \prod_{i=1}^{w} q_{ai} \tag{4.2}$$

and the odds ratio is given by

$$\frac{P(\mathbf{A} \mid B)}{P(\mathbf{A} \mid R)} = \prod_{i=1}^{w} \frac{p_{ai}}{q_{ai}} \tag{4.3}$$

The *score* for the sequence *A* can be defined as the log of the odds ratio, or

$$S = \log_2 \frac{P(\mathbf{A} \mid B)}{P(\mathbf{A} \mid R)} = \sum_{i=1}^{w} \log_2 (p_{ai} / q_{ai}) \equiv \sum_{i=1}^{w} s_i. \tag{4.4}$$

where $s_i = \log_2(p_{ai}/q_{ai})$ is the contribution to the sequence score of the base at position *i*. (Note that we are taking the log to base 2 because eventually we will be referring to "information" in bits. Note that $\log_2(x) = \ln(x)/\ln(2)$.)

The P_{ai} may be used as elements in a positional weight matrix. For the contrasting iid model, the q_{ai} are independent of position (identically distributed), so from here on, we just refer to q_{ai} as q_a. If the base composition is 50% G+C, then $q_a = 0.25$ for each base for duplex DNA (i.e., $q_a = 0.25$ for all a in {A,C,G,T}).

Let's consider a more complex example than those we have considered previously. *Escherichia coli* promoters have been recognized to contain conserved hexamer sequences at approximately -10 and -35 relative to the first position in the transcript.

The consensus for the -10 hexamer is often represented as TATAAT. Table 4.2 lists nine examples of promoter segments for *E. coli;* these are part of a larger data set given already.

Clearly, it is hard to pick out the -10 hexamer in all cases using the consensus given above, so we represent the -10 hexamer as a positional weight matrix using compiled data for a subset of all *E. coli* promoters. First, we create a matrix containing values for counts of each letter at each position (rows i = 1, ... , 4 corresponding to A, C, G, and T, and columns corresponding to positions

Table 9.2: A sample of *E. coli* promoter sequences. These sequences have been aligned relative to the transcriptional start site at position +1 (boldface large letter). Sequences from *-40* to +11 are shown. Close matches to consensus -35 and -10 hexamers are underlined. See also Appendix C.3 for additional examples and sources of the data.

	−35 −10 −1
ORF83P1	CTCTGCTGGCATTCACAAATGCGCAGGGGTAAAACGTTTCCTGTAGCACCG
ada	GTTGGTTTTTGCGTGATGGTGACCGGGCAGCCTAAAGGCTATCCTTAACCA
*amn*P4	TTCACATTCTGTGACATACTATCGGATGTGCGGTAATTGTATGGAACAGG
araFGH	CTCTCCTATGGAGAATTAATTTCTCGCTAAAACTATGTCAACACAGTCACT
aro G	CCCCGTTTACACATTCTGACGGAAGATATAGATTGGAAGTATTGCATTCAC
atpI	TATTGTTTGAAATCACGGGGGCGCACCGTATAATTTGACCGCTTTTTGATG
cai T	AATCACAGAATACAGCTTATTGAATACCCATTATGAGTTAGCCATTAACGC
*clpAp*1	TTATTGACGTGTTACAAAAATTCTTTTCTTATGATGTAGAACGTGCAACGC
crrP2-I	GTGGTGAGCTTGCTGGCGATGAACGTGCTACACTTCTGTTGCTGGGGATGG

as before):

$$\begin{bmatrix} 9 & 214 & 63 & 142 & 118 & 8 \\ 22 & 7 & 26 & 31 & 52 & 13 \\ 18 & 2 & 29 & 38 & 28 & 5 \\ 193 & 19 & 124 & 31 & 43 & 216 \end{bmatrix}$$

The sums of the entries in each column are all 242, which is the number of promoter sequences in this particular data set. The numerical entries indicate an overwhelming preference for A at position 2 and a less pronounced preference for A at position 5.

We can also represent the data by a matrix containing the relative frequencies of each letter at each position:

$$\begin{bmatrix} 0.04 & 0.88 & 0.26 & 0.59 & 0.49 & 0.03 \\ 0.09 & 0.03 & 0.11 & 0.13 & 0.21 & 0.05 \\ 0.07 & 0.01 & 0.12 & 0.16 & 0.12 & 0.02 \\ 0.80 & 0.08 & 0.51 & 0.13 & 0.18 & 0.89 \end{bmatrix}$$

Entries in this matrix correspond to entries in the previous one, each divided by 242. Note that the sums of the entries in each column should be 1.0, and they are, to within rounding error.

We can also represent the data by a third matrix (used for scoring), containing the corresponding values for $\log_2(p_{ai}/q_a)$. Such a matrix is known as a *position-specific scoring matrix* (*PSSM*—articulated as "possom"). If we make the reasonable approximation for E. *coli* DNA that $p_a = 0.25$ for all a, the resulting PSSM is:

$$\begin{bmatrix} -2.75 & \boxed{1.82} & 0.06 & \boxed{1.23} & \boxed{0.96} & -2.92 \\ -1.46 & -3.11 & -1.22 & -0.96 & -0.22 & -2.22 \\ -1.75 & -4.92 & -1.06 & -0.67 & -1.11 & -3.60 \\ \boxed{1.67} & -1.67 & \boxed{1.04} & -0.96 & -0.49 & \boxed{1.84} \end{bmatrix}$$

Now let's see how to score a particular sequence string. Take the sequence ACTATAATCG for example. If we start parsing this from the beginning using windows of width 6, we score hexamers ACTATA, CTATAA, TATAAT, . . . until we come to the end. The scores s_i at each position for TATAAT (matches with the consensus) are boxed in the matrix above. The score for TATAAT is $S = 1.67 + 1.82 + 1.04 + 1.23 + 0.96 + 1.84 = 8.56$. From (4.4), it is clear that $\mathbb{P}(\mathbf{A}|\mathcal{B})/\mathbb{P}(\mathbf{A}|\mathcal{R}) = 2^S$, so for TATAAT, $\mathbb{P}(\mathbf{A}|\mathcal{B})/\mathbb{P}(\mathbf{A}|\mathcal{R}) = 377$.

In contrast, for ACTATA, $S = -2.75 - 3.11 + 1.04 + 1.23 - 0.49 - 2.92 = -9.75$. Suppose that only one base occurred at each position in every example in the training set (e.g., only T appeared at position 1, only A at position 2, etc.). Then $\log2(p_{ai}/q_a)$ would have been $\log_2(1/0.25) = \log_2 4 = 2$ at each position. The score for the six positions would then have been 12, and $\mathbb{P}(\mathbf{A}|\mathcal{B})/\mathbb{P}(\mathbf{A}|\mathcal{R}) = 2^{12} = 4096$. This is the same ratio of probabilities that was implicit when we were looking for an invariant 6 by endonuclease recognition site in DNA.

A positional weight matrix is a *probabilistic* description of a protein-binding site or other signal, with the *underlying* assumption that the state or letter at a given position is unaffected by the state or letter at the previous position. We have emphasized the utility of such matrices for scoring potential sites within a string containing other DNA.

In addition, a PWM (elements representing probabilities at each position) is a probabilistic model that allows us to simulate sites, should we wish to do so. When we wanted to simulate a string of iid letters, we used a single vector of probabilities to generate a letter for each site in a particular simulated sequence.

To simulate a binding site of length *w,* we use consecutively *w* probability vectors, each corresponding to a column in the PWM. This is done in one of the exercises at the end of this chapter.

Ignored Issues

We briefly touch on some practical matters that we have ignored up to now. The first is aligning the sequences. For *E. coli* promoters, this was simplified because the alignment could be approximately fixed relative to the transcriptional start site. (The +1 position is determined experimentally.) For short patterns or signals embedded in extensive other DNA, alignment is a more difficult proposition (i.e., finding statistically meaningful short local alignments in long sequences).

Having footprinting data helps to restrict the lengths of the sequences to be aligned. A second problem is defining the extent of the site (the value of *w*). Here, footprinting data are also extremely helpful. The *information* content $I(X_i)$, mentioned in under head Shaunan's Entropy, can be used to determine the size of the sites. The approach is to calculate $I(X_i)$ across the sequence containing the site and to select those positions where $I(X_i)$ is significantly above the background level.

A third problem is that only a limited sample of the total population of sites is available for use as a training set. This means that the elements p_{ai} of the positional weight matrix are only estimates of the population values. To correct for a small sample size, we may compute p_{ai} as

$$P_{ai} = \frac{n_{ai} + 1}{N + 4}, \tag{4.5}$$

where n_{ai} is the number of sites having letter a at position *i,* and *N* is the total number of sites in the training set. This is particularly necessary when $n_{ai} = 0$ since the $\log_2(p_{ai}/q_a)$ term would then be undefined. This equation provides a correction for small sample sizes by adding a "pseudocount" of 1 for each base at each position.

The 4 in the denominator reflects the fact that a pseudocount of 1 has been supplied for each of {A,C,G,T}. If there were no observations at all ($n_{ai} = N = 0$), the result would be $p_{ai} = 1/4$, which is what we would have predicted with no prior knowledge of the sites (assuming equal probabilities of each base at each site).

Finally, note that, for a signal of length *w, a* positional weight matrix contains 4*w* parameters. For a matrix of probabilities, only three entries in each column are independent since the sum of the four entries must equal 1.0. Therefore, only 3*w* parameters are independent. If there were only six members in the training set, there would on average be only two observations available for estimating each parameter (6w observations÷ 3w independent parameters). This is such a small sample size that we would not expect the parameters to be reliably estimated.

The number of bona fide sites available depends on wet-lab experiments, and this number may not be large enough to provide both training and validation sets of adequate size. Sometimes methods such as *cross-validation* (leave one out) are employed to test the predictive performance of PWMs when the number of sites is too limited to provide both training and validation sets of adequate size.

In Computational Example elsewhere in this chaptere, we *reinforce* these concepts using the site to which transcription factor GATA-1 binds. GATA-1 is a *transcription* factor that regulates transcription in hematopoietic cells (cells that give rise to *blood cells*, such as red blood cells or erythrocytes). The string representing its binding site has w = 6, and it is represented by the consensus (A/T) GATA (A/G) . Examples of GATA-1 binding sites are listed in the TRANSFAC database (http://transfac.gbf.de/TRANSFAC/).

(The database may list the binding site embedded in a string of other sequence or may list either of two strands. We need to perform alignment and record the complements of the sequence given, as appropriate.) We chose GATA-1 binding sites for this example because there is a reasonably large number of listed binding sites for estimating the parameters (49 human sites). These site sequences are listed in Table 4.3.

Table 4.3: Binding sites for hematopoietic transcription factor GATA-1 from *H. sapiens*.

TTATAG	AGATAA	TGATTA
AGATAT	TGATAA	AGATAA
AGATAG	AGATAA	AGATAG
ATATCT	AGATAG	TGATAT
AGATAG	TGATAG	AGATAA
AGATAG	TGATCA	TCAGAG
AGATAG	TTATCA	AAGTAG
TGATAA	AGATGG	AGATTA
AGATAA	TGATAT	TGATAG
AGATAA	AGATAG	TGATAG
CGATAG	TGATAA	AGATAC
AGAGTT	GGATAC	TGATTG
TGATAA	AGATAA	AGATTA
TGATAA	CGATAA	AGAATA
AGATGG	TGATAG	AGATAA
AGATAG	AGATAA	AGATTA
AGATTG		

Computational Example 4.1: PWM representation of GATA-1 sites

We begin by converting the data in Table 4.3 into a numerical represent-ation, as we have done in earlier chapters. The data are stored in the matrix gata.

The first three entries are:

```
> gata
      [,1]  [,2]  [,3]  [,4]  [,5]  [,6]
[1,]  4     4     1     4     1     3
[2,]  1     3     1     4     1     4
[3,]  1     3     1     4     1     3
```

We can produce a PWM in terms of probabilities of bases at each site directly from gata. However, we ultimately want to generate scores according to (4.4), so we also create a PSSM having elements of the form log2(p_{ai}/q_a). This means that we need the probabilities of each base for genomic DNA. Since these sites are from human DNA, we use the human base composition, 41% G+C, to produce the "background" probability distribution based upon the iid model. Vector bg has four elements corresponding to the probabilities of A, C, G, and T, respectively:

```
> bg<-c(0.295,0.205,0.205,0.295)
> bg
[1] 0.295    0.205    0.205    0.295
```

Now we write a function to compute the PWM, given a matrix of sites and the background probability distribution as input:

```
makepwm<-function(x,bg){
# x = matrix of N aligned sites coded numerically
# bg = vector (1×4) of background base frequencies
  L<-length(x[1,]) # Number of positions in each site
  N<-length(x [,1]) # Number of sites
  pwm<-matrix(rep(1,4*L),nrow=4)
# pwm initialized to 1 for each matrix element (pseudocounts)
  for (j in 1:L){
  for (i in 1:N){
  k <- x[i,j]
  pwm[k, j] <- pwm[k, j]+1
  N <- N+4 # Denominator for small sample correction
  pwm<-pwm/N # PWM in terms of probabilities
  log2pwm<-matrix (rep(0,4*L),nrow=4,ncol=L)
# Initialize PWM in terms of log(base 2) of p/q
```

```
f or(i in 1:4){
log2pwm [i,]<-log2 (pwm[i,]/bg[i])
# Scores for each [nucleotide, position], base 2
return (pwm, log2pwm)
}
```

The "guts" of this function are the lines

```
for (i in 1:N){
  k<-x[i,j]
  pwm [k,j]<-pwm[k,j]+1
  }
```

Here we have used the numeric coding of the sequence to identify k, which in the second line after the f or also specifies the numeric row number corresponding to each base in object pwm. The vectorization in R avoids a set of commands such as

```
if(x[i,j]==1)...pwm[1,j]<-pwm[1,j]+1...elseif(x[i,j]==2)...
```

Notice that we are returning two objects from this function. This produces an R "list" object, which we call tmp. Objects in the list tmp are extracted by entering tmp followed immediately by the object name prefixed with the $ sign:

```
> tmp<-makepwm(gata,bg)
> tmp$pwm
      [,1]     [,2]     [,3]     [,4]     [,5]     [,6]
[1,] 0.566040 0.037736 0.924530 0.037736 0.698110 0.433960
[2,] 0.056604 0.037736 0.018868 0.018868 0.075472 0.056604
[3,] 0.037736 0.849060 0.037736 0.056604 0.056604 0.396230
[4,] 0.339620 0.075472 0.018868 0.886790 0.169810 0.113210
> tmp$log2pwm
      [,1]     [,2]     [,3]     [,4]     [,5]     [,6]
[1,] 0.94018  -2.9667  1.6480   -2.9667  1.24270  0.55685
[2,] -1.85670 -2.4416  -3.4416  -3.4416  -1.44160 -1.85670
[3,] -2.44160 2.0502   -2.4416  -1.8567  -1.85670 0.95070
[4,] 0.20322  -1.9667  -3.9667  1.5879   -0.79678 -1.38170
```

(You should check for yourself to see what happens if you enter object name *tmp* alone.)

Now we use `log2pwm` to generate scores for all of the input sites in *gata*. To do this, we need a scoring function, which is shown below. This calculates the score for the first *L* positions of any input sequence, where *L* is the site length, here represented as length `(log2pwm [1,])`. We invoke this repeatedly for the set of GATA-1 sites, and later we use it in another function to scan along a long input sequence.

```
calcscore<-function(seq,log2pwm){
# seq is a vector representing input DNA numerically
# log2pwm is a PWM (4xL) with elements as log base 2
   score <- 0
for (j in 1:length(log2pwm[1,])){
   score<-score+log2pwm[seq[j],j]}
return (score)
}
```

Now we calculate scores for all of the GATA-1 sites. First, we extract *log2pwm* from *tmp* as a separate object:

```
> log2pwm<-tmp$log2pwm
```

Then we loop *calcscore* over all elements in *gata* and store the resulting scores in *gata. score:*

```
> gata.score<-rep(0,length(gata[,1]))
> for(i in 1:length(gata[,1])){
```

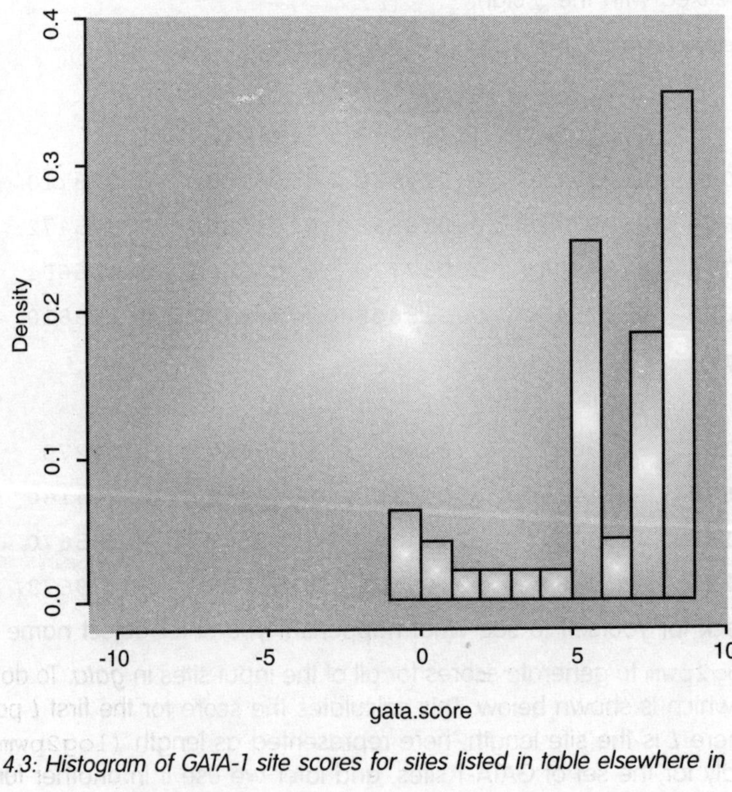

Figure 4.3: Histogram of GATA-1 site scores for sites listed in table elsewhere in this chapter.

```
+  gata.score[i]<-calcscore(gata[i,],log2pwm)
+  }
>
>  signif(gata.score,2)
[1]  3.67  6.09  8.42  -0.61 8.42  8.42  8.42  7.29  8.03  8.03
[11] 5.62  0.60  7.29  7.29  5.32  8.42  6.38  8.03  7.29  8.03
[21] 8.42  7.68  4.60  0.59  5.32  5.35  8.42  7.29  2.23  8.03
[31] 5.23  7.68  8.03  5.25  8.03  8.42  5.35  8.03  -0.25 -0.69
[41] 5.99  7.68  7.68  5.61  5.64  5.99  1.43  8.03  5.99
```

(We use the signif () function to limit the output to two significant figures). We examine the score distribution graphically:

```
>  hist(gata.score,xlim=c(-12,12),ylim=c(0,0.4),
+  nclass=10,  prob=T)
```

The result is shown in Figure 4.3. The score distribution is very broad and asymmetric. Notice that some of the *GATA-1* sites have scores below 0. What would you have expected the average score to be for a set of *background* sequences composed of iid letters? We return to this question in the example after the last section of this chapter. We might wonder whether some of the *GATA-1* sequences in our training set have been misidentified.

Probability Distribution

A PWM represents a signal under the assumption that the identities of letters at a given position are independent of the identities of letters at the previous position. The probability distributions in general differ at every position in the site (i.e., the letters are not identically distributed). We can remove the condition of independence by using a Markov chain description.

We have already used this while describing DNA sequences having dinucleotide frequencies that departed from the iid predictions. That application assumed that identities of bases at various positions were not independent but *were* identically distributed at all positions (i.e., the same transition matrix was employed at every position).

The comparison and contrast of the approach stated already in this book with the one used here is illustrated in Figure 4.4. When we were representing a string with identical distributions of each letter, we started with an initial probability distribution π and produced the probability distributions at each subsequent position by multiplication using a single transition matrix, as diagrammed in Figure 4.4 A.

Only one transition matrix was required because we were considering *homogeneous* chains. The Markov chain illustrated in the diagram in Figure 4.4 A is conventionally represented as shown in Figure 4.5 To represent the signals, the π vector is identified with the vector of probabilities of each state at the first position, and subsequent probabilities are produced by multiplication by a succession of (w-1) transition matrices.

Multiplication of the vector of first position probabilities by matrix 2 produces the probability

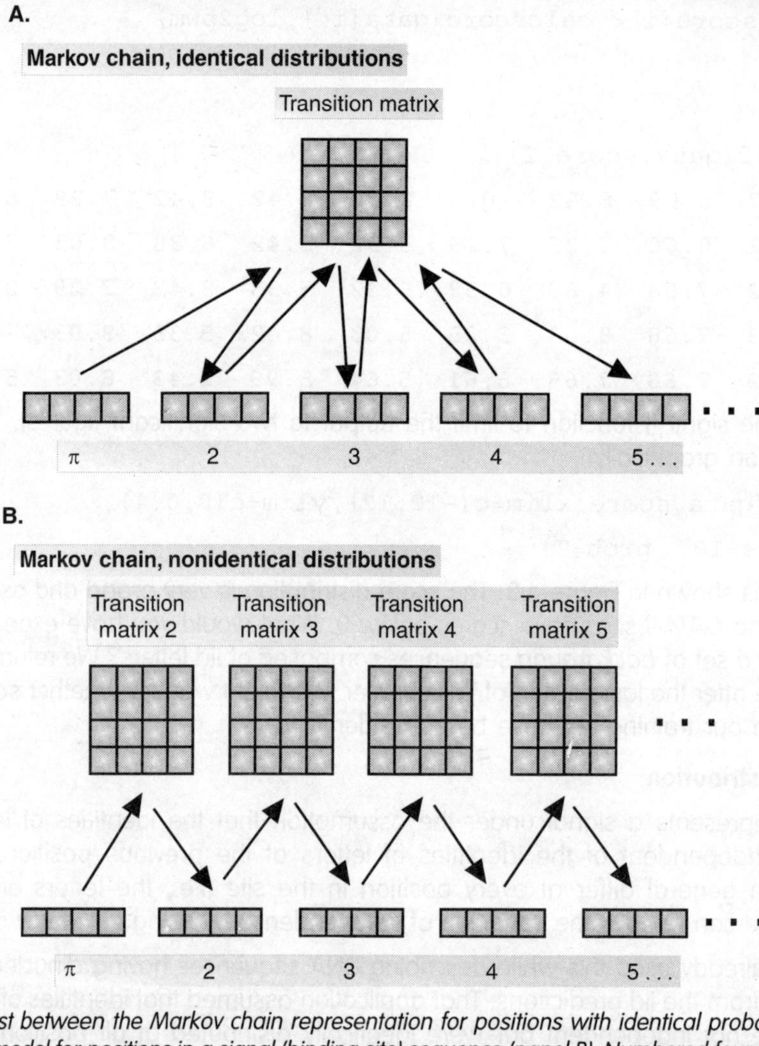

Figure 4.4: Contrast between the Markov chain representation for positions with identical probability distributions (panel A) and the model for positions in a signal (binding site) sequence (panel B). Numbered four-element horizontal rows are the vectors of probabilities for A, C, G, and T at each position. These vectors are transformed to vectors at the next position by matrix multiplication by a transition matrix. In panel A, a single transition matrix is employed. In panel B. a different transition matrix is required for each successive position.

distribution at position 2 given the distribution at position 1, multiplication of the probability distribution at position 2 by matrix 3 produces the probability distribution at position 3 given the distribution at position 2, and so on.

It should be evident to you that the initial vector and the set of transition matrices are sufficient to allow *simulation* of the binding site type to which they correspond. It is important to think about the *data structure* and the number of parameters. For PWMs, the *probabilistic* data were summarized in a matrix that could be produced by binding together column vectors corresponding to the independent probability distribution at each site.

For the Markov chain representation, we can stack the individual transition matrices to form

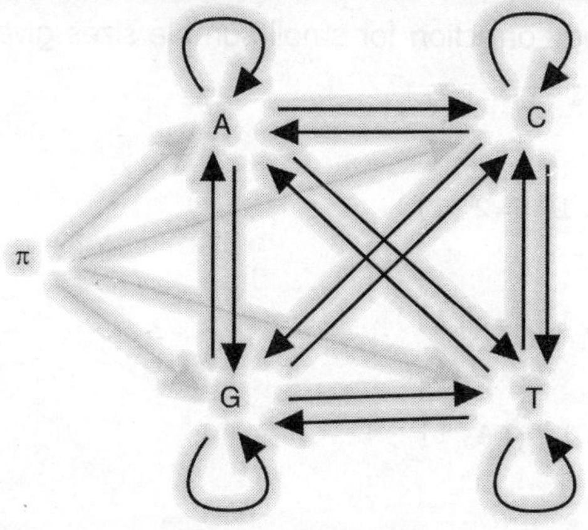

Figure 4.5: Ndicate allowable transitions from one state to another, and associated with each arrow is a probability for that transition. The initial state is drawn from the initial probability distribution p. In any chain generated by the model, state* n + 1 *is assigned based upon the identity of the state at* n *and the probabilities describing the model. The 16 dark arrows correspond to the elements in the transition matrix and the lighter arrows to the entries in the initial probability distribution, Figure 4.4 A.*

a three-dimensional *array*, or three-dimensional matrix. The R statistics package accommodates this type of data construct. With the Markov chain description, the number of parameters has increased compared with the positional weight matrix representation.

Remember that each row of a transition matrix consists of the probabilities that the next letter is A, C, G, or T, given the letter corresponding to the row label. Three of these probabilities are independent, so $4 \times 3 = 12$ of the entries in each transition matrix are independent. There are three independent probabilities in the vector for position 1, and there are $(w - 1)$ transition matrices.

Therefore, the number of parameters in this probabilistic description is $3 + 12(w - 1)$, or for $w = 6$ a total of 63 parameters (in contrast with 18 parameters for a conventional PWM representation, $w = 6$).

Example 4.2: Markov chain representation of GATA-1 sites

Step 1: Preprocess the data

We find it convenient to use logical operators, so we first convert the alphabetical characters into numeric values: `A=1`, `C=2`, `G=3`, and `T=4`. A space is placed between each character. This can be done with Perl or with any text editor. The data may be put in a file called `gata1N.txt`.

Step 2: Import the data

Read the data into a matrix object defined under R. If you do not use a UNIX text editor, you may find that formatting elements are picked up by `read.table`. Save your file as *text only* before using `read.table`.

```
> gata<-read.table("gata1N.txt",header=F)
```

Step 3: Make the probability vectors

Produce the vector representing the probability distribution at position 1. This can be done by simple base counts and using the correction for small sample sizes given in the last section.

```
> length(gata[gata[,1.]==1,1])
[1] 29
> length(gata[gata[,1]==2,1])
[1] 2
> length(gata[gata[,1]==3,1])
[1] 1
> length(gata[gata[,1]==4,1] )
[1] 17
> vectorn<-c(29,2,1,17)
> vectorn
[1] 29 2 1 17
```

Now, applying the small sample correction, we estimate the probabilities from the frequencies and check that the probabilities sum to unity, as they should:

```
> vectori [] <-(vectorn [] +1) / (49+4)
> vectorl
[1] 0.56603774 0.05660377 0.03773585 0.33962264
> sum(vectorl [ 1 )
[1] 1
```

`vector1` represents the probability distribution for the first position of this collection of `GATA-1` binding sites.

Step 4: Produce the transition matrices

We indicate how this is done for `matrix2` (used to create the probability distribution of sites at position 2 given `vector1`). In general, we need to read down each column of gata for columns 2 through 6, evaluating the number of times that 1, 2, 3, or 4 appears in a column (except the first) given that each instance is preceded by 1, 2, 3, or 4 in the previous column.

We first do this by hand, to see how it works, and then by using an R function. (You may wish to revisit Section 2.6, where we used R to implement a first-order Markov process that generated a sequence having predetermined dinucleotide frequencies.)

[a.] Initialize a matrix to hold counts, `matrix0`.

```
> matrix0<-matrix(nrow=4,ncol=4,rep(0,16))
> matrix0
     [,1] [,2] [,3] [,4]
[1,]    0    0    0    0
[2,]    0    0    0    0
```

```
[3,]      0      0      0      0
[4,]      0      0      0      0
```

[b.] Record the following numerical counts in `matrixn (4×4)`: `matrixn[1,1]` is the number of times that a 1 in the first column of the gata matrix is followed by a 1 in the second column for the whole list of binding sites. By inspection of Table elsewhere in this chapter (in numerical from), we see that this happens one time (row 40 of gata), so the [1,1] element is 1.

`matrixn [1,2]` is the number of times that a 1 in the first column of the gata matrix is followed by a 2 in the second column. By inspection of Table elsewhere in this chapter, we see that there are no instances of this at all, so we record 0 for the [1,2] element of `matrixn`. We do one more entry by hand. `matrixn` [4, 3] is the number of times that a 4 in the first column of the gata matrix is followed by a 3 in the second column.

There are 14 occurrences of 4 followed by 3, so we record 14 in matrixn element [4,3]. Once we have calculated all of the matrix elements in matrixn, we need to convert the numerical counts to probabilities to create the transition matrix, matrixp. Recall that, for transition matrices, the four probabilities in each row must sum to 1.0.

So if the elements in the ath row of matrixn sum to *N*, then the probability value at each element of the transition matrix, matrixp, is $(n_{ai} + 1)/(N + 4)$, $i = 1, 2, 3, 4$, if we apply the small sample correction that we used earlier. Clearly, all of these operations are much too tedious for hand calculation; therefore, we write a function in R to perform the job for us:

```
transmatp<-f unction (sites,col,matrix0){
#sites = numeric matrix of n binding sites, w positions
#col = column that transition matrix produces
#matrix0 = matrix of counts for n = col, initialized to 0
matrixn<-matrix0
for(i in 1:length(sites[,1])){
   j<-sites[i,(col-1)]
   matrixn [j,sites[i,col]]<-matrixn[j,sites[i,col]]+1
#Change counts to probabilities
matrixp<-matrixn
matrixp<-matrixp+l #Adds 1 to every element
for(i in 1:4){
   matrixp[i,]<-matrixp[i,]/sum(matrixp[i,])
   #Denominator=sum(matrixn[i,])+4
   }
return (matrixp, matrixn)
> tmp<-transmatp(gata,2,matrix0)
```

The result of this calculation is shown as an R list object below:

```
> tmp$matrixp
          [,1]        [,2]        [,3]        [,4]
[1,] 0.06060606  0.03030303  0.8484848  0.06060606
[2,] 0.16666667  0.16666667  0.5000000  0.16666667
[3,] 0.20000000  0.20000000  0.4000000  0.20000000
[4,] 0.04761905  0.09523810  0.7142857  0.14285714
>    tmp$matrixn
      [,1]  [,2]  [,3]  [,4]
[1,]  1     0     27    1
[2,]  0     0     2     0
[3,]  0     0     1     0
[4,]  0     1     14    2
> matrix2<-tmp$matrixp #Identical to $matrixp above
```

`matrix2` is the transition matrix that generates the probability distribution for the second position given the probability distribution for the first position given in vectors. For completeness, we present all of the other matrices calculated for the remaining positions. These are computed in the same way as for `matrix2`.

```
> matrix3
          [,1]       [,2]       [,3]       [,4]
[1,]  0.2000000  0.2000000  0.4000000  0.2000000
[2,]  0.4000000  0.2000000  0.2000000  0.2000000
[3,]  0.9375000  0.0208333  0.0208333  0.0208333
[4,]  0.5714286  0.1428571  0.1428571  0.1428571
> matrix4
          [,1]       [,2]       [,3]       [,4]
[1,]  0.0384615  0.0192308  0.0576923  0.8846154
[2,]  0.2500000  0.2500000  0.2500000  0.2500000
[3,]  0.2000000  0.2000000  0.2000000  0.4000000
[4,]  0.2500000  0.2500000  0.2500000  0.2500000
> matrix5
          [,1]       [,2]       [,3]       [,4]
[1,]  0.2000000  0.2000000  0.2000000  0.4000000
[2,]  0.2500000  0.2500000  0.2500000  0.2500000
```

```
[3,]  0.3333333  0.1666667  0.1666667  0.3333333
[4,]  0.7200000  0.0800000  0.0600000  0.1400000
> matrix6
         [,1]       [,2]       [,3]       [,4]
[1,]  0.4000000  0.07500000 0.4250000  0.1000000
[2,]  0.4285714  0.14285714 0.1428571  0.2857143
[3,]  0.1666667  0.16666667 0.5000000  0.1666667
[4,]  0.5000000  0.08333333 0.2500000  0.1666667
```

These five matrices, together with vectors, provide the Markov chain representation of the `GATA-1` transcription factor binding sites listed in Table 4.3 If we wanted, we could use the initial probability distribution vectors and these transition matrices to simulate `GATA-1` binding sites in a manner similar to the simulation described already.

One more point needs to be made before we move on: How do we *score* a sequence given a representation of the type shown above? Scoring proceeds in a fashion analogous to the approach used for the PWMs. Convert the probabilities in the initial probability distribution into $\log_2(p_{ai}/q_a)$ using q_a for the genome as a whole (which means that we are comparing sites to an iid model).

Convert the probabilities in the transition matrices into $\log_e (p_{ij}/q_{ij})$ values using $q_{ij} = qj$ (also assuming an iid model). Let's take a specific example using the variable names and notation employed in the example above. Suppose the sequence to be scored is TGATAA.

The score s_1 for the first position is $\log_2(p_{al}/q_{d},$ selecting the vector element corresponding to position 1 in the initial probability distribution vector, vectors. Since we have coded A, C, G, and T, respectively, as 1, 2, 3, and 4, the element that we require is vectors [4] (corresponding to T). The score is s1 = $\log_2$(vector1[4]/q_T).

The score for the second position is taken from the transition matrix corresponding to position 2, matrix2. The element of matrix2 whose row corresponds to the base that appears at position 1 (T) and whose column corresponds to the base that appears at position 2(G) provides the information needed to compute s_2 = $\log_2$(matrix2[4, 3]/q_G). Scores at the other positions are calculated similarly.

Shannon's Entropy

Shannon's entropy is a concept drawn from information theory and signal processing. It measures the degree of uncertainty associated with a set of possible outcomes. Given a discrete random variable X with J outcomes $x_1, x_2, \dots, x_J$ having probabilities $p(x_1), \dots, p(x_J)$, respectively, Shannon's entropy is conventionally defined as

$$H(X) = -\sum_{j=1}^{J} p(x_j)\log_2 p(x_j), \tag{4.6}$$

where $p(x) \log_2 p(x) = 0$ if $p(x) = 0$.

Now suppose we are looking at a sequence $X_1, X_{2....}$ of iid bases. The entropy of the ith position ism

$$H(X_i) = -\sum_{a\in(A,C,G,T)} p(a)\log_2 p(a). \tag{4.7}$$

If the outcomes are equally probable, $p(a)$ = 1/4 and then $H(X_i) = \log_2 4 = 2$. The uncertainty in this case can be represented by two bits. One bit accounts for the possibility that a base is a purine (1) or pyrimidine (0), and the other bit specifies which purine (1 or 0, given that the first bit was 1) or which pyrimidine (1 or 0, given that the first bit was 0) is present. If we know that the base at a position is actually an A, for example, then $p(a) = 1$ if $a = A$, and all other $p(a) = 0$. In this case, $H(X_i) = 0$ *(i.e.,* the uncertainty is zero).

The *information* is a measure of how much the entropy is reduced after a "signal" (in this case, resulting from natural selection at a position within a region of DNA) has been received:

$$I(X_i) = H_{before} - H_{after}. \tag{4.8}$$

If the bases at a position had initially been distributed with equal probabilities, and over the course of time had evolutionarily been fixed to be an A, then using the calculation above, we see that the information at that position is now $I(X_i)$ = 2 bits.

Another measure of information content (which becomes identical to Shannon's entropy when the probability distribution is uniform) is the *relative entropy,* or Kullback-Leibler distance. We start with a set of aligned sites, where, as before, the probability of finding base *a* at position *i* is given by p_{ai}. Let q_a represent the background distribution of bases in a genome or in a random model of a genome. The relative entropy is then defined as

$$H(p_i \,\|\, q) = \sum_{a\in(A,C,G,T)} p_{ai}\log_2(p_{ai}/q_a) \tag{4.9}$$

The ll* indicates that a *distribution pi is* being examined relative to a distribution *q*. Using the log-odds scoring scheme that we employed above, we can calculate the expectation for the score at any position in the sequence:

$$\mathbf{E}(S_i) = \sum_a p_{ai}s_a = \sum_a p_{ai}\log_2(p_{ai}/q_a). \tag{4.10}$$

In this expression, the P_{ai} correspond to the distribution at position *i*. We see that the expected value for the score at any position is the same as the relative entropy at that position. This provides an intuitive view of the nature of relative entropy.

Why have we introduced relative entropy and information content? The major reason is to help define the *extent* of signals. Remember that an experimental measure of the extent of a binding site comes from footprinting experiments. But how can we decide the extent of a signal in probabilistic terms given a set of sequences containing that signal?

One approach is to plot information content as a function of position along the set of sequences. The binding site boundary is taken as that position where the relative entropy exceeds a particular threshold (e.g., the value of relative entropy exceeded by no more than 5% of the positions in iid sequences).

The informational perspective has led to a depiction of signals as sequence logos (Schneider

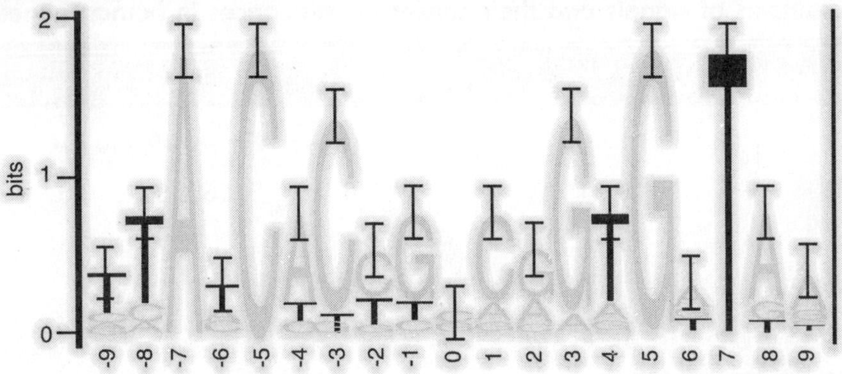

Figure 4.6: [This figure also appears in the color insert.] DNA sequence logo of lambda operator sites. The illustration includes one more position to the right and left of the sequences shown in Table 4.1. Heights of each letter A, C, G, *or* T *(in bits) represent the contribution of each letter to the information encoded at each position.*

and Stephens, 1990). A sequence logo is a graphical representation of a signal in which the total height at each position corresponds to the relative entropy. The height of each *letter* at each position is calculated by multiplying the relative entropy at that position by the frequency of the corresponding letter.

A logo representing the lambda operator sites is shown in Figure 4.6. Logos are better visual representations than consensus sequences because the logos indicate both the amount of relative entropy at each position (the height of the stack of four letters) and the relative contribution of each base (the relative height of the letter) at that position.

Molecular Signals in Eukaryotic Cells

Signals are important components of eukaryotic genes. Examples of promoter signals, transcriptional signals, and mRNA processing signals are provided in Table 4.4. As stated already, the consensus sequences by writing down the actual base that is overwhelmingly preferred at each position. In Table 4.4, we expand the representation by using single-letter codes to describe different *combinations* of bases.

For example, (A/T) is represented by W, and (A/G) is represented by R. With this notation, the GATA-1 consensus can be more flexibly represented as WGATAR. Lowercase letters are sometimes used to depict a weak preference for a particular base, with all others appearing at lower frequencies. The last two positions in the 3' splice signal are designated by this notation.

These types of designations are not appropriate for gene finding because we need to be able to generate a score. Positional weight matrices (also called profiles in this context) or Markov chains are effective representations of these signals.

Regulatory sequences control the timing and levels of gene expression by binding specialized transcription factors. The regulatory regions of genes can extend tens of thousands of base pairs upstream of the transcriptional start site, and some sequences (enhancers) can be found either upstream or downstream of the promoter (sometimes within introns or downstream of the polyadenylation site).

Table 4.4: Examples of signals and their consensus sequences in human genes.

Site	Consensus																
	-3	-2	-1	0	+1	+2	+3	+4	+5	+6	+7	+8	+9	+10	+11		
TATA box	S	T	W	T	W	W	A	W	R	S	S	S	S	S	S		
	-4	-3	-2	-1	0	+1	+2	+3	+4								
CAAT box	R	R	C	C	A	R	K	S	R								
	-2	-1	0	+1	+2	+3	+4	+5									
Cap site	K	C	W	S	Y	S	S	S									
	-9	-8	-7	-6	-5	-4	-3	-2	-1	0	+1	+2	+3				
Start site	S	S	S	S	S	C	R	M	C	A	T	G	R				
	-3	-2	-1	0	+1	+2	+3	+4	+5								
5' splice signal	N	A	G	G	T	R	A	G	W								
	-15	-14	-13	-12	-11	-10	-9	-8	-7	-6	-5	-4	-3	-2	-1	0	+1
3' splice signal	Y	Y	Y	Y	Y	Y	Y	Y	Y	Y	Y	N	Y	A	G	g	t

Weaker preferences are indicated by lowercase letters. Locations with preferences for particular combinations of bases are indicated by IUPAC-IUB codes. In addition to A, C, G, T, they are R = A or G (purine), Y = T or C (pyrimidine), S=CorG,W= A or T, K = G or T, M = A or C, and N = any base.

Probabilistic descriptions can be used as an aid to locating these binding sites (e.g.,the GATA-1 binding site). Numbers and locations for k-tuples corresponding to signals such as GATA-1 may provide similar information about signal content.

However, many binding sites are quite short, and therefore they are likely to occur by chance in long lengths of sequence. Our probabilistic description has not taken into account the *distributions* of the sites along the DNA although typical values for locations of GC boxes, CAAT boxes, and TATA boxes relative to the transcriptional start sites are known.

The spacing (or alternatively density) of candidate transcription factor binding sites is additional information that can be used for assessing whether they have been correctly identified.

This sort of information can be incorporated in *hidden Markov models* (not discussed in this book). Clearly, signals are only one set of features that describe eukaryotic genes, but they are an important component of gene-finding tools.

Scoring Technique

We have described how to represent signals in probabilistic terms and how to produce scores for instances of any signal. The scores can be used to classify any candidate string into one of two categories: sites or nonsites. There are two types of errors that can result from this procedure. Let the *null hypothesis* $\mathcal{H}$ be that the sequence to be tested is a site. A *Type I error* is one that classifies a site as a nonsite (i.e., a false negative, rejecting $\mathcal{H}$ when it is true). A *Type II error* is one that classifies a nonsite as a site (a false positive, failing to reject 7-l when it is false):

$\mathcal{H}$ is:	Assigned class is: True	False
True	correct	Type I error
False	Type II	error correct

The performance of a classification method is often described in terms of *sensitivity (Sn,* the proportion of actual features detected) and *specificity (Sp,* the proportion of predicted features that are real). In terms of Type I and Type II errors, we have

$$Sn = 1 - \mathrm{P} \quad \text{(Type I error)},$$

$$Sp = 1 - \mathrm{P} \ \text{(Type II error)}.$$

If #TP is the number of true positive predictions, #FP is the number of false positive predictions, #TN is the number of true negative predictions, and #FN is the number of false negative predictions, then

$$Sn \approx \frac{\#TP}{\#TP + \#FN}, \quad Sp \approx \frac{\#TN}{\#TN + \#FP} \tag{4.11}$$

Given a training set of sites, scoring produces a distribution of scores for these sites, shown as an idealized distribution, A, in Figure elsewhere in this chapter. Given a set of sequences that are not sites, the same scoring method produces another distribution, B, that in general overlaps the first one. We can place a *cutoff score* C on the graph and classify any sequence with score $S \geq C$ as a site and classify any sequence having $S < C$ as a nonsite.

The area of distribution A to the left of C represents the fraction of false negative assignments, and the area of distribution B to the right of C is the fraction of false positive assignments. We may move the cutoff lower in an effort to avoid "losing" actual sites, but we do so at the expense of a greater number of false-positive assignments. We can move the cutoff to the right to reduce the false-positive assignments, but then we lose sensitivity.

In actual practice, we may be scanning an entire genome for a limited number of sites (e.g., one site per 10^3 to 10^4 bp). In this case, the number of nonsites examined is equal to $(G - w) - n$, where G is the number of base pairs in the genome, *w* is the length of the site, and *n is* the number of actual sites. Ordinarily, *G >> n, so* many more nonsites than sites are scored.

The result is that if we set the cutoff midway between the scores corresponding to the maxima of the two curves, the *number* of false positive assignments will vastly exceed the *number* of actual sites. For this sort of application, we might accept a lower sensitivity to achieve higher specificity. If, however, we are only analyzing the upstream region of a particular gene of interest, then we might move the cutoff to the left to increase the sensitivity at the expense of specificity.

We make the trade-offs necessary for the problem at hand. We illustrate these principles in

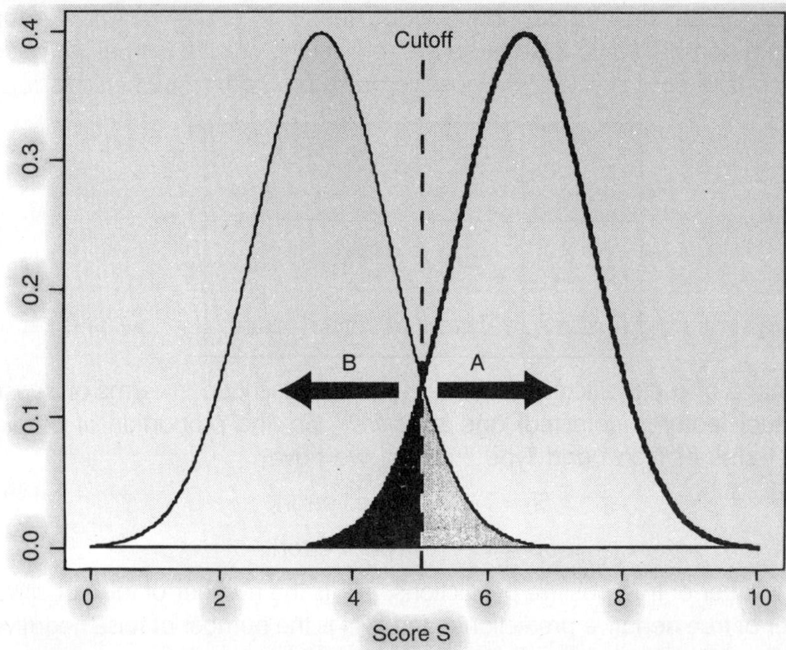

Figure 4.7: Idealized illustration of the classification ot objects of types A and B based upon their score distributions. An arbitrary cutoff score midway between the maximum scores for the two object types is indicated. Objects B having scores larger than the cutoff score (light shaded area) are incorrectly classified as A, while objects A having scores below the cutoff score (dark shaded area) are incorrectly classified as B. If objects are classified to answer the question of whether they are objects of type A, then the light shaded area corresponds to false positive assignments and the dark shaded area corresponds to false negative assignments. The cutoff score can be adjusted up or down (arrows) to minimize false positive assignments (with correspondingly greater false negative errors) or to minimize false negative assignments (with correspondingly greater false positive errors).

Computational Example 4.3. Another approach to assessing a classification scheme uses the *false discovery rate (FDR),* defined to be the expected proportion of false positive features among those features called positive. This is given by

$$FDR \approx \frac{\#FP}{\#FP + \#TP}.$$

For details on the calculation of the FDR, see, for example, Storey and Tib-shirani (2003).

Computational Example 4.3: Classification of sites using GATA-1 PWM

We use the PWM computed in Computational Example 9.1 to illustrate the classification problem. Remember that we had a training set comprising 49 sites and that these represent a sample of all possible sites. First, we employ our PWM to simulate 5000 sites and score them to determine the site score distribution implied by our PWM.

Next, we simulate a set *of* 5000 "back-ground" sequences *of* the same length as GATA-1 sites (6 bp) by using the iid model, and we determine the score distribution *of* this set. Finally, we examine FP and FN error rates at different cutoff scores.

Step 1: Simulating GATA-1 sites

Our R function makepwm produced a matrix *of* probabilities at each position as one *of* its two returned objects. This is just what we need to simulate the sites. The function to do this is:

```
simmotif<-function(pwm){
# pwm is a PWM matrix of probabilities (4xL)
L<-length(pwm[1,]) #Number of positions in the motif
motif<-rep(O,L) #Create and initialize motif vector
dna<-c(1,2,3,4) #Numeric codes for A, C, G, T
for (j in 1:L){
   motif[j]<-sample(dna,1,p=pwm[,j])
   }
return(motif)
}
```

We have used sample before, so this function should be clear to you. The only wrinkle is that for each cycle of the for loop, a different probability distribution is used, corresponding to each column of pwm. The need for these probability distributions was anticipated when we devised the makepwm function. To simulate *N* sites, we make an *N* ×6 matrix and then run simmotif *N* times. Remember that the pwm argument that we need is in tmp as tmp$pwm.

```
> N<-5000
> gata.motifs<-matrix(nrow=N,ncol=6)
> for(i in 1:N){
+       gata. motifs [i, ] <-simmotif (tmp$pwm)
+       }
```

We check some of the simulated sites to make sure that they look reasonable (compared with the gata object).

```
> gata.motifs[1:5,]
     [,1] [,2] [,3] [,4] [,5] [,6]
[1,] 4    3    1    4    1    3
[2,] 1    3    1    4    1    3
[3,] 4    3    1    4    1    3
[4,] 1    3    1    4    4    3
[5,] 1    4    1    4    3    3
```

Step 2: Scoring simulated sites

We now wish to generate the score distribution for the simulated sites. What would we predict for this distribution, based on the training set scores? We can employ the calcscore function operating on each row of gata. motif s to obtain the scores:

```
> gata.motifs.score<-rep(O,N)
#vector to hold the results of computation
> for(i in 1:N){
+ gata. motifs. score [i] <-
+ calcscore(gata.motifs[i,],log2pwm)
}
```

We can compare this set of scores to the training set by looking at the means:

```
> mean(gata.motifs.score)
[1] 5.094539
> mean(gata.score)
[1] 6.07064
```

At first, it may seem strange that the simulated motif scores have a smaller average than those of the training set, but things become clearer when we examine the score distributions. We produce a histogram of gataf motifs . score in the usual manner, with the result shown in Figure 4.8 (solid lines: code supplied later).

Notice that our simulated sites have a score distribution that is skewed left (as is the training set distribution), and note that the central part of the distribution has been filled in. This accounts for the lower mean value for the scores. Remember that the training set is a sample of the set of actual sites.

As an exercise, you can test how much individual samples vary by repeatedly sampling 49 of the simulated set of 5000 motifs, recomputing the corresponding PWMs, and plotting the resulting scores for each sampling.

Step 3: Simulating the background

We now simulate sequence strings of length N according to the iid background model. The function to do this is:

```
simbg<-function(bg,L){
#bg is a vector of probabilities for A, C, G, T (1x4)
#L = length of sites to be simulated
seq<-rep(0,L)
dna<-c(1,2,3,4)#Numeric codes for DNA
seq<-sample(dna,L,replace=T,p=bg)
return(seq)
}
```

We apply this function 5000 times in the same manner as we did with simmotif, and examine the resulting score distribution.

```
> for(i in 1:N){
+ back.sim[i,]<-simbg(bg,6)
+ }
> back.sim.score<-rep(0,N)
> for(i in 1:N){
+ back.sim.score[i]<-calcscore(back.sim[i,],log2pwm)
}
```

Now we examine the score distribution for the background and compare it with the distribution for the simulated motifs:

```
> hist(back.sim.score, prob=T,xlim=c(20,20),
+ ylim=c(0,0.25),lty=2)
> hist(gata.motifs.score, xlim=c(-20,20),
+ prob=T,lty=1, add=T)
```

The distribution of the background scores is more nearly symmetric, and the mean is at -6.56. In Computational Example 4.2, we asked what we might have expected the mean score for the

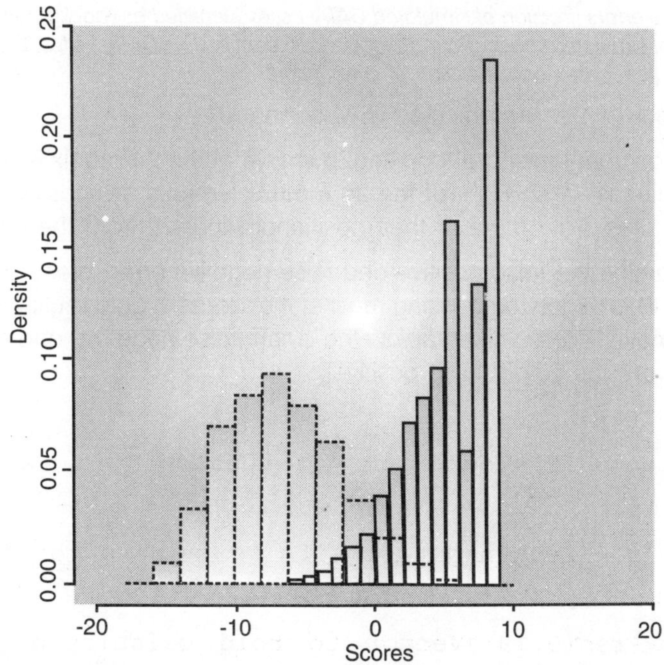

Figure 4.8: Score distributions for GATA-1 sites (solid lines) simulated from the PWM corresponding to sites in Table elsewhere in this chapter and for "background" sequences (broken lines) simulated from an iid model using the base composition of human DNA.

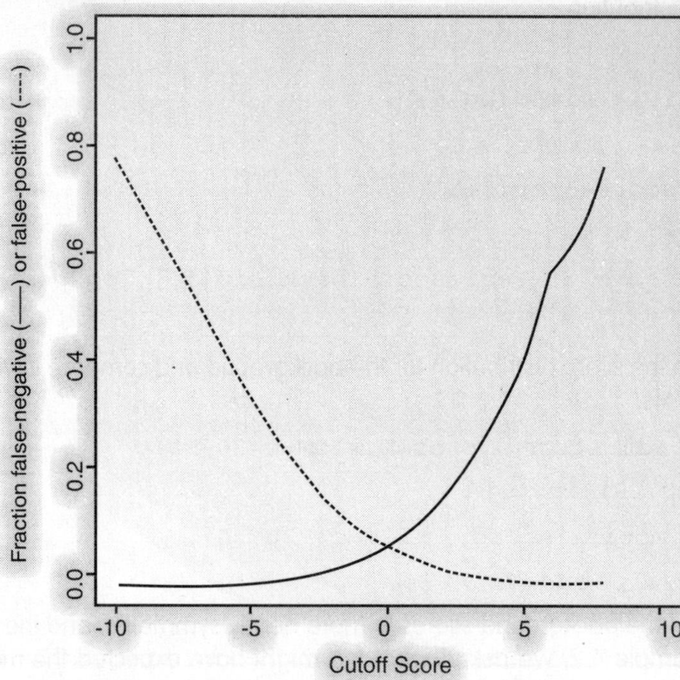

Figure 4.9: False negative errors (fraction of simulated GATA-1 sites identified as "background": solid line) and false positive errors (fraction of simulated "back-ground" sequences identified as GATA-1 sites: broken line) for simulated motifs and background sequences as a function of cutoff score.

background distribution to be. In particular, why is it not 0?

The answer comes from looking at tmp$log2pwm: 16 of the 24 matrix elements are negative, and their average value is -2.32. Eight of the 24 matrix elements are positive, and their average value is 1.14. The iid sites draw more of their positional scores s_i from the negative values.

Now we can examine the false positive and false negative error rates. To do this, we specify a set of cutoffs from -10 to +8 (encompassing most of the scores in gata.motifs. score) and calculate the fraction of gata.motifsf score values *below* the cutoff (false negative) and the fraction of back. sim. score values *above* the cutoff (false positive).

```
cutoffs<-c(-10:8)
false.neg<-rep(0,19)#Vector to hold calculated values
for(i in 1:19){
   false.neg[i]<-length(gata.motifs.score[gata.motifs.score[]
         <cutoffs[i]])/N}
   false.pos<-rep(0,19)#Vector to hold calculated values
         for(i in 1:19){
   false.pos[i]<-length(back.sim.score[back.sim.score[]
         >cutoffs[i]])/N}
```

```
> plot(cutoffs,false.neg,type="1",xlim=c(-10,10), ylim=c(0,1))
> points(cutoffs,false.pos,type="1",lty=2)
```

The plot in Figure 4.9 shows the expected opposite behavior of the false positive and false negative error rates. If we were to choose a cutoff score of 0.0, we would have a false negative rate of 6.6% and a false positive error rate of 6.8%. This latter value is deadly for scanning long sequences.

For example, if we were to examine 100,000 base pairs upstream of a particular gene, we would incorrectly predict approximately 6800 GATA-1 sites (assuming that human DNA is adequately represented by the iid model).

As mentioned above, to avoid the "noise" associated with the false positive error, we might choose to increase the cutoff and accept the concomitant false negative errors. Ultimately, we might choose a more sophisticated probabilistic model that includes the expected site distribution (e.g., a hidden Markov model), but this is beyond the scope of this chapter.

Chapter 5

Measuring Antibodies

Protein and antibody *microarray* measurements can be performed in a number of different ways and formats. Reverse arrays wherein many samples, typically cell lysates or tissue sections, are probed simultaneously by a single antibody, are a powerful approach to characterize patient-derived, primary tumor samples, especially when *employed* in *conjunction* with laser capture microdissection. Reverse arrays have also been developed to study signal *transduction* pathways.

Arrays of antigens for the *detection* of *auto-antibodies* have had the biggest impact so far in clinical environments. Two-color approaches that simultaneously compare the *responses* of large antibody arrays to chemically labeled reference and diseased samples continue to be powerful approaches to disease marker discovery. Closest to the enzyme-linked *immunosorbent* assay (ELISA) tests widely used in research and clinical laboratories are antibody *microarrays* using multiplexed sandwich assays.

In this approach, different capture antibodies are *spatially* arrayed on a surface and a variety of detection schemes can be employed, as has been reviewed recently. All achieve multiplexing capability by applying mixtures of detection antibodies to an array of *immobilized* capture antibodies. In this sense, antibody *microarrays* are simply *multiplexed sandwich assays.*

Compared with a single assay in a microtiter plate, the concentrations and amounts of reagents involved are, however, quite different, and it is believed that for this reason, microarray measurements generally fall into the ambient analyte regime, meaning that the measurement does not deplete the *analyte* in the assay volume.

Because quantitative assays require *calibration* curves, we have developed a slide holder that divides a 1-inch × 3-inch slide into 48 separate wells (4 rows by 12 columns). The well-to-well spacing, 4.5 mm, is the same as is used in the standard 384-well microtiter plates. Each well contains the same 12 ×12 antibody microarray with 144 spots (triplicate spots for 48 antibodies).

We use eight wells per slide to simultaneously generate eight-point calibration curves for all analytes, using mixtures of purified antigens. A single slide holder contains four slides with 48 identical antibody microarrays, and on each slide four dilutions of ten samples are measured against an eight-point calibration curve. Unique to our platform is the use of *resonance* light-scattering (RLS) particles, in essence 78-nm diameter colloidal gold particles coated with an anti-biotin antibody, for the detection of the *biotinylated* primary *antibodies* by signal *generation* through light *scattering*.

As we have shown recently, RLS can detect as few as 400 binding events on the surface of a 24-plex mouse antibody array. The light-scattering intensity of these *colloidal* gold particles is such that single particles can be seen under a light *microscope*. RLS staining has also been shown to *dramatically* increase the number of detectable gene messages in microarray gene *expression* profiling experiments.

The high sensitivity of the stain is based on the *light-scattering* properties of colloidal gold particles, and others have used colloidal gold for the *ultrasensitive*, almost single-molecule detection of antigens after PCR amplification of DNA *barcodes* on *nanoparticles*. As will be illustrated below, RLS staining of *biotinylated* reagents is a fast, simple, and robust *procedure* that yields highly sensitive *multiplexed* sandwich assays.

We routinely use RLS particles with two antibody arrays for human *plasma* proteins with 43 and 45 *analytes* respectively, as well as a 24-analyte antibody array for mouse plasma proteins. In addition, we are currently evaluating RLS staining procedures for microarray assays that study protein-DNA *interactions* as well as profiling of proteases using PNA-encoded probes.

The technical challenges of array measurements are numerous. With the RLS particles, we have found a very sensitive and robust staining method. Manufacturing of antibody arrays is a *process* that needs a strong commitment to details and tight process control, but good results can be obtained. Assay *optimization*, and availability and *reliability* of the *reagents*, will continue to be a challenge.

In our experience, a maximum of 50 analytes can be multiplexed before cross-reactivity issues and antigen-independent antibody cross-talk will severely limit and affect the *dynamic* range and sensitivity of many sandwich assays on the array. There seems to be an intrinsic *barrier* for the degree of multiplexing. Needless to say, keeping a collection of 50 capture antibodies, 50 antigens, and 50 detection antibodies fully functional on a day-to-day basis is a *formidable* task.

Smaller sets of reagents can be used over extended periods with good results, but clearly better antibody reagents are urgently needed. Because of the multiplexed nature of these measurements, the *diluent* and sample preparation methods cannot be optimized for best performance of all analytes. As has recently been shown, the choice of anticoagulant seems to have an effect on the results for specific *analytes*, and therefore compromises have to be made with respect to sample preparation and assay optimization.

REQUIREMENTS

Procedure

1. GAPS II or UltraGAPS amino-silane coated glass microarray substrates are purchased from Corning, Inc..
2. A Genemachine Accent printer from Genomic Solutions (Ann Arbor, MI) is equipped with

a *humidity* control unit and an in-house built cooling unit for the print plate. A larger blotting pad is also installed.

3. Four SMP3 steel quill pins (Telechem, Sunnyvale, CA) are typically used in parallel.
4. Monoclonal and affinity purified *polyclonal* antibodies to be *immobilized* on the *microarrays* are purchased from a variety of vendors and used without further purification. Antibodies are stored according to the suppliers' recommendations, typically in liquid form at 4°C.
5. Capture antibodies are diluted to a final concentration of 0.25 mg/mL into a printing buffer that is optimized to the hydrophobicity of the surface as well as the printing conditions. The 5X printing buffer is available from Invitrogen (Carlsbad, CA).
6. Casein blocking solution (1%) for blocking of antibody arrays in preparation for storage is purchased from Pierce, and filtered through a 0.22-μm filter at room temperature. The filtrate is stored at 4°C.

Sampling

Complete, ethylenediamine tetraacetic acid (EDTA)-free protease inhibitor cocktail tablets are purchased from Roche, Inc.

Biotinylated Detection Method

1. Biotinylated detection antibodies are purchased from a variety of vendors and used without further purification at concentrations between 0.013 and 0.5 μg/mL, as determined during assay optimization. *Detection* antibodies are stored according to the suppliers' *recommendations,* typically in liquid form at 4°C.
2. Other detection antibodies are *biotinylated* using NHS-PEO4-biotin (Pierce) according to *protocols* provided by the *manufacturer.*
3. For the standard curves, antigens are purchased from a variety of vendors and used without further purification. *Aliquots* of the *antigens* are flash frozen in liquid nitrogen and stored at -80°C.
4. Antigens are diluted into casein blocking solution. Dilutions for calibration curves are freshly *prepared* for each assay from flash-frozen stock solutions. *Casein blocking* solution (1%) is purchased from Pierce.
5. *Bovine serum albumin, sodium chloride,* and Tween-20 are purchased from Sigma. 6. Goat IgG is purchased from Jackson ImmunoResearch Laboratories, Inc.
7. Phosphate-buffered saline (PBS) stock solution (10X PBS) is purchased from Invitrogen.
8. Complete, EDTA-free protease inhibitor *cocktail* tablets are purchased from Roche, Inc.
9. RLS particles and archiving solution are purchased from Invitrogen (Carlsbad, CA).
10. A slide holder to hold four *microarray* substrates was developed in house. The slide holder has 192 wells with 4.5-mm spacing separating each of the four microarray substrates into 48 *physically* separated 3 mm × 3 min areas, similar to the footprint of a standard 384-well microtiter plate. The slide holder is *commercially* available through GNF's Department of Engineering.

11. A 384-plate washer (Tecan, Research Triangle Park, NC) was modified to allow for automated and simultaneous washing of all 192 wells in the slide holder. *Generally*, three cycles of 400 µL wash volume per well in the overflow wash mode are used. The wash solution consists of 0.005% Tween-20 in 1X PBS.
12. During all incubation steps the slide holder is shaken on a Tecan orbital shaker at 1500 rpm, changing directions (clockwise to counter-clockwise) every 5 min.

Image and Data Analysis

1. A 16-bit charge-coupled device (CCD) camera-based scanner optimized for the detection of *resonance* light scattering is available from Invitrogen (Carlsbad, CA).
2. Images are processed with ArrayVision, Version 8.0 (Imaging Research, St. Catharines, Canada). The software is provided with the scanner.
3. For *quantification* of the images, an Excel macro was developed. Sample files will be made available upon request.

PROCEDURES

Antibody microarray measurements as described here are conceptually simply *multiplexed* sandwich assays in *microarray* format. Rather than obtaining a concentration readout for a single protein of interest, tens of proteins are measured simultaneously with the same small amount of sample. In our format, 15 µL of sample are sufficient for the quantification of up to 45 different analytes. However, in practice, microarray measurements are quite complex.

This starts with the manufacturing of microarrays that requires tight environmental and process control, and great care must be taken in optimizing printing conditions to the surface properties of the microarray substrate before acceptable results can be obtained. Because of the multiplexed nature of the assay, compromises must be made in how samples can be handled.

The sandwich microarray assay itself roughly follows the scheme of an ELISA plate assay, with a sample incubation period, a detection antibody incubation period, and a staining step, with additional blocking and washing steps in between. Staining with RLS particles follows a simple procedure that results in a highly sensitive and generic signal generation method for *biotinylated* reagents frequently used in microarray applications, such as *protein, antibody, oligonucleotide,* and *cDNA microarrays.*

In contrast with plate assays, microarray measurements require an image to be taken and analyzed. We will also describe how we calibrate the assays and analyze the actual protein concentrations. Last but not least, the process of developing and optimizing a panel of assays into an antibody microarray is-discussed under the head optimization, and likely constitutes the most challenging aspect of antibody microarray measurements.

Processing of 16 microarray slides in four slide holders (with a total of 768 microarrays), from mounting the slides, performing the assay, to archiving all slides, requires approx 5 h.

Preparing Print Plate

1. To avoid transferring particulate matter to the print plate, capture antibody stock solutions

are spun for 30 min at 30,000g (at 4°C).

2. *Capture* antibody stock solutions are diluted with 1X PBS and 5X printing buffer to a final antibody *concentration* of 0.25 mg/mL.
3. 8 µL of each antibody printing solution is transferred to a 384-Well Round Bottom Low Volume plate (Corning, part number 3676).
4. After all samples have been added, the print plate is sealed with an aluminum adhesive lid (Beckman) and spun at 500g for 30 s to remove air *bubbles* and to settle the samples to the bottom of the wells.
5. The print plate is stored in the cold-room for at least 24 h before the printing operation begins.

Printing of Microarrays

1. The microarrayer is situated in a dedicated humidity- and temperature-controlled room that is supplied with high-efficiency particulate arresting (HEPA)-filtered air generating a near clean-room environment. The room is set to 70% relative humidity and 20°C. The humidity control on the microarrayer is set to 75% relative humidity.
2. The array printer is equipped with a temperature-controlled stage used to keep the print plate at a constant temperature of approx 15.5°C throughout the printing process. The temperature for the sample plates is set just slightly above the dew point to prevent evaporation and *concentration* of the samples to be printed as well as to prevent *condensation*.
3. SMP3 pins deposit approx 0.6 nL of solution in a single spot. 50 spots are *preprinted* onto a blotting pad to remove excess material from the outside of the pin. Antibodies are printed in triplicates with a spot-to-spot distance of 200 µm. 48 identical subarrays with 4.5-mm array-to-array spacing are printed with four pins in parallel.

   Each pin is refilled after 72 spots have been printed onto two microarray slides (two rows of 12 subarrays with triplicate spots per subarray). The print plate is allowed to equilibrate to the stage temperature while basic printer maintenance, such as removing dust, changing rinse and cleaning solutions, and cleaning blotting pad, is performed.
4. Print pins are cleaned and inspected under microscope.
5. The sonication station is filled with fresh, deionized water (Millipore).
6. The print pins are inserted into the proper locations in the print head.
7. In the final step, the array substrates are mounted on the print deck, the cover is removed from the print plate, and the printing program is started.

Postprinting Process

1. Microarrays are kept on printer for at least 1 h after printer has finished spotting the arrays.
2. To block printed slides with casein solution as preparation for storage, the antibody microarray slides are removed one by one from the printing deck, and gently and slowly

lowered with a pair of forceps into a slide-mailer filled to three-quarters with blocking solution.

3. After 1 h room temperature incubation, remove slides from the slide-mailer one by one and place each slide in its own 50-mL centrifuge tube (Falcon tube).
4. To remove the excess blocking reagent from the slides, centrifuge the slides for 1 min at 285g.
5. Remove slides from the 50-mL centrifuge tubes and place them into a clean *slidemailer*. Store slides at room temperature in a *desiccator* until they are needed.

Sampling

1. Our collaborators collected up to 150 μL of blood from mice via retroorbital eye bleeds into Sarstedt Microvette 200 K-EDTA RB blood tubes (Fisher Scientific) or into *heparin-coated* collection tubes. Blood cells were separated by *centrifugation*, and *plasma* samples were flash frozen in liquid nitrogen. Similarly prepared human samples were provided to us by a number of collaborators in frozen form.
2. Samples are stored as frozen aliquots at -80°C.
3. After *thawing*, the samples are *spun* for 30 min at 30,000g at 4°C to remove precipitates and *aggregates*.
4. Complete, EDTA-free protease inhibitor *cocktail* is applied at a concentration of one tablet per 50 mL of diluent used throughout the assays.
5. For samples that are obtained in liquid form, protease inhibitor cocktail is added as soon as possible before flash freezing aliquots of samples.

Sandwich Microarray Assays

Slide Preparation

1. Four preblocked slides are mounted in the slide holder.
2. The slides are washed once with 1X PBS/0.005% Tween-20 buffer.
3. Excess wash solution is removed by inverting the slide holder and spinning the slide holder briefly (285g, 30 s).

Incubating Sample

1. Dilutions of plasma samples and of standard mixtures for calibration curves are prepared in casein in a 384-well plate. For the most concentrated dilution, unprocessed plasma or freshly prepared mixtures of purified antigens are diluted 1:1 with 12.5 μL casein blocking solution. Subsequent dilutions are serially diluted 10-fold into casein using 2.4 μL of the first dilution.
2. Twenty *microliters* of plasma and standard samples are applied to each antibody array of the slide holder with a multi-channel *pipettor* or a *laboratory* robot.
3. After an incubation time of 1 h, all arrays are washed with IX PBS/0.005% Tween-20 buffer.

4. Excess wash solution is removed by inverting the slide hoider and spinning it briefly (285g, 30 s).
5. During all incubations, the solutions are mixed *continuously* by placing the slide holder on an orbital shaker (Tecan, Research Triangle Park, NC) in a climate-controlled room (1500 rpm, 20°C, 70% relative humidity).

Incubation With Detection Antibodies

1. A mixture of biotinylated detection antibodies diluted into casein containing 0.5 mg/mL goat IgG is applied to the microarrays for 1 h with *vigorous* mixing followed by a wash step.
2. Excess wash solution is removed by centrifugation as described previously.

Staining With RLS Particles

1. Prior to staining with RLS particles, slides are re-blocked with casein solution for 10 min.
2. In a final 1-h incubation step, RLS gold particles are applied to the arrays at a final dilution of 0.25 OD_{600} in casein containing 300 mM NaCl and 0.5 mg/mL goat IgG.
3. Excess material is removed by washing.
4. The slide holder is disassembled, the slides are dipped twice into 50 mL deionized water, and are spun dry (400g, 1 min) in Falcon tubes before coating with RLS archiving solution (Invitrogen, Carlsbad, CA).

Archiving

1. To enhance the light-scattering properties of the RLS particles as well as to reduce light-scattering effects from surface scratches and dust particles, a polymer coating is applied to the slides.
2. Slides are coated with archiving solution by very slowly dipping each slide individually into a slide mailer three-quarters filled with the archiving solution.
3. Excess material is allowed to drip off the slide. Each slide is placed in a separate 50-mL Falcon tube, and excess archiving solution is removed by centrifugation (400g, 45 *s*).

Data Analysis

1. Archived slides are typically scanned the next morning on a 16-bit CCD camera-based *scanner* optimized for the detection of RLS light scattering. Four slides at a time are loaded into the holder of the scanner.

A resolution of 10 μm is chosen. The *exposure* time is adjusted with the standard curves so as to not saturate the 16-bit detector. Typical exposure times are 0.3 s.

2. After imaging, the TIFF file is loaded into Array Vision. A mask with the array layout is applied to the image. The orientation and the dimensions of the mask are adjusted. Proper alignment of the mask with the spots of all subarrays is checked manually. Spots affected by artifacts such as dust and *streaks* are manually *flagged* and excluded from the analysis before signals are integrated. Raw median-trimmed mean (MTM) values (without subtraction of the local background) are exported to an Excel worksheet.

3. Raw MTM values are copied into the RAW data table in the analysis Excel file. Within Excel, a Visual Basic program will convert the raw data, using inputs describing the array layout, *concentrations* of *antigen* standards, and dilution factors, into concentration readouts in pg/mL for each analyte and for each dilution.
4. Analyte names, location on the array, and the concentrations of antigens used for the calibration arrays, as well as the array layout and sample dilutions, are defined in the SETUP worksheet.
5. The automatically fitted standard curves for each analyte are displayed in the MANUAL worksheet. These eightpoint calibration curves are inspected visually, and the fit can be modified by resetting and restricting the fitting parameters and removing obvious outlier data points.

   Below the calibration curve, a table lists the raw signal and the concentration value for each of the four dilutions measured for each of the 10 samples that are being analyzed per antibody microarray slide.

   The program automatically chooses the lowest dilution that gives a signal in the dynamic range of the assay. Alternatively, a dilution can be selected manually, which often allows for a more accurate comparison across many samples, especially for assays where matrix effects result in dramatically different concentration readouts depending on which dilution is used.
6. After clicking on STORE DILUTION, the results are exported to a SUMMARY data sheet that lists the concentration values for all analytes and samples of one microarray slide. For multislide experiments, these SUMMARY data sheets are combined and processed further.

Optimization

Reagents Required

1. To identify microarray-compatible sandwich pairs, several commercially available antibodies are obtained against each analyte of interest and tested in all possible combinations, with each antibody acting as capture and also as detection antibody.
2. All unmodified antibodies are arrayed out as capture antibodies and incubated with their individual biotinylated variants at 0.5 µg/mL in the absence or presence of increasing concentrations of single antigens (100 pg/mL, 10 ng/mL, 1 µg/mL).
3. Antibodies associated with high antigen-independent background signals are discarded. The most sensitive sandwich pair remaining is selected for further testing and for the final panel assembly.

Assembly of Multiplexed Immunoassays

1. For the assembly of multiplexed immunoassays, up to 45 capture antibodies targeting distinct antigens are arrayed and incubated with a mixture of the 45 corresponding detection antibodies in the absence or presence of increasing concentrations of individual antigens (e.g., 100 pg/mL, 10 ng/mL, 1 µg/mL). This experimental setup will help evaluate the concentration-dependent cross-reactivity of each analyte with each individual sandwich assay on the array.

2. In iterative rounds of testing and reshuffling, individual immunoassays are assessed for their potential to work together in assemblies of 45. Generally cross-reactive and grossly incompatible immunoassays are discarded. The remaining immunoassays are grouped into separate subpanels with a view to minimize antigen-dependent cross-reactivity. In this phase of assay optimization, generic antibody concentrations are used for spotting and detection.
3. To evaluate the antigen-independent antibody cross-talk in the final panel, each detection antibody is incubated individually at a concentration of 0.5 μg/mL in the absence of antigens with the entire capture antibody array.
4. To evaluate the antigen-dependent on-feature cross-reactivity in the final panel, the detection antibody mix is incubated with increasing concentrations of individual antigens.
5. Individual antigen titrations against the entire panel are used to adjust antigen concentrations to cover the dose-response curve of the immunoassay in an eight-point threefold dilution series.
6. Multiple standard curves are recorded with antigen and detection antibody mixtures, and the performance of individual immunoassays in the multiplexed format is compared to the assay performance using individual analytes with respect to background signal in the absence of antigen and with respect to the dynamic range of each assay.

   To reduce the effect of the increased antigen-independent back-ground in the fully multiplexed assay, the concentrations of all detection antibodies are stepwise lowered as far as possible.
7. Finally, test measurements are performed with plasma samples to fine-tune the standard concentrations and panel composition as needed.
8. To a limited extent, the assay sensitivity can be enhanced by increasing the concetration of the capture antibody solution used in the printing process.

POINTS TO REMEMBER

1. Although 48 antibody samples are printed in triplicates on 12 × 12 arrays, 3 samples are reserved for a biotinylated antibody to mark the array outline for easier image analysis.
2. Many alternative substrates are commercially available, but the spotting buffer must be adjusted to the hydrophobicity of the surface and to the temperature and the humidity of the printing environment in order to obtain good-quality spots.

   In the printing process, tight humidity control is essential to prevent concentrated antibody solutions from clogging the capillary pins as well as for good spot consistency. The antibody solutions need to be free of particles, and near clean-room conditions are recommended. We compared the performance of multiplexed sandwich assays on a variety of commercially available slides (data not shown), including plain amino-silane-coated slides for noncovalent attachment, aldehyde- and amine-reactive slides for covalent attachment, as well as a number of slides coated in-house with an amine-reactive vinyl axazolone-polymer featuring different mixtures and types of copolymers.

In terms of sensitivity of the assays, we saw little difference between different substrates; proper adjustment of the printing conditions to the hydrophobicity of the surface in order to obtain good spot morphology appeared in our hands more important than the choice of attachment chemistry. Physical adsorption to amino-silane-coated slides appears to perform sufficiently well for fulllength antibodies.

3. The printer is situated in a temperature- and humidity-controlled room with HEPA-filtered air and "sticky" mats.at the entrance to provide near cleanroom conditions.
4. Despite several attempts, we have not been able to establish assays using antibody sera or unpurified antibodies as capture antibodies.
5. Care must be taken to not overly biotinylate the protein, as this may cause high background signal. Conditions are adjusted so that approximately two to four biotin molecules are conjugated to a single antibody molecule.
6. Depending on the applications, other matrices may be more advantageous.
7. The actual wash volume per well is 800 μL because the wash head was modified and features only 192 outlets. Shorter, gentler wash protocols should be tested for a given application.
8. Shaking the slide holder during incubation periods results in a more uniform back-ground. To efficiently mix an assay volume of only 20 μ.L, the shaker must be operated at > 1200 rpm.
9. Higher capture antibody concentrations will increase signal intensity, but non-specific background and smearing of spots because of over-saturation of the surface may result. Capture antibodies can be spotted at concentrations as high as 1 mg/mL.
10. Print plates can be prepared up to 2 d prior to printing.
11. Depending on the print buffer and the hydrophobicity of the microarray substrate used, other environmental conditions may produce equally good results.
12. The effective temperature in the well is closer to 18°C. With the current settings, antibody solutions will not dry up or be diluted by condensation during print runs as long as 24 h.
13. SMP3 pins will deliver approx 180 spots using the current print buffer and conditions before the capillaries need to be refilled. For most consistent spot sizes, at least 50 spots should be preprinted and no more than 100 spots should be printed onto the microarray substrates.

    The spot diameter and shape will depend on the printing conditions, printing buffer, as well as the hydrophobicity of the microarray substrate. Using the reagents and protocols described, antibody spots are approx 120 μm in diameter.
14. SMP3 print tips are cleaned by dipping them into cleaning solution (VPI 10, V&P Scientific) followed by several rounds of drying with filtered nitrogen gas and dipping into deionized water and finally isopropanol. Pins are inspected under a microscope for dirt and mechanical damage and wear. If necessary, the cleaning procedure is repeated several times or the pin is replaced.

15. If the slides are lowered too quickly, the samples will tend to smear across the slide.
16. Blocked antibody arrays can be stored dry in a desiccator at room temperature for more than 2 mo without loss in sensitivity and performance.
17. In all sample addition steps, care must be taken not to touch the bottom of the array. Solutions are added to the walls of the well and the slide holder is briefly spun to move the solution droplets to the bottom of the well and to remove air bubbles.
18. To reduce assay variations, wash solutions are completely removed from the slide holder by centrifugation after each wash step.
19. Extra care must be taken to avoid particulate matter such as dust from settling onto the slides. Contamination of the slide surface by particulate matter will cause streaks and light-scattering artifacts. The archiving solution needs to be applied as uniformly as possible.
20. Although slides are typically dried at room temperature overnight before scanning, they can also be scanned after quick drying in an oven. Fast-drying alternative archiving solutions are available from Invitrogen. Archiving also creates a permanent record of the assay, and slides can be re-scanned several times. Slides can be recoated with archiving solution.
21. Calibration curves for each analyte are generated by fitting 24 data points (8 concentrations × 3 replicate spots) to a four-parameter logistic curve using $1/Y^2$ weighting. Unknown sample concentrations are calculated using the corresponding signal values, the curve parameters, and the dilution factors.

    The concentrations calculated for each of the three replicate spots on an array are compared to each other; if one of the spots gives a value 1.5-fold higher or lower than the median of the three, it is rejected. For the rest of the spots, an average is calculated. The rejection criterion was introduced in order to reduce the errors caused by randomly occurring spots of poor morphology, and it is applied to the measurement of samples as well as calibration standards.
22. Of the four dilutions, usually only one or two are within the dynamic range of the assay. To obtain a single concentration value, the program automatically chooses the lowest dilution that gives a raw signal at least twice the background.
23. For the development of our antibody arrays, over 350 antibodies and more than 120 protein standard preparations for over 100 antigens were tested.
24. These test arrays can contain sandwich pairs for several different analytes in parallel.
25. The successful antibody pairs can represent (a) combinations of monoclonal antibodies for antigen capture with monoclonal or polyclonal antibodies for detection, (b) polyclonal antibodies for capture with the identical polyclonal antibody for detection, or (c) soluble antigen receptor domains for capture with either monoclonal or polyclonal antibodies for detection. This early reagent selection phase screens for optimal assay reagents only on the level of the individual immunoassay. Individual detection antibodies are tested against a panel of capture antibodies targeting the identical antigen in the presence or absence of this same antigen. The cross-talk between individual immunoassays is not accounted for at this stage.

26. The antigen-independent on-feature background will increase with the number of detection antibodies in the mix. These nonspecific antibody-antibody interactions effectively limit the dynamic range as well as the sensitivity of the arrayed assays, and need to be minimized. This can be accomplished by establishing the lowest possible concentration of each detection antibody that maintains acceptable performance of each individual assay.

    The experiment to evaluate antigen-independent cross-talk of each detection antibody against the final capture antibody array is hereby used as a guide for which detection antibody is most critical in the optimization process. In the case that individual sandwich assays detect very abundant analytes or when very high-affinity antibodies are available, the detection antibody concentration, and in principle the capture antibody concentration as well, can be lowered substantially.

    Following this strategy, the sensitivity of the overall panel, and especially the sensitivity of problematic individual assays operating close to the detection limit, can be optimized in an iterative fashion at the cost of the sensitivity of other less critical assays. Detection antibody concentrations used in our panels range from 0.0 13 to 0.5 μg/mL.

Chapter 6

Genome Sequencing in Parasites

Continued funding of parasite genome sequencing projects has led to the successful *cataloging* of several hundred thousand expressed sequence tags (ESTs) from many different organisms. ESTs (short stretches of DNA sequence derived from cloned *complementary* DNAs [cDNAs]) have proved to be an invaluable starting resource for gene discovery projects in many laboratories across the world and have *contributed* to our understanding of diverse *biological* processes.

However, despite the obvious advantages of scanning large databases containing individual EST sequences for specific gene identification purposes, there is limited information to be obtained with this "*piecemeal*" approach for subsequent interpretation of gene expression studies in the modern *genomic* age.

For example, it would require years of gene sequence analysis and several rather complicated experimental gene expression techniques (yeast 2 *hybrid* protocols, *glutathione-S-transferase* (GST)-capture assays, etc.) to dissect the related components of a signaling cascade in *Brugia malayi* from the starting point of a putative *microfilarial* chemokine receptor identified via the filarial EST databases (http://circuit.neb.com/fgn/filgenl.html).

Additionally, if one were interested in the gene transcription pathways associated with sexual dimorphism in schistosome biology, one would find it difficult to *ascertain* this information via traditional experimental techniques (subtractive cDNA libraries, differential display, etc.) or querying the existing *schistosome* EST databases (http://verjol8.iq.usp.br/schisto/; http://www.nhm.ac.uk/hosted_ sites/schisto/; http://schistosoma.chgc.sh.cn).

Clearly, a more global approach to the study of gene expression is needed to relate ESTs to other gene products in complicated biological systems. With the advent and refinement of cDNA microarray technology, expression analysis and functional relationship studies can now be easily *initiated,* which, as a result, has led to a revolution in the way gene discovery projects are performed in the 21st century.

cDNA microarray platforms useful for gene expression studies are a vital component of functional genomics, as protein microarrays will be a component of functional proteomics. Functional genomics is a catch-all term used to describe a research area in which EST or genomic sequence information is used to decipher the *functional* meaning of *genomes.* A cDNA microarray containing different EST clones allows for a parallel, *simultaneous,* high-throughput screening strategy from very small amounts of starting material. Essentially, cDNAs identified through parasite EST sequencing efforts can be arrayed in spatially defined grid patterns.

High densities of cDNA "spots" can be achieved on *chemically* modified glass *microscope* slides (although membranes can be used, the advantages of glass over membrane are discussed elsewhere) through the use of a *programmable* robot. These cDNA microarrays are then used as a probe (by *convention-probes* are the spotted DNA on the array and target is the cDNA generated from each RNA pool) to assess the relative abundance of gene *transcripts* between two separate RNA target pools using Southern *hybridization* procedures.

In addition to single bimodal types of *experimental assays,* cDNA microarrays can also be used to study the longitudinal changes in parasite gene transcription in response to a *physiological* stress, drug, or immunological pressure. This chapter describes the procedures essential for the successful design of cDNA microarray experiments for use in *parasitological*-related gene expression studies.

Techniques and protocols that have been used successfully in studies on schistosome-host *immunobiology* are included, although modifications to these *protocols* used by other investigators have been mentioned where applicable. The following techniques have been modified and adapted from protocols published within the cDNA *microarray* community and are cited as such. Intentionally excluded are the line details of data mining and bioinformatics, as this in itself would require a separate *discussion* and is an ever-evolving *mathematical* and *statistical* discipline.

REQUIREMENTS

Extraction of cDNA Probe

1. Oligonucleotide polymerase chain reaction (PCR) primers to amplify cDNA probes (from any commercial source).
2. Standard PCR reagents and *thermal cycler* (capable of handling 96-well PCR plates).
3. Standard 96-well PCR plates.
4. Self-adhering PE foil (Macherey-Nagel 740676).
5. NucleoFast 96-well filter plates (Macherey-Nagel 7435000.4).
6. Vacuum manifold (e.g., Mill ipore/MultiScreen, Qiagen/QIAvac 96, Promega/VacMan 96, BioRad/Aurum).
7. Centrifuge alignment frame (Millipore MACF09604).
8. 96-well V-bottom plates (e.g., Nunc 442587).
9. 384-well V-bottom plates (e.g., Genetix X7022).
10. Bar-coded y-amino propyl silane-coated slides (Corning 40003).

11. 4X Spotting buffer: 600 mM sodium phosphate, 0.04% sodium dodecyl sulfate (SDS). 48.9 mL 600 mM Na_2HPO_4, 1.1 mL 600 mM NaH_2PO4, 200 μL 10% SDS.
12. Printing tool (and pins).
13. Robotic printer: MicroGrid II, BioRobotics (biorobotics.co.uk); GeneTac G3, Genomics Solutions (also incorporating GeneMachines and Cartesian Technologies); Genetix "Q" Array (genetix.co.uk); LabMan HDMS (labman.co.uk).
14. UV stratalinker (Stratagene).
15. 80°C oven.

Extraction of RNA

1. Trizol (Invitrogen 15596-018).
2. Rotor stator homogenizer (Tekmar TISSUMIZER or equivalent).
3. Chloroform.
4. Ethanol (EtOH).
5. RNeasy column chromatography kits (Qiagen: midi kit 75142, maxi kit 75162).
6. 3 M sodium acetate, pH 5.2.
7. 1 M Tris-HCl, pH 7.4.
8. RNase-free H_2O.
9. Filter barrier pipet tips.

Methods of cDNA Labeling

1. Superscript II reverse transcriptase, SSRT 11 (10,000 UmL) (Invitrogen 18064-014).
2. Oligo-dTV primer *(1 μM* scale synthesis) 5'-TTT TTT TTT TTT TTT TTT TTV-3' (V designates any other deoxynucleotide except dT).
3. Microcon-YM30 *concentrators* (30,000 MW cutoff) (Amicon 42410).
4. *Deoxynucleotides* (100 mM each) (*Pharmacia*: dATP 272-050, dCTP 272-060, dGTP 272-070, and dTTP 272-080).
5. 5-(3-Aminoallyl)-2'-dUTP (1 mg) (Sigma A0410).
6. Cy3 mono-reactive dye pack (5 mg) (Amersham PA23001). Resuspend individual Cy dye tubes (5 tubes/pack) in 10 μL dimethyl sulfoxide (DMSO). Aliquot 1.25 μL to eight separate tubes. Evaporate dyes in Speed-Vac in dark. Store samples under vacuum in a dessicator at 4°C in dark.
7. Cy5 monoreactive dye pack (5 mg) (Amersham PA25001). Prepare as above prior to use.
8. 0.5 M EDTA, pH 8.0.
9. 1 M NaOH.
10. 20X saline sodium citrate (SSC).
11. DMSO.

12. 10% SDS.
13. 50X amino-allyl (aa)- dUTP/dNTP mix (2:3 ratio of aa-dUTP/dTTP): 10 μL dATP (100 mM stock), 10 μL dCTP (100 *mM* stock),10 μL dGTP (100 mM stock), 6 μL dTTP (100 mM stock), 4 μL aa-dUTP (100 mM stock).
14. Master mix for cDNA labeling (for 50 rxns): 300 μL SSRT II 1st strand cDNA buffer (supplied with SSRT II enzyme), 150 μL 0.1 *M* dithiorthreitol (DTT) (supplied with SSRT II enzyme), 150 μL RNAse-free H_2O, 30 μL (50X) amino-allyl (aa)-dUTP/dNTP mix (2:3 ratio of aa-dUTP/dTTP).
15. 100 mM sodium bicarbonate ($NaHCO_3$).
16. 100 mM sodium carbonate anhydrous (Na_2CO_3). *Mix 100 mM* $NaHCO_3$ and 100 mM Na_2CO_3 together until final pH 8.5-9.0 is achieved.
17. 4 *M* hydroxylamine (Sigma H9876).
18. Qiaquick PCR purification kit (Qiagen 28104).
19. $pd(A)_{40-60}$ (Amersham 27-7988-01) (reconstituted at 8 mg/mL).
20. Yeast transfer RNA (tRNA) (Sigma R87.59) (reconstituted at 4 mg/mL).
21. Bovine serum albumin (BSA) (Fraction V: Sigma).
22. 50X Denhardt's reagent.
23. Deionized formamide.
24. Sodium pyrophosphate.
25. 1 *M* Tris-HC1 (pH 7.4).
26. Isopropanol.
27. Hybridization chambers (ArrayltTM AHC-1 or equivalent).
28. Lifter-slips (Erie Scientific: size depends on cDNA array dimensions)/coverslips (size depends on cDNA array dimensions).
29. Coplin jars.
30. Slide swing handle rack (RA Lamb E102).
31. Cot-1 DNA; mouse and human (Invitrogen).

Scanning

1. Scanner: e.g., GenePix 4000B, Axon (http://www.axon.com); ScanArray 3000, Packard BioChip Technologies/GSI Lumonics. (http://www.dssimage.com/microarraySystems.html); Avalanche, Molecular Dynamics (Amersham Biosciences) (http://www.mdyn.com); DNAscope, GeneFocus (http://www.genefocus.com).
2. Spot-finding software: e.g., GenePix Pro, Axon (http://www.axon.com); Microarray Suite, Scanalytics (http://www.scanalytics.com).
3. Data-mining software: e.g., Spotfire DecisionSite for Functional Genomics, Spotfire, (http://www.spotfire.com); Genespring. Silicon Genetics (http://wwNA·. sit icon genetics.com); Vector Xpression 3, Informax Inc. (http://www.informaxinc.com/).

PROCEDURE

Amplification of cDNA Probe

cDNA probes to be *robotically* printed onto chemically modified glass slides are *generated* according to standard PCR protocols. Briefly, a 100-μL reaction amplified for 35 cycles generally is sufficient to generate adequate probe material to be spotted onto several hundred glass slides. For ease in *processing* multiple samples, 96-well PCR plates are recommended for the *amplification*. The source of template used for each unique parasite cDNA typically is a phage suspension or a lysed *bacterial colony* (obtained from cDNA libraries or cloned individually).

Usually, these clones have been sequence verified. There is no need to isolate purified DNA (containing parasite-specific cDNA) before proceeding with the PCR reactions unless problems arise with the amplification. PCR primers used in the amplification are either complementary to cDNA vector sequences, flanking the parasite-specific inserts in the starting library (e.g., T7/T3, M 13R/M 13F), or are gene-specific.

Analyze a 3-μL sample from each PCR reaction on a F /c agarose gel to verify size and number of products. The remaining PCR reactions may be stored at -80°C until purification can be performed.

Purification

Before the parasite-specific cDNA probes are suitable for printing, they must be purified away from all contaminating salts, dNTPs, and primers. An easy and *efficient* way to achieve high-throughput *purification* of multiple samples is via the use of NucleoFast° 96-well filter plates (Macherey-Nagel).

1. Transfer each 100-μL PCR reaction directly onto the membrane of the NucleoFast 96-well filter-plate placed on top of an appropriate *vacuum manifold* (with waste collection plate).
2. To remove contaminants by *ultrafiltration*, apply a vacuum of -400 to -600 mbar (reduction of *atmospheric* pressure), typically for 10-15 min (NO modifications to this protocol are necessary if mineral oil was used during the PCR reactions, although the actual *filtration* time may be longer).
3. Dispense 100 μL nuclease-free double-distilled H_2O (ddH_2O) to each well of the 96-well filter plate while *applying* the vacuum until the water has passed through the membrane. Repeat this wash two additional times and completely *ventilate* the *manifold* for 60-90 s.
4. To recover the DNA, add 100 μL nuclease-free ddH_2O (ensure no vacuum is applied) and transfer filter plate to a suitable microplate shaker with moderate shaking for 10 min. Gently remove the water from the filter plate wells with a *multichannel pipet* and *dispense* into the supplied *elution* 96-well plates.

This will deliver the PCR product at the *original concentration*. A precipitation step may additionally be included at this point if the final concentration of DNA is required to be higher. It is advisable to check purification procedure by *analyzing* PCR products (approx 1 μL) on a 1% *agarose gel*. Purified products can be stored at -80°C in 96-well plates (source plate) that have been sealed with self-adhering PE foil and the lid safely taped to prevent *evaporation*.

Printing

1. A small aliquot (15 μL) of the total purified PCR product (from source plate) should be transferred to a new V-bottomed 96- or 384-well plate (print plate).
2. Add 5 μL (4X) spotting buffer to each well and mix thoroughly.
3. Print PCR products onto y-amino propyl silane-coated slides according to *robotic arrayer* instructions. Be sure to include appropriate controls (which can be added to empty wells of print plate) including, but not all-inclusive: genomic DNA (positive control), yeast tRNA (negative control), buffer only (negative control), plasmid and phage DNA from which PCR products were amplified (negative controls).
4. After printing, slides should be allowed to dry at room temperature, baked for 2 h at 80°C and UV cross-linked (600-mJ dose) to facilitate maximal fixation of PCR products. At this stage, if the slides used for printing are not pre-bar coded, then generally it is useful to etch *pertinent* information into each glass slide with a diamond scribe (while salt residue still defines the border of the array).

   This information should include slide number, print series, and borders of the printed array. Marking each slide in this manner also serves to distinguish the side of glass that contains DNA.
5. After marking glass slides, denature immobilized PCR products by *submerging* in a boiling dH_2O bath (turn heating unit off immediately after slides have been submerged) for 2 min.
6. After denaturing DNA, transfer slides to a 95% EtOH bath and fix for 1 min. Spindry slides for 5 min at 1000g in a benchtop centrifuge. Slides can now be stored in a dust-free *microscope* slide box at room temperature and are ready for *hybridization*.

Extraction of Total RNA

As in all applications that involve the isolation of RNA, care must be taken to avoid introduction of ubiquitous RNAses. Either purchase RNAse/DNAse free *reagents* (preferred) or treat all solutions (except those containing Tris) thoroughly with *diethyl-pyrocarbonate* (DEPC) prior to use. Change gloves frequently, and use filter tips and *disposable plastics* when isolating RNA from parasite material.

Generally, it is best to work rapidly during the *isolation* of RNA (the RNA isolation procedure should be performed in 1 d). The method outlined below produces total RNA free from *contaminating* carbohydrates and protein and involves the use of both TRIZOL Reagent (Invitrogen) and Qiagen RNeasy affinity columns (Qiagen). RNA isolated in other fashions (although suitable for Northern blotting, RNAse protection, or reverse *transcriptase* [RT]-PCR assays) generally is not suitable for cDNA microarray applications and can often lead to high background because of biological *contaminants*.

Extremely low amounts of genomic DNA may be isolated during the procedure described here. If this presents a problem (there are reports that Superscript reverse *transcriptase* II can prime off genomic DNA, leading to false positives: *see* Genisphere's Web site http://www.genisphere.com). Qiagen provides a protocol for genomic DNA digestion on a column, or alternatively, one can follow any protocol that uses the enzyme RNAse-free DNAse.

TRIZOL Extraction

According to the manufacturer's instructions, add a sufficient volume of TRIZOL to parasite material (1 mL of TRIZOL per 10^7 cells or I mL TRIZOL per 100 mg *tissue*. If starting from parasite-infected cell cultures (e.g., *Toxoplasma* gondii-infected fibroblasts or *Plasmodium* sp.-infected red blood cells), then refer to commonly used *techniques* to enrich for parasite material (if desired).

Keep in mind that there is a trade-off between obtaining increased quantities of purified parasite material (enriching away from host material) vs the effect this enrichment has on parasite *transcriptional* processes. Each *researcher* will have his or her own *protocols* for obtuining parasite material (fresh material is best, especially if old parasite material was stored improperly), as well as his or her own favorite method of *disrupting/homogenizing* samples. For multicellular parasites (*helminths, nematodes*, or other large *metazoans*), a *homogenizer* works quite well for this important application.

1. Using a cleaned homogenizer (clean with 50 mL ddH_2O, 50 mL 100% EtOH, and 7 mL TRIZOL between each parasite sample), thoroughly disrupt parasite material (10-15 s). Make sure the tip of the homogenizer stays below the surface of TRIZOL to avoid foaming of solution. After complete *homogenization/disruption* of sample material, let the sample sit at room temperature for 5 min.
2. Add 0.2 mL *chloroform* per 1 mL TRIZOL and *vigorously* shake samples for at least 1 min. Let samples incubate at room temperature for 2 min to ensure the complete *dissociation* of *nucleoprotein* complexes.
3. Spin samples to separate phases (4°C at 12,000g for 15 min).
4. Carefully transfer *aqueous supernatant* (containing the RNA) to a new RNAsefree tube without disturbing the interface (containing protein and other cellular debris) and proceed with Qiagen RNeasy affinity purification.

Purification of Quiagen

The second phase of the total RNA purification procedure is to use a Qiagen affinity column to *reversibly* bind the *nucleic acid* from the *aqueous* solution obtained above (see Qiagen RNeasy *handbook* for detailed principles and procedures of product).

1. Add an equal volume of 70% EtOH to the aqueous supernatant obtained from the TRIZOL extraction (dripwise while *vortexing* gently to avoid local RNA precipitation).
2. Transfer this solution to a Qiagen affinity column. If the volume of solution is greater than the capacity of the Qiagen column, add the sample in successive aliquots. Spin for 5 min at room temperature (3000-5000g) to facilitate RNA binding.
3. Vortex the collected flow-through (FT) and *re-apply* to Qiagen column. Spin again for 5 min at room temperature (3000-5000g). This step ensures that all RNA in the sample is bound to the column.
4. Discard the FT and stepwise wash the column as indicated in the Qiagen handbook. Volumes of washing *buffers* will depend on Qiagen column used (midi or maxi column).
5. After the last wash in buffer RPE (supplied in the Qiagen kit) and ensuring that the *Qiagen*

column is completely dry (additional 5 min spin after removal of last FT), carefully transfer the column to a new conical tube (supplied in the Qiagen kit).

Elute RNA from the column with three successive equal volumes of RNAse-free ddH_2O (also supplied in the kit). All three samples containing the eluted RNA can be collected in the same collection tube.

6. Add 1/10 vol of 3 M sodium acetate (pH 5.2-5.3) and 2.5 vol of 100% EtOH to the RNA sample. Vortex this mixture and incubate on ice for at least 10 min (alternatively, place at -20°C overnight to facilitate precipitation of RNA). Spin precipitated RNA at 12,000g for 15 min (4°C). You may have to transfer sample to several 1.5-mL *micro-centrifuge* tubes.
7. Wash pellets twice with 75% EtOH by vigorously vortexing sample during each wash. Dry pellets completely in Speed-Vac.
8. Dissolve RNA pellet(s) from each sample in RNAse-free ddH_2O by adding directly to the top of the dried sample and placing at 65°C for 10 min (it may be necessary to incubate longer at 65°C). Do not pipet the sample up and down (RNA may stick to the inside of the pipet tip). After the sample has completely dissolved, combine all aliquots of each sample (if necessary) and transfer to a new microcentrifuge tube.

   Spin out any insoluble debris (3 min at max speed in microcentrifuge) and quantitate each RNA sample by spectrophotometry. *Aliquot* RNA (10-μg aliquots) into RNAse-free microcentrifuge tubes and store samples at -80°C. It is *advisable* to check the integrity of the RNA sample by *electrophoresis* through a *denaturing* agarose gel.

Labeling of cDNA With Amino-Allyl dUTP

Although there are alternative cDNA-labeling methods used in cDNA microarray applications (see http://www.nhgri.nih.gov/DIR/Microarray/main.html and http://www.genisphere.com for *information* on two additional techniques), the protocol listed below describes a procedure that uses amino-allyl dUTP and the fluorescent dyes Cy3 and Cy5.

This eDNA-labeling protocol has been slightly modified from one that is available at http://www.microarrays.org (the original protocol was developed at Rosetta Inpharmatics, Kirkland, WA), with the user supplying all the reagents (alternatively, kits that use a similar procedure can be purchased from *commercial* sources-e.g., Fairplay system from Stratagene).

As little as 8 μg of total RNA can be used to generate fluorescently modified cDNA molecules of sufficient quantity and intensity to be used in cDNA microarray *hybridizations*. The protocol detailed below uses 10 μg of total RNA as the starting input material for each separate cDNA reaction (Cy3 and Cy5). *Empirical* testing of input RNA is *recommended* for optimal results (generally, more input RNA leads to generation of cDNA with higher signal *intensity*).

Labeling Technique

1. Combine 14.5 μL total RNA (10 μg) with 1 μL oligo-dTV (5 μg/μL) primer in a 0.2-mL RNAse/DNAse-free PCR tube. 10 μg from each sample where gene expression is being compared is necessary for each reaction (e.g., 10 μg from adult female *S. mansoni* RNA and 10 μg from adult male *S. mansoni* RNA).
2. Heat RNA/oligo-dTV mixture at 70°C for 10 min (it is useful to do all incubation steps in a

PCR thermal cycler). Immediately place samples on ice and leave for 5 min. This step effectively removes secondary structures associated with total RNA mixtures.

3. Add 12.6 μL of the master mix containing the *amino-allyl* dUTP to each denatured RNA/ oligo-dTV sample. Place reactions at 42°C to prewarm samples prior to adding 1 μL Superscript Reverse Transcriptase II (SSRTII). Incubate reactions for 1 h, then add another 1-μL aliquot of SSRTII to each and continue incubating for another hour.
4. Remove cDNA reactions from 42°C and add 10 μL 0.5 M EDTA (chelates SSRTII) to each, followed by addition of 10 L of 1 N NaOH (hydrolyzes any remaining RNA). Place samples at 65°C for 15 min and then immediately on ice. Neutralize sample by adding 25 μL of *1 M* Tris-HCl (7.5).
5. To each neutralized reaction, add 450 μL H_2O and transfer contents into separate Microcon-YM30 concentrators (one for each cDNA reaction). Spin at 13,000g in a microcentrifuge for 8 min. When reservoir volume gets to about 50 μL, add another 400 μL $_{H20.}$ Repeat wash with a third 400=μL vol of H_2O (three total washes). These three washes effectively eliminate all unincorporated dNTPs and free Tris (Tris will interfere with Cy dye conjugation reaction).
6. When volume of sample after third wash reaches approx 10 μL, invert the Microcon-YM30 concentrators and place into a new 1.5-mL microcentrifuge tube. Elute washed cDNA samples from filter by spinning at 3000g for 4 min.
7. Dry each cDNA sample (will not see dried pellet) in Speed-Vac (time varies depending on temperature).
8. Resuspend each cDNA pellet in 4.5 μL of DNAse-free ddH_2O.
9. Resuspend an aliquot of dried Cy dye (one cDNA sample will be labeled with Cy3 and the other will be labeled with Cy5) in 4.5 μL of 0. 1 M carbonate/bicLufbonate buffer (pH 8.5-9.0).
10. Mix cDNA with the appropriate Cy dye and incubate in the dark for 1 h at room temperature (Cy dyes will conjugate to the amino-allyl dUTP that has been incorporated during the cDNA reactions).
11. To stop *conjugation* reaction, add 4.5 μL of 4 *M* hydroxyla.nine to each cDNA sample (excess amino groups effectively quench free Cy dye) and incubate in the dark for 15 min.
12. Add 35 μL of 0.1 M sodium acetate (pH 5.2) to each conjugation reaction and combine the two samples in which gene expression will be compared. Transfer combined sample to new 1.5-mL microcentrifuge tube.
13. Add 500 μL PB buffer from QIAGEN PCR clean-up kit to each combined Cy3/ Cy5-labeled cDNA sample and apply to QIAquick column. Spin samples for 30-60 s, at maximum speed in microcentrifuge. Reapply FT to column and spin again. Discard FT (all labeled cDNA should be bound to column), and apply 750 μL PE wash buffer (ensure *ethanol* has been added to PE buffer as per *Qiagen* instructions) to column.
14. *Spin columns* again for 30-60 s, at maximum speed in microcentrifuge. Discard FT. As Cy3 and Cy5 dyes can weakly bind to Qiagen column, it is advisable to perform two additional PE washes (three total). After the third wash, spin for 1 min to completely dry column.

15. Place dried column (containing washed, bound, labeled cDNA) in new 1.5-mL collection tube and add 100 μL elution buffer (EB) to membrane. Let sit at room temperature for 1 min (in dark), then spin for 1 min at 13,000g to elute cDNA. Add same volume of EB and repeat elution (collect both fractions in same tube).
16. Bring the 200 μL eluted, labeled cDNA target to 400 μL with DNase-free H_2O and apply to Microcon-YM30 concentrator. Concentrate sample to approx 20-30 μL using a microcentrifuge. Add 2.2 μL yeast tRNA (4 mg/ml...' and 2.2 μL poly d(A) (8 mg/mL) and dry the remaining solution using a Speed-Vac. Resuspend the pellet in 40 μL prewarmed (48°C) filter-sterilized hybridization buffer containing 40% deionized formamide, 5X SSC, 5X Denhart's reagent, 1 mm Na-pyrophosphate, 50 mM Tris, pH 7.4, 0.1% SDS.
17. Denature labeled cDNA target for 2 min at 95°C and spin for 3 min at maximum speed in *microcentrifuge* to pellet any *insoluble* material that may interfere with *hybridization*. The cDNA target (minus any insoluble pellet) is now ready for hybridization to cDNA *microarray*.

Hybridization of cDNA

cDNA microarrays can be prehybridized for up to 1 h prior to addition of labeled cDNA target. This is an additional cautionary step to prevent *nonspecific* hybridization events from occurring (may not be needed in all cases).

1. Place slide containing cDNA elements to he used for hybridization in a 50-mL conical tube containing a solution made up of prehybridization buffer (I% BSA/0.1 % SDS/5X SSC/50% formamide). Place *conical* (containing slide) in 42-48°C H_2O bath or *hybridization* oven, with *gentle* shaking for 40-50 min.
2. Remove slide from *prehybridization* conical and briefly wash by immediate immersion and shaking in a *Coplin jar* containing dH_2O. Repeat wash by transferring slide to another Coplin jar containing an additional *aliquot* of dH_2O. Finally, submerge slide in a third Coplin jar containing 100% *isopropanol*.
3. After removal from isopropanol, immediately transfer slides to slide rack and spin dry (do not let air dry) in a benchtop centrifuge (10 min at 1000g). Prehybridized slides now can be stored in a clean 50-mL conical tube. When handling slides, touch only sides of glass and always use powder-free gloves.

   As an alternative to standard glass coverslips, Lifter-slips (coverslips with Teflon-coated raised edges) are commonly used in cDNA microarray hybridizations (Erie Scientific). Lifterslips facilitate equal distribution of labeled target cDNA across all elements of cDNA microarray.
4. Place the Lifter-slip directly on top of cDNA microarray glass slide with the Teflon-coated raised edges toward the cDNA. Carefully pipet 37 μL of the spun, denatured, labeled cDNA target (leave 3 μL of the target behind so that any insoluble pellet is not disturbed) between Lifter-slip and glass slide and capillary action will draw the liquid to sufficiently cover a fabricated DNA microarray with dimensions of 25 mm × 60 mm.
5. Place array into hybridization chamber (make sure chamber has a few drops of ddH_2O or

3X SSC in each end to maintain humidity), clamp hybridization lid down tightly, and incubate whole apparatus overnight (14-18 h) at 42-50°C.

Washing

After overnight hybridization, slides must be carefully removed from hybridization chambers and washed to remove unbound target DNAs. The washing procedure listed below is a good place to start, however, optimization of washing times, detergent/salt concentrations, and temperature may be necessary.

Three Coplin jars are used in this procedure, but any suitable washing tray may be utilized. If using Coplin jars, only two slides should be processed at one time before changing the wash solutions.

1. Remove cDNA microarray from chamber by carefully picking up one end of the glass slide (coverslip still attached) with forceps and place into Coplin jar 1 containing wash solution one (0.5X SSC/0.1% SDS). Carefully shake slide with forceps until coverslip falls off. Remove coverslip from Coplin jar and continue shaking cDNA microarray slide with forceps for 4 min.
2. Remove cDNA microarray from Coplin jar 1, touch end of slide to paper towel to remove excess wash solution (do not allow slide to dry completely), and transfer slide to Coplin jar 2 (containing wash solution two, 0.5X SSC/0.01 % SDS). Again, shake slide with forceps for 4 min.
3. Finally, remove cDNA microarray from Coplin jar 2, quickly touch end of slide to paper towel, and transfer slide to Coplin jar 3 (containing wash solution three, 0.06X SSC). Wash slide by vigorous shaking for 4 min as before.
4. Remove cDNA microarray from Coplin jar 3, immediately submerge in 100% isopropanol, and spin dry using a benchtop centrifuge (5 min at 1000g). You can now store the cDNA microarray in a dark slide box until ready to acquire fluorescent information via a scanner (processed cDNA arrays can be stored for at least a week in the dark before scanning, without losing data).

Data Mining and Image Analysis Issues

Fluorescent data is acquired through the use of a cDNA microarray scanner (review of some commercially available scanners is available) and appropriate image analysis software. Data mining and filtering (usually the most timeintensive aspect of cDNA microarray experiments) can be performed usinglimited software packages (Excel', Microsoft) or quite statistically thorough software packages such as GeneSpring (Silicon Genetics). The specific details about using such equipment and software are beyond the scope of this chapter.

Because of the relatively high cost of scanners and software, anyone performing cDNA microarray experiments should consult appropriate references to ascertain which scanner, image analysis software, and data-mining packages are the most suitable for each laboratory's intended experimental setup. It is also recommended that at least three independent hybridization experiments be performed (analytical replication) with each pair of RNA samples to account for intrinsic variables associated with this gene expression platform.

Control experiments are also recommended to assess RNA quality, cDNA labeling, and cDNA microarray hybridization conditions. One control experiment includes performing a "self vs self" hybridization (the two RNA samples are identical, but labeled with different dyes). As both RNA species are the same, this control reveals how tightly a cDNA array estimates that the two samples from the same source produce equivalent signals.

Deviation away from normalized gene expression ratio of 1 indicates that the array may be failing in some areas (general or specific). If only some portion of the gene expression ratios is far from 1, then their position on the array can be reassessed (background, signal, etc.) before proceeding with more complicated (and valuable) experimental hybridizations. Another control experiment (very good for bimodal comparisons) involves "dye swapping"-switching the Cy dyes used to label the starting cDNA samples.

This control experiment provides additional confidence in gene expression between two samples (the same genes should be differentially expressed even when dyes are switched). Finally, independent confirmation of gene expression, using traditional biological techniques (Northern blotting, RT-PCR, real-time PCR, *in situ* hybridization, protection assays, etc.), should be sought on all differentially expressed cDNAs prior to any detailed advanced investigation.

POINTS TO REMEMBER

1. Alternative spotting buffers used by other investigators include SSC (IX through 5X final concentration) or 50% DMSO (with this spotting buffer, generally less evaporation of samples in print plates occur during printing run). Some investigators also add betaine (N,N,N-trimethylglycine: Sigma) to their spotting solution (typically 3X SSC) at a final concentration of 1.5 M.

   This additional spotting compound contributes to high spot homogeneity and reduced background, although slides may need to be baked for longer periods of time (overnight) at 80°C to ensure complete fixation of deposited DNA spots.

2. Typical PCR conditions for amplifying *S. mansoni* cDNA EST inserts from phage suspension or bacteria solution: initial denaturation at 9.5°C for 5 min; then 35 cycles of 95°C for 30 s, 59°C for 30 s, 72°C for 4 min; final 72°C extension for 5 min; hold at 4°C. Phage suspension was prepared as follows: single plaque cored into 500 µL SM buffer and stored frozen (4 pL of this solution was used as template material for PCR).

   Bacterial suspension was prepared as follows: 10 µL from a 1 mL overnight culture was added to 100 µL dH2O and heated for 5 min at 95°C. This suspension was stored frozen (4 µL of this solution was used as template material for PCR).

   Size of amplified product will undoubtedly vary among ESTs. Generally, 500- to 2000-bp ESTs are suitable for arraying (more detailed examination of PCR product size on hybridization kinetics and signal intensity has recently been examined).

3. Template DNA generally does not need to be purified prior to PCR amplification. However, in case of particularly difficult PCR amplifications, a template purification step may be necessary. Refer to Sambrook et al. for general plasmid and phage DNA isolation protocols or access http://www.nhgri.nih.gov/DIR/Micro-array/main.html or http://www.microa

rays.org for typical microarray template purification procedures.

4. Centrifugation may replace the use of a vacuum manifold if necessary. In this case, centrifuge 96-well filter plates containing PCR products at 4500g for 10-15 min in substitution of vacuum application.

5. It is critical to shake moderately (not vigorously) to elute DNA samples from the membrane and also to prevent cross-contamination from well to well. Evaluate the rotation speed before elution with a water/dye combination. If you wish to obtain PCR products at a higher concentration, simply elute with less volume. DNA in a range of 0.15-0.25 mg/mL is a good starting concentration for robotic printing (concentration estimates can be made from gel electrophoresis).

   Alternatively, a precipitation step may be performed using isopropanol, according to protocols at http://www.microarrays.org. Excessively concentrated DNA may clog pins or produce "comet tails" (caused by localized reattachnnent of cxcess ;pot material to the slide surface during postprinting processing), whereas DNA below this concentration leads to weak signal and a decrease in the dynamic range for detecting differentially expressed genes.

   Refer to Yue et al, http://www.nhgri. nih.gov/DIR/Microarray/main.html, for methods of determining precise starting DNA concentration in source plate (typically a fluorescent assay that can be performed in 96-well plates).

   Additionally, SYBR GREEN II RNA gel stain (Molecular Probes) can be used as quick test to determine how efficiently the DNA has been deposited onto each microarray prior to hybridization (protocol available at http://www.probes.com/handbook/boxes/0681.html).

6. A review of the specifications and features of some commercially available robotic printers or arrayers is provided by D. Bowtell. Additionally, an alternative to printing PCR products onto y-amino propyl silane-coated slides is to print onto poly-L-lysine-coated slides. Both coating agents essentially perform the same function-provide a hydrophobic and positive charge critical for spotting and positioning of DNA.

   Methods and reagents to coat Cold Seal hrand microscope slides (this particular glass has minimal intrinsic autofluorescence-Erie Scientific) can be found at http://www.nhgri.nih.gov/DIR/Microarray/main.html. Although coating microscope slides in house with poly-L-lysine is more cost-effective than purchasing either poly-L-lysine or y-amino propyl silane-coated slides, batch-to-batch variation in slide coating often can occur if care is not taken in quality control.

   There are many varieties of printing tools and pins (split pins are commonly used), and the pros and cons of each make are beyond the scope of this chapter. Consult an experienced company (e.g., Biorobotics or TeleChem) for details of each tool and pin type.

7. In cases of high background, an extra slide-blocking step can be performed with succinic anhydride (Corning; www.corning.com, provides instructions with their y-amino propyl silane-coated slides, or alternatively, instructions can be found at http://www.nhgri.nih.gov/DIR/Microarray/main.html).

   Succinic anhydride converts the residual amines on the surface of the precoated slide into carboxylic moieties (reduces the positive charge of non-DNA-containing areas and, therefore, limits the areas in which cDNA can bind nonspecifically). New techniques have improved this blocking procedure by effectively limiting the amount of DNA solubilization that sometimes occurs during succinic anhydride treatment.

8. The correct volume of TRIZOL reagent added to initial parasite sample should be empirically determined for optimal yields of RNA.
9. In some cases (multicellular, multitissue parasites such as adult nematodes and helminths, etc.), a more vigorous shaking step is suggested (3 min instead of just l min).
10. If volume of material is too large to spin in a microcentrifuge, use a Sorvall RC-5B (or equivalent) centrifuge and spin at 6,000-10,000g increasing the time to 30 min (Sorvall HB-6 rotor). This centrifugation step will require RNAse-free 15- or 30-mL Corex tubes (treated with 0.1 M HCl in DEPC-H_20 for at least 8 h and subsequently autoclaved).
11. Occasionally, during this step some precipitation of nucleic acid (when starting with approx 1 g of material) is observed. This does not appear to interfere with the subsequent binding of RNA to the Qiagen column as long as the sample is thoroughly vortexed prior to addition to column.
12. The three successive elution volumes are based on the size (midi or maxi) of the Qiagen column used. Suggested elution volumes for each column are indicated in the supplied manufacturer's handbook.
13. As RNA isolated in this manner is free of contaminating carbohydrate molecules (carbohydrates bound to RNA often makes RNA difficult to resuspend), complete Speed-Vac drying of the sample generally does not inhibit the subsequent solubility of each sample.
14. RNA is typically diluted 1/100 in 0.01 M Tris-HCl (7.5) for estimations of purity (OD_{260}:OD_{280} ratio). An OD_{260}:OD_{280} ratio of 2.0-2.3 (in 0.01 M Tris-HCl) generally indicates highly pure RNA free of contamination. For accurate RNA quantitation, measure samples in RNAse-free H_2O.
15. The cDNA master mix in this protocol used a 2:3 ratio of aa-dUTP to dTTP. This ratio works quite well for schistosome RNA. However, empirical determination of the correct ratio for each parasite species is essential for optimal labeling efficiency (this ratio should be a good starting point for optimization experiments).
16. Depending on the microcentrifuge being used as well as fine differences in microcon membrane construction, the time spent washing the cDNA samples may vary. Try to avoid drying the microcon membrane completely, as some difficulty may be encountered when eluting the cDNA (although according to the manufacturer's instructions, 10 μL of H_2O can be applied to dried membranes, allowing for recovery of cDNA sample material).
17. It is important not to let the Cy dyes sit in an aqueous buffer prior to conjugation, only rehydrate the dyes directly prior to addition of suspended cDNA sample. Small residual amounts of DMSO do not affect conjugation reaction.
18. Strong color will be observed in the FT following the first wash. This material represents

unconjugated dyes. Some color may also be noticed on the Qiagen membrane after spin drying of the samples (after the third wash).

This material likely represents trapped Cy3/Cy5-labeled SSRT II (free amino groups on the enzyme capable of being modified by the dyes during the conjugation reaction) and/or labeled cDNA targets. With 10 μg input RNA, one does not consistently observe color on the membrane prior to elution of cDNA or in the eluent.

19. It can be useful to include either human or mouse Cot-I DNA (1 μg Cot-1 DNA/ 10 μg input RNA) during this concentration step as an additional blocking agent in schistosome cDNA microarray hybridizations. Cot- I DNA is 50-3 10 by in size and enriched for repetitive DNA sequences such as Alu and Kpn (human Cot-1) or Bl, B2, and L1 (mouse Cot-1) family members.

    Including one of these blocking agents prevents undesired hybridization between conserved repetitive DNA sequences in the labeled cDNA and probe DNA by competitive hybridization. Other blocking agents can be added during hybridization, such as yeast tRNA and poly d(A). Yeast tRNA reduces undesired nonspecific DNA hybridization, while poly d(A) that is 40-60 dATP bases in length promotes specific hybridization between the labeled cDNA and probe DNA. Poly d(A) reduces hybridization of the polyA sequences in the probe DNA to the polyT tract in the labeled cDNA.

20. This hybridization buffer is one that works well for schistosome cDNA arrays, although each user should empirically test different components (blocking agents) and formulations (formamide vs aqueous based hybridization buffers) to ensure optimal hybridization kinetics. An aqueous hybridization buffer may also be used consisting of 3.5X SSC/0.275% SDS.

    In this case, the hybridization is carried out at 65°C overnight instead of 42-50°C. However, dehydration of labeled cDNA sample material between coverslip and cDNA microarray may occasionally occur when using an aqueous hybridization buffer. Using either hybridization buffer, do not place labeled cDNA target on ice after heating at 95°C, as SDS will precipitate.

21. A glass coverslip/Lifter-slip can be cleaned of dust and debris by using compressed air. Ensure that the compressed-air canister does not also spray oil droplets or other potential contaminants onto the glass coverslip. Alternatively, the covers lip/Lifterslip may be cleaned using EtOH. Care should be taken not to introduce air bubbles between glass coverslips and cDNA microarrays.

    If air bubbles are observed, then carefully push bubbles to the side of the coverslip with clean forceps prior to hybridization. Small air bubbles will usually dissipate during the hybridization.

Chapter 7

Position Effect Variegation (*PEV*)

Position effects describe the observed alteration in *protein-coding* gene expression that may accompany a change in genomic position of a given gene. A position effect may result from chromosomal translocation or other genomic rearrangements. Recent advances in chromatin studies in several different species including yeast, *Drosophila,* and mouse have contributed significantly to better understanding of human *diseases* resulting from *abnormal epigenetic* effects.

Molecular models attempting to explain position effects in humans have been proposed; however, none of them adequately addresses a variety of mechanisms. According to the *noncontact* models, the *cis*- or *trans-regulatory* elements, or locus control regions, are physically separated from the target gene and act either at the RNA level, by protein interactions, or by *mediation* of boundary elements, termed insulators.

On the contrary, the contact models invoke *spatial-temporal* modifications of chromatin structure (e.g., active chromatin hub). In both models, the conserved nongenic sequences (CNGs) may play an important role in genomic regulation of gene expression. The recent introduction of new techniques including *tagging* and recovery of associated proteins (RNA-TRAP) and capturing chromatin conformation (CCC or 3C), has provided powerful tools to investigate position effects in humans.

Chemical modifications to DNA or histories that alter the structure of chromatin without changing the DNA coding sequence are described as epigenetic. Change in histone code, DNA *methylation*, or genes packaging in the nucleus can result in altered regulation of gene expression that has been referred to as a position effect.

Early studies of heterochromatin in *Drosophila* have led to the discovery of the *phenomenon* of position effect variegation (PEV), a juxtaposition of a euchromatic gene in a novel *cisacting heterochromatin* environment wherein the gene becomes *epigenetically* silenced. Subsequently, PEV has been described also in yeast and *mammals*.

The molecular mechanisms for PEV have remained elusive, but recent progress in chromatin studies has begun to illuminate molecular aspects of this *phenomenon*. It has been *speculated* that the spreading of *heterochromatic* packaging protects the cell against the invading *retrotransposons* and *transposable* elements.

In addition to well known features of *heterochromatin,* e.g., *histone deacetylation, histone methylation* (histone H3 on lysine 9, H3-mK9), and heterochromatin-associated protein HPI, small RNAs, RNAi, and siRNAs have been shown in yeast to play important roles in heterochromatin assembly through the RNA-induced initiation of a transcriptional gene silencing complex.

The infrequently described trans-regulation is probably best represented in *Drosophila* by the transvection phenomenon, a pairing-dependent interallelic compleinentation, or the ability of one *locus* to affect the expression of the allele, on the homologous chromosome.

In recent years, transvection has been reported in *Drosophila* at a rapidly growing number of loci, and transvection-like phenomena have been described in a number of different *organisms, including plants, fungi,* and *mammals.*

Despite the near completion of the human genome sequence, little is known about genomic aspects of gene regulation in humans. Kleinjan and van Heyningen reviewed position effects of chromosomal rearrangements in humans and proposed a few potential mechanisms: (1) separation of the gene from its enhancer or promoter region (locus control region); (2) *juxtaposition* with an enhancer element from another gene; (3) removal of the long-range *insulator* or *boundary* elements; (4) competition with another *enhancer*; and (5) PEV.

A useful model for studying such regulation is provided by genomic rearrangements resulting in diseases, in which chromosome breakpoints do not disrupt the causative gene but instead map outside the intact gene.

BALANCE EFFECT

Many of the balanced chromosome aberration breakpoints associated with abnormal phenotype in humans have been mapped up to 400 kb both *upstream* and *downstream* from the *causative* gene. Remarkably, a few cases with putative *cis-acting* regulatory elements have been described as distant as 1 Mb upstream from the target gene.

Interestingly, no chromosome breakpoints have been found in the genomic regions approx 400-800 kb from the *disease-causative* gene. This statistically *significant bimodal* distribution of breakpoints suggests the role of a chromatin structure, rather than the noncontact mechanisms.

Limb Mal Formation

Preaxial polydactyly (PPD) (MIM 190605) is one of the most common limb *malformation* in humans. Belloni et al. and Roessler et al. reported a position effect approx 250 kb upstream of the *SHH* gene, responsible for PPD. By analyzing a chromosome *breakpoint* of an apparently balanced *reciprocal* translocation t(5;7)(q11;q36) in a patient with PPD and a *transgenic insertion* site in a polydactylous mouse mutant sasquatch (Ssq) (*supernumerarypreaxial* digits of the hind feet), Lettice et al.identified an evolutionary conserved non-coding regulatory element termed ZRS.

The ZRS regulatory site is located approx 1-Mb 5' from the sonic hedgehog gene *SHH, in* intron 5 of the *LMBR1* gene. Using a *cis-traps* genetic test in mice, Lettice et al. and Sagai et al. demonstrated *cis-* and not trans-interactions between the *Ssq* locus and the *Shh* gene. These

Table 7.1: Long-Range Position Effects in Humans.

Gene affected	*Maximal distance of breakpoint from gene*		*Rearrangement type*
	Upstream	*Downstream*	
SOX	~900 kb	1.3 Mb	Translocation
POU3F4	~900 kb		
FKHL7	~1.2 Mb		
MAF	~1 Mb		
SHH	~1 Mb		
DACH	~1 Mb		
OTX2			
PLP1			
NSDI			

findings further suggested the presence of a long-range cis-regulating enhancer of *SHH,* capable of causing *congenital abnormalities.*

Supporting these findings, *Lettice* et al. identified four different point mutations in the ZRS in unrelated Dutch, Belgian, and Cuban families with PPG that did not have coding mutations in *SHH.* Further genomic comparison using the computer program VISTA among human, mouse, and teleost fish medaka revealed the mammals-fishes-conserved-sequences MFCS1 (corresponding to ZRS), MFCS2, and MFCS3 in and around *LMBR1.*

Interestingly, these elements are located only approx 100 kb from *Shh* in medaka fish. When deleted in mouse, MCFS1 resulted in a complete loss of *Shh* expression and truncation of the mouse limb.

CIS-Regulatory Elements

Evolutionarily conserved long-range cis-regulatory elements have been found also in introns of other genes. Three chromosome translocation breakpoints were mapped within introns 6, 11, and 12 of *MRPS22* in patients with *blepharophimosis/ptosis/epicanthus inversus syndrome* (BPES). These breakpoints are located approx 170 kb upstream to the BPES-causitive FOXL2.

In addition, four microdeletions with a 126-kb overlap were identified more distal to these breakpoints, 230 kb upstream to *FOXL2.* Interestingly, the shortest region of deletion overlap contains several CNGs with putative transcription-factor binding sites that represent potential candidate cis-regulatory elements *(21a).* Moreover, *translocation* breakpoints have been identified within the last intron of *PAXNEB,* more than 150 kb downstream of *PAX6,* the gene responsible for aniridia.

Haploinsufficiency of SOX9

Haploinsufficiency of *SOX9* on 17q24.3 causes campomelic dysplasia (CD), (MIM 114290), a clinically distinct syndrome characterized by semilethal skeletal *malformation* syndrome with or without XY sex reversal. The two mechanisms responsible for *SOX9* haploinsufficiency are *intragenic* mutations and chromosome rearrangements including *translocations, inversions,* and *deletions.*

The analysis of 12 patients with CD and apparently balanced chromosome rearrangements showed the breakpoints scattered up to approx 900 kb upstream of *SOX9,* whereas the gene itself was intact. Recently, Pop et al. reported an approx 1.5-Mb *microdeletion* located approx 380 kb upstream of *SOX9.*

Huang et al. described a chromosome *duplication* dup(17)(q24.124.3) associated with XX female-to-male sex reversal and *Bishop* et al. reported XX female-to-male sex reversal in a transgenic *Odd Sex* mouse with a 134-kb insertional deletion, resulting from a recombinant construct insertion 0.98 Mb upstream of *SOX9.* The authors proposed that a Dct promoter of the inserted construct interacted with *gonad-specific* enhancer elements, yielding sex reversal. Thus, both the *chondrogenic* and *gonadal* functions of *SOX9* appear to be *precisely* regulated by *elements* located as far as 1 Mb away from the gene itself.

Detailed DNA analysis of the genomic region *extending* up to 1 Mb proximal to *SOX9* failed to uncover any protein-coding genes, suggesting that the chromosomal rearrangements remove one or more cis-regulatory elements from an extended *SOX9* region. Recently, four independent 17q chromosome breakpoints were mapped 786-809, 899, 900, and 932 kb upstream of *SOX9.*

In addition, in a prenatal identification of acampomelic CD with male-to-female sex reversal in a fetus with *a de novo* balanced complex karyotype, the 17q breakpoint mapped approx 1.3 Mb downstream of *SOX9.* This is perhaps the longest-range position effect described in the field of human genetics, and the first report of CD with the chromosome breakpoint mapping 3' to *SOX9.*

Using a regulatory potential computer program in conjunction with the analysis of the rearrangement breakpoints, highly conserved genomic sequences that potentially represent cis-regulatory elements, were identified both upstream (1.1 kb *SOX9 evolutionarily* conserved regulatory element 1 [SOX9cre 1]) and downstream of *SOX9.*

These were shown to co-localize with *SOX9* in the interphase nucleus despite being located 1.1 Mb upstream and 1.3 Mb downstream to it, respectively. This suggests that the murine SOX9cre1 element has been modified by the inserted Dct promoter in *Odd Sex* mice and that in humans SOX9crel is responsible for the isolated Robin sequence phenotype cosegregating with a balanced t(2; 17)(g24.1;24.3) translocation reported recently by Jamshidi et al.. The 17q breakpoint in this family maps very close and proximal to SOX9cre1.

It is also likely that some patients with isolated Robin sequence have mutations involving SOX9cre 1 or other regulatory elements of *SOX9.* Thus, the mild *campomelic dysplasia* phenotype found in patients with translocation breakpoints near *SOX9* may be the result of slightly modified expression of chondrocyte-specific genes, secondary to altered *SOX9* regulation as a result of spatial dissociation from the distal SOX9crel.

BOUNDARY ELEMENTS

Insulators, or boundary elements, are DNA elements that have the potential to block the function of enhancers on gene promoters as well as spreading of chromatin silencing. Almost all reported insulators have been shown to require the regulatory *chromatin insulator* protein CTCF. CTCF also binds to several sites in the unmethylated *imprinted-control* regions that are essential for *blocking* the enhancers.

Recently, Yu et al. demonstrated that a post-translational modification, poly(ADP-ribosyl)ation of CTCF, *regulates* its activity as a *transcriptional* insulator. Insulators are considered a very *promising* and useful tool in gene therapy *applications* for ensuring high-level and stable expression of *transgenes*. The structure and function of insulators have been studied best on the chicken β-*globin* locus and in the *Igf2* and *1119* region on human chromosome 11p15.5.

HS4 Insulator Element

Pioneering work by Felsenfeld and *colleagues* on the chicken β-*globin locus* has led to the identification of an HS4 insulator element and demonstrated the important role of CTCF protein in boundary elements. They were also able to show the blocking activity of *insulators* in an in vitro construct. Yusufzai et al. showed that, together with CTCF, the nucleolar protein nucleophosmin tethers the insulator to the nucleolus and, thus, may generate *chromatin* loop structures *separating* promoters from their *enhancer*.

Recent chromatin studies revealed that, similar to observations made in *yeast, histone* modifications (acetylation and H3K4 *methylation*) in human appear to be pivotal for the insulator function.

Deregulation of Expression

Beckwith-Wiedemann syndrome (BWS) is caused by deregulation of expression of imprinted genes in chromosome 11p15.5. BWS is characterized by prenatal overgrowth, *midline abdominal wall defects, macroglossia,* and *embryonal tumors. H19* and *Igf2* genes implicated in the BWS are expressed from the maternal or the paternal chromosome, respectively. Webber et al. described the unmethylated imprinted-control region (ICR) that acts as a chromatin boundary between differently imprinted *Igf2* and *H19*.

Two independent groups showed that the binding of the enhancer-blocking protein CTCF to the ICR depends on its *methylation* status. The binding of CTCF to ICR on *maternal chromosome* averts interaction between the *Igf2* promoter and its enhancer, leading to *Igf2* silencing. On the *paternal allele, spreading* of methylation of the ICR inhibits *H19* and eliminates the CTFC-ICR assembly, resulting in normal *Igf2* expression.

Thus, *DNA methylation*- and CTCF-dependent enhancer blocking mediate genomic imprinting of *H19* and *Igf2* loci. Supporting this notion, Prawitt et al. *(54a)* described a 2.2-kb familial microdeletion involving three CTCF-target sites in the H19/IGF2-imprinting center and proposed that it is causitive for BWS and Wilms tumor.

Autosomal Dominant Disorder

CTG trinucleotide repeat expansion in the DM 1 locus on chromosome 19q 13.32 affects the expression of two adjacent genes *DMPK* and *SIX5* and leads to myotonic dystrophy, an *autosomal* dominant disorder characterized by *myotonia, muscular dystrophy, cataracts, hypogonadism, frontal balding,* and *electrocardiogram* changes. Filippova et al. demonstrated that CTCF binding sites flank the CTG repeat and act as insulator element between *DMPK* and *SIX5*.

Similar to the *H19* and *Igf2* genes, methylation of these sites eliminates binding of CTCF. Thus, methylation of the DM I locus in *myotonic dystrophy* disrupts the insulator function. The role of CTCF in the insulator function has been proposed also in establishing a regulatable epigenetic switch for X chromosome *inactivation* and in cancer.

COMPARATIVE ANALYSIS

Comparative genomics studies of several different species enabled identification of conserved non-coding DNA sequences that appear to have important biological functions. Several computer programs have been developed to perform such comparative analyses. In addition to the anticipated protein-coding genes and non-coding RNA genes, the multispecies comparison of human chromosome 21 sequence revealed the presence of a large number of highly conserved DNA sequence blocks of unknown function.

Intriguingly, these potentially functional CNGs, or multispecies conserved sequences, are significantly more conserved than protein-coding genes. In addition, it has been estimated that in the human genome there are approx 60,000 CNGs, approximately twice as many as the number of coding genes and that CNGs may constitute 0.3-1 % of the human genome. It is *tempting* to speculate that these elements may play a role in long-distance functional or structural gene regulation (e.g., in *chromatin folding/ loop* formation by matri x scaffold attachment regions).

α-Globin Gene

Recently, by studying the *α-globin* gene, Tufarelli et al. described a novel mechanism of position effect-related gene regulation, "anti sense-mediated cis-acting methylation utilizing non-coding RNA (ncRNA)," that is similar to the *XIST/TSIX-mediated* X chromosome inactivation. It has been estimated that at least 20% of human genes have antisense transcripts.

Several non-coding transcripts have been found associated with human diseases including B-cell lymphoma, lung cancer, prostate cancer, cartilage-hair hypoplasia, spinocerebellar ataxia type 8, DiGeorge syndrome, *autism*, and *schizophrenia* (69). ncRNAs-based gene regulation may be an important mechanism of cell function particularly in more complex organisms; however, the molecular mechanisms still await *elucidation.*

MORPHOLOGY OF CHROMATIN

Distal enhancers have been hypothesized to interact with gene promoters through a set of proteins that act by bending DNA to bring the enhancers within physical *proximity* of their respective target gene, thus, regulating its function (*chromatin* loop contact model). After analyzing the β-*globin* locus using the 3C technique, Tolhuis et al. proposed an "*active chromatin hub* (ACH)" model, wherein active genes and DNAse I hypersensitive sites are in close spatial proximity forming a three-dimensional (3D) hub of *hyperaccessible* chromatin and the remaining *fragments* loop out.

Interestingly, this structure was found only in gene-expressing erythroid tissues and was linearized in nonexpressing brain tissue. Moreover, during erythroid differentiation, *cis-regulatory* elements of the β-*globin* locus form a nuclear *compartmentalization* with RNA polymerase II representing ACH. Subsequently, the same group proved that multiple interactions between *cis-*

regulatory elements and the target gene are required to maintain the ACH structure. Supporting this notion, Horike et al. *demonstrated* that formation of a *silent-chromatin* loop is a novel mechanism underlying *MECP2* gene regulation in the Rett *syndrome*.

Formation of Chromatin Loop

By attaching chromatin to the nuclear matrix, *matrix-scaffold* attachment regions (M/SARs) are thought to participate in chromatin loop *formation* and represent initiation sites for *chromosome condensation*.

Recently, Ioudinkova et al. hybridized a DNAse I digested nuclear matrix containing the M/SAR elements to a 60mer oligonucleotide arrays covering the chicken α*-globin* locus and identified a co-localization of 40-kb α*-globin* chromatin loop attachment region with the nuclear M/SAR and the previously described CTCF-dependent enhancer blocker elements.

Compartmentalization

Transcriptionally active chromatin is believed to be *markedly compartmentalized* in chromosome territories. Active *loci* are located *predominantly* at or near the surface of compact *chromatin domains*, depositing newly synthesized RNA directly into the *interchromatin* space. Tanabe et al. showed *evolutionary* conservation of chromosome territory arrangements in cell nuclei from higher primates.

Using chromatin precipitation, sucrose gradient sedimentation, and array comparative genomic hybridization (aCGH) techniques, Gilbert et al. showed that gene-rich domains are enriched in open chromatin fibers and suggested that domains of open chromatin may create an environment that *facilitates* transcriptional activation and could provide an evolutionary *constraint* to maintain clusters of genes together along chromosomes.

A growing number of human diseases have been shown to result from abnormal chromatin structure. *Dysfunctions* in proper chromatin *remodeling* can affect epigenetic interactions and in turn lead to gene expression *multi-system* disorders and *neoplasias*.

The importance of chromatin-nuclear envelope interactions has been emphasized further in laminopathies, a new group of genetic diseases, wherein mutations in *lamin A/C*, a constituent of the nuclear envelope, can presumably deregulate chromatin-nuclear envelope interactions that *manifest* with a broad *spectrum* of disorders.

These include Emery-Dreifuss and limb-girdle muscular *dystrophies* (skeletal muscles), Charcot-Marie-Tooth *neuropathy* (*peripheral nerves*), *dilated cardiomyopathy* (heart), *Dunnigan familial partial lipodystrophy* (adipose tissue), and *mandibuloacral dysplasia* (skeletal system).

The best known example of disease resulting from *abnormal chromatin* structure or function is probably facioscapulohumeral *muscular dystrophy* (FSHD). FSHD has been included also in a novel group of genetic diseases associated with *transcriptional derepression*.

Muscular Dystrophy

FSHD is the third most common *muscular dystrophy* and usually *manifests* in the second decade of life with *progressive* wasting of facial, *upper arm*, and *shoulder muscles*. FSHD is caused by interstitial deletion involving *polymorphic heterochromatic* 3.3 *kb D4Z4* repeats in 4q35. The number of repeats varies between 11 and 100 copies in normal individuals. In patients affected with FSHD, the number of *D4Z4* repeats is lower than 11 (<35 kb) and is associated with derepression and altered methylation of *ANT1, FRG1,* and *FRG2* and hypomethylation of *D4Z4*. An element within *D4Z4* has been proposed to bind a multi- protein complex and mediate transcriptional repression of the 4q35 genes, *ANTI, FRG 1,* and *FRG2.* Consequently, an epigenetic PEV-like mechanism with *D4Z4* acting as a silencer element has been proposed.

Recently, four models attempting to explain the pathomechanism of FSHD have been postulated. In a cis-spreading scenario, interstitial deletion of *D4Z4* leads to local chromatin *relaxation* with spreading of upregulation.

The *insulator* mechanism predicts inefficiency of the contracted *D4Z4* boundary element to the distally located heterochromatin and in the chromatin cis-loop model, the number of *D4Z4* repeats determines the 3D chromatin structure and, thus, direct interactions between *D4Z4* and the upstream genes. The fourth model supplements the third one and implies the perturbation of 4qter with nuclear lamina and subsequent misbalance of chromatin and transcription factors.

MOLECULAR TECHNIQUES

Recently, a number of molecular techniques that enable insight into chromatin structure and function have been developed. Dekker et al. described a methodology that can estimate the spatial organization and frequency of interactions between two genomic regions in organisms from bacteria to human. After *formaldehyde*-induced *crosslinking, EcoRI* digestion, intramolecular ligation, and reverse crosslinking, the frequency of different loci being connected is assessed by quantitative *polymerase* chain reaction (PCR) with primers specific to gene promoter and other *investigated loci.*

Contact Model

To identify the locus control regions that directly interacts with the investigated gene (contact model), Carter et al. developed a RNA fluorescence *in situ* hybridization method called RNA-TRAP. The tissue to be analyzed is attached to poly-L-lysine coated glass slides and fixed with formaldehyde. Intron-specific oligonucleotide probe labeled with digoxigenin is hybridized to the expressed pre-mRNA, followed by addition of *anti-digoxigenin* antibody labeled with horse radish and subsequently *biotin-tyramide* that on contact with horse radish, covalently attaches to neighbor proteins (*tyrosines*).

Such prepared cells are removed from slides, fragmented by *sonication,* and purified by affinity *chromatography* on a *streptavidin-agarose* column. Finally, similar to the 3C technique, the gene-interacting DNA fragment can be identified by PCR. Both 3C and RNA-TRAP techniques have proven to be sensitive and specific and have been used successfully for assessment of the distances between long-range gene enhancers and promoters, the influence of 3D organization of a gene locus for its *regulation,* and *dynamic* organization of genes into shared nuclear compartments.

Chromatin Participation

The combination of aCGH with *chromatin* precipitation (ChIP) techniques enabled the development of a novel useful *methodology,* termed ChIP-on-chip or ChIP-chip. Like in the two above techniques, in the first step, the DNA and proteins are crosslinked using *formaldehyde*. Subsequently, the specimen is sonicated and immunoprecipitated with an antibody against the investigated protein. The separated DNA-protein complex is *cohybridized* with the reference DNA using aCGH with target DNA of interest to determine the localization of the associated DNA.

Optical Study

Bussiek et al. applied scanning force microscopy to analyze DNA loop formation on nucleosomes. Another tool that proved to be useful in decoding the human chromatin structure on the genomic level is electron *immunomicroscopy*.

McArthur et al. showed that real-time PCR enabled *quantification* of the sensitivity of *chromatin* to digestion by DNase I from small amounts of *tissue* samples. Such an approach provides the potential to achieve accurate and detailed mapping of chromatin structure.

CONCLUDING REMARKS

Studies of position effects contributed significantly to a better understanding of gene *regulation*; however, we are only beginning to understand the *epigenetic* effects of *chromatin*. 3D chromatin *dynamics*, evolutionarily conserved *nongenic* sequences, and boundary elements (*insulators*) appear to be the major factors responsible for position effect in humans.

Chapter 8

Marker Chromosomes

The presence of a small *supernumerary* marker chromosome (SMC) in a karyotype creates a *diagnostic dilemma*, because the resulting *duplications/triplications* may cause *abnormal* development, depending on the location and size of the extra material. The most common SMC is the inv dup(15), the effect of which varies with size of *triplication* as well as the parent of origin. inv dup(22) is associated with the highly variable cat eye *syndrome*.

Both are thought to be caused by U-type *recombination* between *neighboring* low-copy repeats (LCRs), resulting in both symmetric and *asymmetric bisatellited dicentric supernumerary chromosomes*. Studies are underway to associate the abnormal features of each syndrome with specific genes in the duplicated regions.

SMCs, detected *prenatally* or postnatally, have presented a diagnostic dilemma since the birth of *cytogenetics* nearly 50 years ago. SMCs, also referred to as extra structurally abnormal chromosomes, are present in addition to the normal chromosome complement, usually not *identifiable* by standard *staining techniques*, and typically smaller than the smallest *autosome*.

In a landmark study of almost 400,000 *amniocenteses*, 0.04% had an SMC of *cytogenetically unidentifiable origin*. The incidence of SMCs at live birth has been variously reported as 0.05% and 0.07%. SMCs can also exist as rings and occasionally may be present in multiple copies per cell. *Mosaicism* has been seen in 59% of all SMC cases, but is higher (69%) for SMCs derived from *nonacrocentric* chromosomes.

In any case, the presence of an SMC may lead to an imbalance for whatever genes are duplicated in the SMC. When studied using fluorescence *in situ* hybridization (FISH), the chromosomal origin of most SMCs can be identified by the presence of chromosome-specific α-satellite DNA (although *commercial* satellite probes to *differentiate* chromosomes 13 from 21 and 14 from 22 are not available).

More recent variants of the FISH technique, such as *spectral karyotyping*, centromere-specific multicolor FISH, and *subcentromeric multicolor*-FISH have allowed finer identification, the latter technique determining the presence of specific *subcentromeric* euchromatin. SMCs derived from chromosome 15 can also be identified by specific staining with DAPI, a DNA-binding dye. In a study of 112 *autosomal* SMCs ascertained both *prenatally* and *postnatally*, 68% were derived from *acrocentric autosomes*, and of those, 51% were derived from chromosome 15 (35% of all SMCs).

The risk of an abnormal *phenotype* associated with the 32% of SMCs that are non acrocentric -derived has been calculated as approx 28%, whereas SMCs derived from *acrocentric autosomes*, excluding chromosome 15, carry a lower risk (approx 7%) of abnormal *phenotype*. Correlations between the SMC origin and phenotype have been attempted.

SMCs derived from chromosome 15 are associated with defined risk and distinct *phenotype*, depending on their genetic composition. As described below, *imprinting* of the *proximal* region of the chromosome 15 long arm also alters gene dosage effects. Specific SMCs derived from chromosome 22 cause a distinct phenotype, the cat eye syndrome (OMIM 115470).

INVERTED DUPLICATION

SMCs containing two copies of a *centromere*, arranged in an apparent mirror image symmetry around a central axis, have been referred to as isodicentric (idic), pseudoisodicentric (psu dic) or, more correctly, inverted duplication (inv dup). The best studied, recurrent inv dup SMCs are derived from chromosomes 15 or 22. These chromosomes are bisatellited because of an *acrocentric* p-arm on each end.

They also contain two centromeres, usually separated by at least several megabases of *euchromatin*. Such chromosomes are often stable, presumably because one centromere is *inactivated* (hence the term pseudoisodicentric). An mv dup SMC associated with an otherwise normal karyotype results in a total of four copies of the excess region, and is therefore a *triplication* or partial *tetrasomy*.

inv dup SMCs have been referred to as isodicentric, but this term is appropriate only if the duplications on each side are symmetrical. inv dup *chromosomes* have been found often to be *asymmetrical*, with one side of the chromosome considerably larger than the other. This results in a total of four copies of some regions and only three copies of others.

inv dup SMCs can also originate from the other *acrocentric chromosomes*, but with less frequency and are associated with a less well-defined *phenotype* than that for chromosomes 15 and 22. *Nonacrocentric* chromosomes are also a source for inv dup SMCs. These may contain no α-satellite DNA and yet are stable, providing an opportunity to address the question of centromeric function in the absence of the DNA normally present at centromere.

These unusual inv dup SMCs are C-band negative, yet have a G-banded primary constriction, which acts as an active *kinetochore* and reacts with CENP-C antibodies. It is presumed that these SMCs activate noncentromeric sequences that function as neocentromeres.

It has long been suggested that inv dup chromosomes are derived from a "U-type" rather than the normal "X-type" exchange between *nonsister* or sister *chromatids* at meiosis I. Mediated by LCRs on chromosomes 15 or 22, a U-type exchange between repeats in opposite *orientation* would lead to a dicentric SMC as well as an *acentric* fragment composed of two copies of the rest of each *chromatid*.

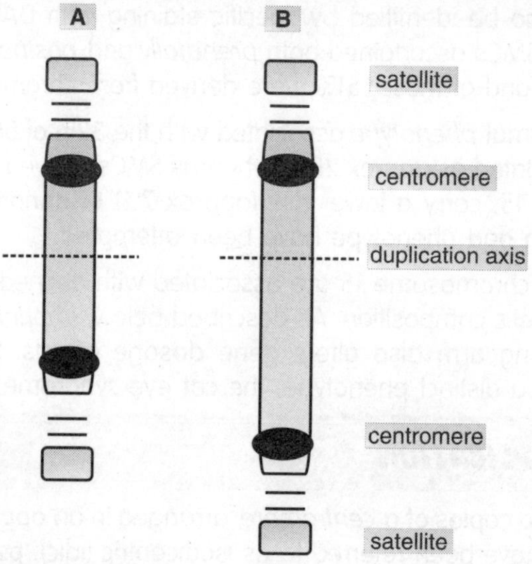

Figure 8.1: Structure of bisatellited and dicentric inv dup chromosomes. These chromosomes can have a symmetrical duplication (A) or be asymmetrical, where one side of the duplication is largerthan the other (B). In an asymmetric inv dup, the region nearest the centromere is present in two extra copies (light gray), whereas the more distal region is present in only one extra copy (dark gray).

The acentric fragment would be lost, whereas the SMC would be retained through *nondisjunction* and *inactivation* of one *centromere*. Each LCR in chromosome 22q11 is composed of complex blocks of repeats. LCR2 and LCR4, in which most rearrangements occur, contain shared repeat blocks in both the same and opposite orientations, which facilitate U-type exchanges.

If a U-type exchange occurs between elements of the same LCR (*allelic*), the resulting inv dup chromosome would essentially be symmetrical. A U-type exchange between similar elements of different LCRs (*nonallelic*) would produce an *asymmetric* SMC. *Asymmetric* inv dup SMCs could also result from a *paracentric* inversion of a region between LCRs, followed by *recombination* within the inversion loop (15). This would similarly result in a dicentric SMC and an acentric fragment that is lost.

INVERTED DUPLICATION OF CHROMOSOME (15)

inv dup chromosomes derived from chromosome 15 account for approx 35% of SMCs, and, after trisomy 21, are the most common *autosomal chromosomal aberration*. Although the phenotype associated with the inv dup itself can be variable, *uniparental disomy* or deletion of the normal chromosome 15 can accompany the inv dup chromosome with additional clinical consequences.

The presence of abnormalities on the "normal" chromosomes 15, the size of the inv dup, and the parental origin of the chromosomal *abnormalities* all affect the *severity* of the outcome. This information is critical in the context of genetic *counseling*, particularly in the setting of prenatal ascertainment of a *de novo* inv dup.

Because of their relatively common frequency, chromosome 15 rearrangements have provided a rich source of information for the study of chromosomal abnormalities. The presence of a set of

LCRs on the proximal long arm of chromosome 15 predisposes this region to a *heterogeneous* group of inter- and *intrachromosomal* rearrangements.

The *nonallelic* copies of these LCRs can misalign during *meiosis*, and the resulting *nonallelic homologous* recombination or unequal crossover event gives rise to structurally abnormal chromosomes. The LCRs are present in the breakpoint regions involved in the rearrangements, and are named in ascending order from the most *centromeric*, as *break point* (BP)1 to BP5.

The chromosomal disorders Prader-Willi *syndrome* (PWS; OMIM 176270) and Angel man syndrome (AS; OMIM 105830) most commonly involve a deletion of the *genetic material* between BP2 and BP3A/3B, with PWS deletions occurring on the *paternally* derived chromosome, and AS deletions occurring on the *maternally* derived chromosome.

Although both syndromes include developmental delay, PWS individuals *typically* have a neonatal failure to thrive and hyperphagic obesity, among other *neuro-developmental* findings, whereas AS individuals typically have *seizures* and *absent* speech. The difference between the phenotypes is owing to the presence of imprinted genes in the deletion region, which are expressed on only one chromosome in a manner dependent on parent of origin.

The paternally expressed genes responsible for PWS are *clustered* toward the *centromeric*

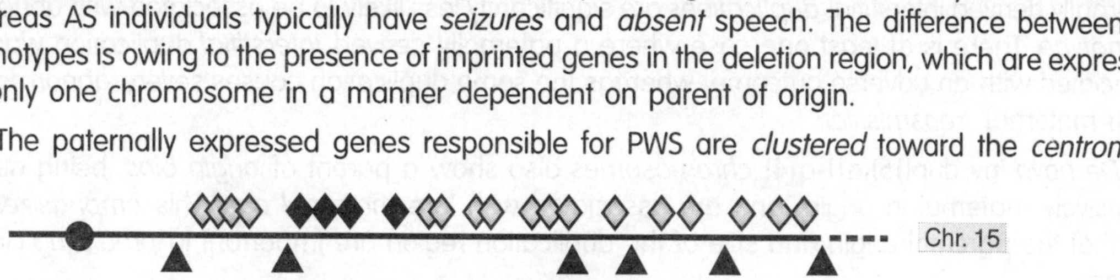

Figure 8.2: Examples of various types of inv dup(15)s. Chromosome 15q11-q 14 contains a set of genes that are expressed only from the paternally inherited allele (black diamonds, Prader-Willi syndrome [PWS] candidate genes), from the maternal allele in a tissue-specific fashion (white diamonds, including the Angelman syndrome [AS] gene UBE3A) or from both alleles (not imprinted, gray diamonds). Rearrangements of chromosome 15q11-q14 generally involve a set of low-copy repeats labeled BP I through BP5; the PWS/AS interstitial deletions typically occur between BP2 and BP3A/3B. inv dup(15) chromosomes can involve any of the breakpoint regions, and can be symmetrical or asymmetrical, as in the bottom example. inv dup(15)s that contain material telomeric to BP2 (shaded gray) are associated with an adverse outcome. The centromere is represented by the black circle toward the left.

end of the BP2-BP3A/3B deletion region. Four of the PWS candidate genes, *MKRN3, NDN* (necdin), *MAGEL2,* and *SNURF/SNRPN,* are protein-coding and expressed in the nervous system during development. Another PWS *candidate* gene encodes a functional RNA, the imprinting center *transcript.* This transcript also contains a set of small *nucleolar* RNAs as well as acting as an antisense RNA for a more *telomeric* gene, *UBE3A.*

In tissues in which the imprinting center transcript is active, this *antisense* regulation imparts paternal *allele-specific* gene *silencing* of *UBE3A. Chromosomal* deletions or other types of mutations that inactivate the maternal allele of *UBE3A* causes AS. In addition to deletions, unequal recombination events can give rise to rare *interstitial* duplications and *triplications,* and to the more common inv dup.

Maternally derived interstitial duplications of 15q11-q14 give rise to a phenotype, characterized by various types of *intellectual* impairment, which is distinct from the deletion syndromes, whereas paternally derived interstitial *duplications* are significantly less likely to be associated with abnormal *phenotype.* There is at least one case where a paternally derived *interstitial* duplication was not associated with an adverse outcome, whereas the same duplication causes severe *abnormalities* upon maternal *transmission.*

De novo inv dup(15)(q11-q14) *chromosomes* also show a parent of *origin bias,* being almost exclusively maternal in origin, and are associated with late *maternal age.* This *emphasizes* the fact that the parental origin and size of the duplication region are important in *predicting* clinical outcome.

The smallest inv dup(15)s contain no duplicated material from the PWS/AS deletion region between BP2 and BP3 and carry a low risk of adverse outcome, despite the presence of at least four protein coding genes in the region *centromeric* to BP2. The largest interstitial and *supernumerary* duplications contain extra copies of the chromosomal material extending distally to the PWS/AS BP3 region, and are associated with dysmorphic features, severe developmental *delay, autism, seizures, strabismus,* and *abnormal dermatoglyphics.*

inv dup(15)s, because of their derivation from an imprinted region, can present an unusual situation in terms of gene activity. A PWS or AS phenotype has been found in individuals who, in addition to the inv dup(15), carry either a deletion or uniparental *disomy* for chromosome 15, leading to the parent of origin phenotypes associated with loss of one parental copy of imprinted genes.

The inv dup(15)s carrying duplicated PWS/AS region material are categorized according to the breakpoints involved, with examples given in Figure elsewhere in this chapter. Several groups have identified the repeat regions involved in inv dup(15)s and *interstitial* duplications, using FISH, somatic cell hybrids and array *comparative* genomic *hybridization* to detect the dosage and position of single genomic sequences.

At high resolution, some inv dup(15)s that were thought to be symmetrical are in fact asymmetrical, and involve recombination between two different BP repeat clusters. The most common of the larger inv dup(15)s involves BP4 and BP5, whereas most of the *remainder* are the result of a *symmetrical* BP3 to BP3 *rearrangement.*

Individuals with this latter class of inv dup(15) carry one *paternally* derived and three *maternally* derived copies of the PWS/AS region, whereas individuals carrying the BP4:BP5 inv dup(15)s carry this genetic material plus additional copies of maternally *inherited* genes distal to the PWS/AS region. Clear phenotypic differences among these latter classes, if they exist, have yet to emerge.

A dosage effect associated with inv dup(15)s is evident, however, with reports of individuals with hexasomy for 15q11-q14 at the severe end of the *phenotypic spectrum*. The high frequency of *chromosomal* rearrangements is undoubtedly related to the presence of the "BP" LCRs, but rearrangements appear to arise from a variety of *recombinational* mechanisms.

Reciprocal recombination events of the PWS/AS deletion region likely cause interstitial duplications involving the same LCRs. Studies using probes located near the chromosome 15 centromere suggest that aberrant U-type exchange resulting in inv dup(15) can occur between sister *chromatids*, or between *homologous* chromosomes either during or after recombination.

The development of informative markers closer to the centromere of chromosome 15 and sequencing of the breakpoints will *facilitate* further *analyses* of the structure and mechanism of origin of these structurally abnormal *chromosomes*.

INVERTED DUPLICATION OF CHROMOSOME (22)

The consequence of inv dup(22) for clinical phenotype is in one way simpler than that for the inv dup(15) since chromosome 22 does not appear to be influenced by genomic imprinting. No convincing sex bias has been seen in chromosome 22 rearrangements, and three cases of maternal uniparental disomy 22 have shown normal phenotype. However, the remarkable *phenotypic* variation of the associated syndrome results in *diagnostic* problems of its own.

inv dup(22)(q11) is associated with cat eye syndrome (CES), which is characterized by *abnormalities* of the eye, heart, anus, kidneys, skeleton, gastrointestinal tract, and face. Patients may show mild mental retardation, but many are within the normal *intelligence* quotient (IQ) range. The syndrome derives its name from ocular *coloboma*, although only about half of CES patients show this feature.

Preauricular skin tags or pits are the most constant feature. The CES phenotype is surprisingly variable, ranging from apparently normal to multiple severe and *lifethreatening malformations*. The incidence of CES has been estimated to be 1/50-150,000 (OMIM 115470). Because some patients are mildly affected, there are *numerous* cases of inheritance of the CES chromosome, sometimes through multiple *generations*.

Although the CES chromosome is often stable, it is not uncommon to detect *mosaicism*. There are also cases where the inv dup(22) has shown *instability* within a family, resulting in secondary *rearrangements*. The parental origin of few inv dup(22)s has been determined. By comparing Q-banded *polymorphisms*, two cases were shown to originate from both maternal *homologs*.

A similar analysis identified a third case of maternal origin, but from the same chromosome 22 homolog. Thus, both intra- and *interchromosomal* recombination can be involved in the production of inv dup(22).

The 22q11 region has a variety of chromosomal rearrangements that are mediated by LCRs . Most prominent is the common deletion associated with disgorge *syndrome/velocardiofacial* syndrome (dGS/VCFS). The majority of dGS/VCFS deletions localize between LCR22-2 and LCR22-

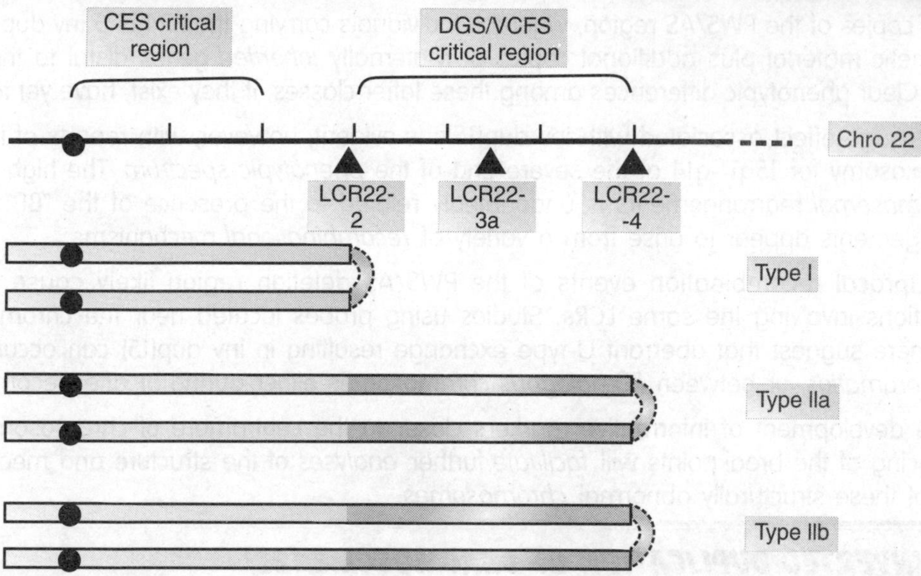

Figure 8.3: Composition of various inv dup(22)s associated with cat-eye syndrome (CES). The inv dup(22) breakpoints coincide with the location of the of low copy repeats associated with DGS/VCFS deletions. Type I and IIb chromosomes are symmetrical duplications, with the latter containing the DGS/VCFS critical region as well as the CES region. Type IIa chromosomes are asymmetrical, with only one extra copy of the DiGeorge syndrome/velocardiofacial syndrome region, but two extra copies of the CES region. There is also one reported case of a CES chromosome with a LCR22-3a breakpoint, although in this case the location of the second breakpoint is not clear (50, case 2).

4, *spanning* approx 3 Mb. A small percentage of deletions occur between LCR22-2 and LCR22-3a, spanning approx 1.5 Mb. CES chromosomes also appear to break at these LCRs.

An inv dup(22) is formed from breakpoints in two chromatids. In a type ICES chromosome, both breakpoints are at LCR22-2, only the CES region (proximal to LCR22-2) is included and the inv dup(22) is symmetrical. In type II CES chromosomes, the dGS/VCFS region is also included in the duplication. Type II chromosomes can be symmetrical, with both breakpoints in LCR22-4 and two additional copies of the dGS/VCFS region (type IIb), or asymmetrical, with breakpoints at LCR22-2 and LCR22-4 and with only one extra copy of the dGS/VCFS region (type IIa).

A case has also been found with at least one breakpoint at LCR22-3a. The CES phenotype results from overexpression of dosage sensitive genes in the CES *critical region.* The critical region for CES was determined by the analysis of an unusual patient with a relatively stable, *supernumerary double* r(22).

This child had all major physical features of CES, although development could not be assessed as the child died at 17 days. The ring contained approx 2 Mb of proximal 22g1 I from the centromere to a region distal to the *A TP6Egene.* A typical inv dup(22) contain at least an additional 1 Mb. The first 1-1.5-Mb from the 22 centromere contain *pericentromeric* repeats that are unlikely to contain active and unique genes.

Thus, the majority, if not all, of the genes in the CES critical region lie within an approx 700-kb interval that contains at least 14 predicted or known genes. The gene or genes causative for CES features has yet to be determined. Candidates identified include a secreted growth factor *predicted* to have adenosine *deaminase* activity *(CECRI),* and a *transcription* factor involved in chromatin remodeling.

A proportion of CES chromosomes have one or two copies of the DGS/VCFS region in additional to four copies of the CES region. Duplications of the DGS/VCFS region have been associated recently with variable symptoms including learning disabilities, *cognitive* and *behavioral abnormalities, palate defects, hearing loss, heart defects, growth deficiency,* developmental and motor delays, and *urogenital anomalies.*

This likely represents only a spectrum of the *microduplication* 22q11.2 syndrome, because the patients were ascertained during testing specific for the DGS/VCFS deletion. In any case, one would assume that the addition of extra copies of the DGS/VCFS region would worsen the phenotype of CES, but this has not been observed.

However, because only 10 CES patients have been characterized in this way, this lack of *correlation* is likely to be because of the extreme variability of the overall syndrome masking more subtle changes brought on by additional copies of the DGS/VCFS region. Only the analysis of a much larger cohort of CES patients will determine whether a correlation exists between *phenotype*

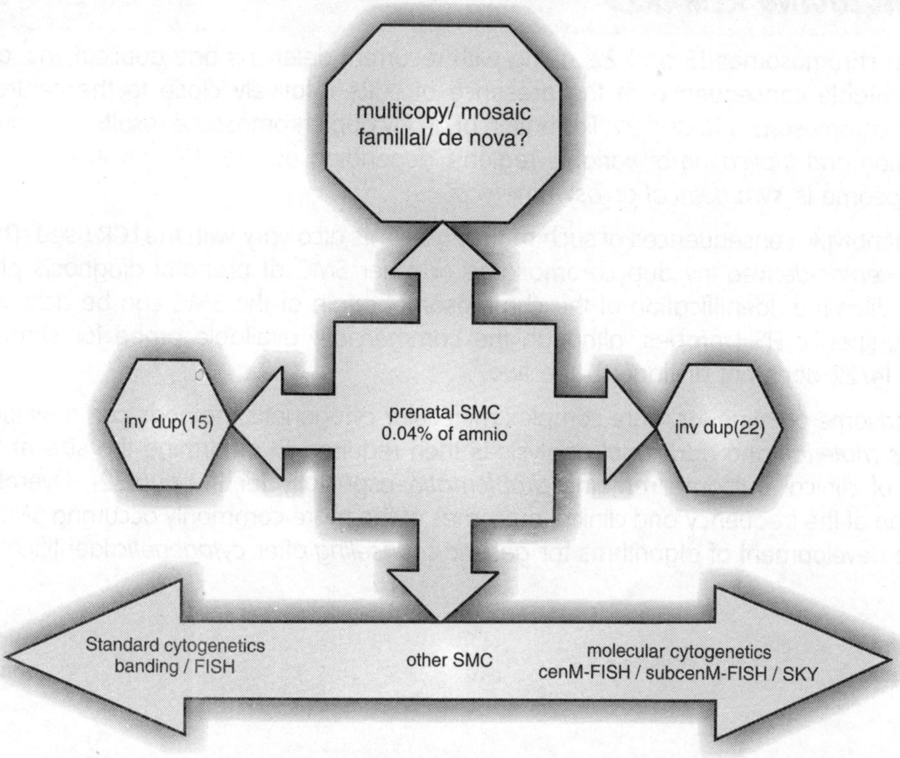

Figure 8.4: Prenatal identification of small marker chromosomes-a diagnostic predicament. Prenatally identified small marker chromosomes (SMCs) involving chromosomes 15 and 22 have partially predictable outcomes. The phenotypic outcome for a fetus with an SMC can depend on its molecular composition, whether the SMC is familial orde novo, *and whether it is present in multiple copies or only in some cells. Overall, molecular and cytogenetic tests have been or are being developed that should unequivocally identify the extra material present in the SMC.*

and the size of the CES chromosome. CES is usually associated with the inv dup(22), however *interstitial* duplications have also been reported.

Again, because of *phenotypic variability* and problems of *ascertaining* mild cases, it is not yet clear whether three copies of the CES region produce less *susceptibility* to *malformations* than four copies. Interestingly, the father and grandfather of the child with the *supernumerary* double r(22) both had a single supernumerary r(22) and appeared normal, stressing the variability of this syndrome even within families.

There is also a class of inv dup(22) associated with a normal phenotype. These chromosomes are often familial, and *presumably* contain no dosage-sensitive genes. The size in comparison to CES chromosomes has not been determined yet, and no commercial probe exists to differentiate the two.

The cytogenetic appearance of these small inv dup(22)s is not always helpful in distinguishing them from CES chromosomes. This makes the prenatal discovery of an inv dup(22) a serious genetic counseling dilemma.

CONCLUDING REMARKS

inv dup chromosomes 15 and 22, along with recurrent deletions and duplications, appear to be the inevitable consequence of the presence of LCRs relatively close to the centromere of *acrocentric* chromosomes 15 and 22. Formation of an inv dup chromosome results in a combination of *duplication* and *triplication* of variable regions, depending on the LCRs involved and whether the chromosome is *symmetrical* or *asymmetrical.*

The phenotypic consequences of such rearrangements also vary with the LCR used. The finding of an *acrocentric-derived* inv dup chromosome or other SMC at prenatal diagnosis presents a diagnostic dilemma. Identification of the chromosomal origin of the SMC can be achieved using centromere-specific FISH probes, although the *commercially* available probe for chromosomes 13/21 and 14/22 does not distinguish the two.

Chromosome painting, or more complex molecular cytogenetic methods can now identify the *extraneous material,* and additional analysis is then required to determine the size of the SMC. *Prediction* of clinical outcome remains *problematic,* especially for inv dup(22). Overall, further investigation of the frequency and clinical outcomes of the more commonly occurring SMCs should enable the development of algorithms for *genetic counseling* after *cytogenetic* identification of an SMC.

Chapter 9

Biochemical Nature of Genome

Organic chemistry is the science of the synthesis and properties of molecules that are constructed from only a few atom types, such as *carbon, hydrogen, nitrogen, oxygen,* and *sulfur,* with carbon atoms constituting the majority of the core of these chemicals.

As these atoms are also the building blocks of naturally occurring peptides or oligonucleotides, this chemistry was termed *organic,* as opposed to other disciplines of chemistry. Such chemicals, commonly referred to as *small molecules,* are valuable as *medicines* to treat diseases ranging from *headache* to cancer.

Small molecules have recently proven to be *extremely* useful tools to explore the functions of the cell at the genome level, giving rise to the new *paradigm* of chemical genomics. The functional dissection of molecular pathways, cells, and organisms by having a *small-molecule* ligand or *modulator* for every gene product, was the vision of Stuart L. Schreiber, one of the *pioneers* of the chemical genomics field.

Several recent reviews and books deal with the implication of chemical genomics towards drug discovery. The chemical genomics paradigm is seen as a logical follow-up of the Human Genome Project, and several public initiatives on chemical genomics have been started on national levels.

Thus, in the United States, the National Human Genome Research Institute has set up a chemical libraries plan that moves the National Institutes of Health into high through put screening and small-molecule development to "determine function and therapeutic potential of genes and to define molecular networks."

In Germany, the Nationale Genomforschungsnetz initiated a chemical genomics platform by *assembling synthetic* compound libraries as probes for protein function. These compounds will help to validate new targets for novel *therapies* more rapidly, and will enable researchers in the public and private sectors to take these targets and move them through the drug-development pipeline. In this overview we would like to focus on some chemical aspects, problems, and solutions to the *application* of small molecules as tools for chemical *genomics.*

MOLECULAR VARIATION

What are small molecules? For the sake of *simplicity*, let us consider only those organic chemicals as small molecules that have a molecular weight of less than 1000 Dalton. Such a definition is by its nature arbitrary, but allows *separating* out other classes of organic molecules such as molecules that are oligomers of smaller building blocks such as, for example, *proteins*, *oligonucleotides*, or *oligosaccharides*.

The number of all possible and different small molecules with a molecular weight below 1000 is assumed to exceed 10^{60}. To be a specific modulator of a target protein, a small molecule has to act as a ligand, binding to its target. As opposed to to large molecules, small molecules have an average interaction area of up to 400 $Å^2$ with their target-which is, for example, the size of a typical enzyme substrate site.

This relatively small interaction area poses a serious problem to the whole concept of chemical genomics-is it at all possible to find a specific small molecule for each gene product? Indeed, it appears more likely that small molecules are *promiscuous* and may interact rather with a range of targets that have similar binding sites. Many of these "*unwanted*" interactions might not result in *undesired* effects, giving way to selective drugs in vivo.

In odd cases, such secondary interactions may cause toxicities or other side effects; in lucky cases, these secondary interactions might be the real reason why a particular small *molecule* is an effective drug. Therefore, the rational design of such dual- or triple-action compounds has emerged as a new *paradigm*.

The functional, biological diversity of small molecules appears to be inherently more limited than that of large molecules. Whereas the diversity of oligomers can be described by *metrics* like *sequence* and *secondary*, *tertiary*, and *quaternary* structure, the diversity of small molecules is harder to capture. The development of *qualitative* and *quantitative* measures for the chemical diversity of small molecules has only started to evolve with the advent of *combinatorial* chemistry, which enabled the synthesis of a large number of small molecules in one experiment.

As opposed to large molecules, the required high functional diversity of small molecules has to be packed into a rather small volume made up of only 20-30 non-hydrogen atoms. To bind efficiently at a protein-ligand site, most of the small-molecule ligand's binding energy must come from sites not *exploited* by the natural protein *ligand*. The limited *molecular* volume has to interact with the target protein with a maximum binding energy per *atom* to achieve both the required *affinity* and *specificity*.

This binding energy has been calculated for maximal free-energy contributions per *non-hydrogen* atom with approx 1.5 kcal/ mol across a wide variety of *macromolecule*-small-molecule interactions. The empirical data also revealed a significant trend to smaller *contributions* per atom as the relative molecular mass of the ligand increases. Thus, small-molecule ligands are binding to their targets with approximately three times more binding energy per unit area than protein *ligands*.

As subnanomolar binding can be achieved with ligands containing as few as 10-20 atoms, the remaining 10-20 atoms could potentially be used to obtain selectivity. The like-lihood of obtaining the desired functional diversity for small molecules is also correlated to the nature of the *biological* targets of interest. Thus, it is more likely to find enzyme *inhibitors* than small molecules that block protein-protein *interactions.*

Functional diversity of small molecules for chemical genomics experiments can therefore be defined in terms of obtaining maximum affinity to a given target protein by a minimum of molecular volume in order to minimize unwanted binding to other proteins. The *realization* of this Max-Min concept in a library of small-molecule compounds poses a *serious* challenge to *organic chemists.*

As the molecular volume of small molecules is rather similar and does not leave enough room for major variations, the only way to obtain both affinity and selectivity is a maximum diverse distribution of atoms in this small volume. This can not be achieved by varying a small number of building blocks like the 20 amino acids for proteins, but requires the whole repertoire of organic chemistry to *assemble* novel chemical scaffolds.

This recent understanding has led not only to a *re-appreciation* of natural-product chemistry but also to a series of initiatives that aim at the development of novel chemistries to obtain chemical diversity.

Pharmaceutical Compounds

A large range of small-molecule physico-chemical properties can be computed and used for diversity selection and drug likeliness. Thus, drug-like molecules should follow the "rules of five" that requires a logP <= 5, molecular weight <= 500, number of hydrogen bond acceptors <= 10, and number of hydrogen bond donors <= 5. Molecules violating more than one of these rules may have problems with bioavailability or the likelihood of having good drug properties.

A polar surface area of <120 Å^2 has been shown to correlate with bio-availability and can be computed in a straightforward way. A new method based on quantum chemistry performs the rapid, automated fragment-wise construction of an approximate charge density (σ-profiles) on the molecular surface-describing better the surface of the ligand a target protein will interact with.

An optimally diverse compound library of drug-like molecules for chemical genomics experiments would be a collection of representatives from a variety of compound *clusters* of similar molecules. A comparison between various simple and more complex structural descriptors and clustering methods has been given by Brown and Martin, using large sets of different molecules.

Interestingly, the simplest method, counting 153 different substructure keys using a priori knowledge about substructure elements from a fragment dictionary, provided a better similarity measure than using more complex two- or three-dimensional criteria. This points to the difficulty of generating general representations of molecules, and it is probably impossible to hope to characterize a large, diverse data set by a small set of representatives.

Calculated octanol-water partition *coefficients,* molar *refractivities,* or dipole moments can be used as additional criteria to assemble a useful library for *biological* testing. Exclusion criteria should also include knowledge on promiscuous or reactive substructures such as, for example, *excluding* compounds that exhibit thiourea, nitro, or a-haloketo groups such as defined by Rishton.

The above principles have been applied with success to select compound libraries for chemical genomic experiments. We would like to give two representative examples for such a successful selection.

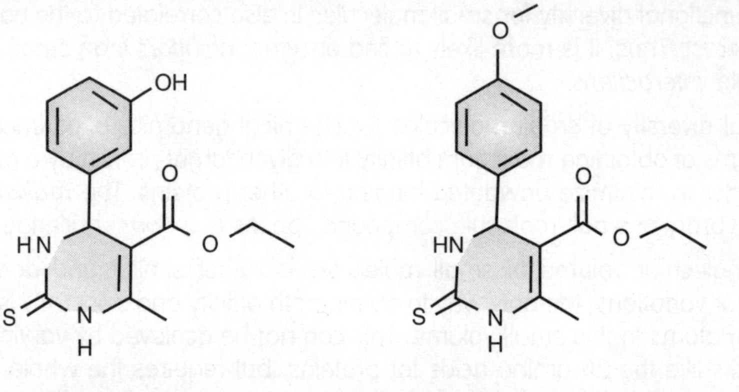

Monastrol DHP1

Monastrol was identified as a novel antimitotic compound from a selected library of 16,320 commercially available compounds. The library was screened by applying a series of phenotype-based screens, selecting 139 compounds that induced phosphonucleolin levels, than subsequently eliminating 52 compounds from these 139 that targeted tubulin and a further 42 antimitotic compounds that target the interphase *cytoskeleton*, leaving overall only 5 compounds with the desired cellular effect.

From these, monastrol arrests *mammalian cells* in mitosis with *mono-astral spindles* surrounded by a ring off chromosomes. The target of monastrol was found to be the mitotic motor protein Eg5, a bipolar kinesin known to be required for *spindle bipolarity*. This cellular activity of *monastrol* is very sensitive to structural changes: the chemically similar DHP2 compound is already completely *inactive*.

In a *conceptually* similar experiment, compounds were selected from commercial vendors to give a screening library of 73,400 diverse, drug-like molecules that comply with the rules of five. Screening of this library in a *permeable* yeast two-hybrid system was accomplished to select molecules that block the protein-protein interaction of cl-DBD-H-Ras(G186) with AD-Raf-1, but not of LexA-DBD-hsRPB7 and AD-hsRPB4.

From this library, 38 active compounds were selected for further screening in mammalian cells to inhibit serum-induced transcriptional activation through SRE and AP-1 sites from the c-fos promoter. MCP 1 was than selected by performing additional cell-based screening to remove unselective or *toxic* compounds. Interestingly, the chemical structure of MCP I resembles other structures that are able to inhibit protein-protein *interactions*, such as, for example, FKB-001, featuring flexible side chains that enable the *molecule* to fold into cavities at the *protein surface*.

On the other hand, both molecules are *accessible* via fast and parallel *synthesis* methods. As opposed to selecting *libraries* of general molecules from historic compound *collections*, for *combinatorial* libraries the problem of a diversity metric is simplified due to the limited number of building blocks that are used for their synthesis.

Using volume, *lipophilicity*, charge, and H-bond donor or acceptor descriptors, it was shown that combinatorial libraries may be designed to exhibit the same diversity as commercial drugs with respect to the density of these structural fragments. *Subsets* from a large virtual library can then be chosen by D-optimal design to assemble an information-rich diverse library. More advanced methods (CATS and TOPAS) have been developed recently to select and synthesize compounds for screening based on *two-dimensional criteria*.

Targeted Compounds

Proteins of a certain family usually exhibit structurally similar binding sites; a prominent example is the *adenosine triphosphate* (ATP)-binding site of the approx 518 kinases identified so far, also called the human kinome. Not surprisingly, compounds that mimic the structural features of ATP are used with success to synthesize more or less selective kinase inhibitors.

These inhibitors can be used both as tool compounds to study the effect of a particular kinase in cell signaling events or as potential drug candidates. A great deal of chemical, *biological*, and *biostructural* information on kinase inhibitors is available and provides a *straightforward* starting point for designing compound libraries that target kinases.

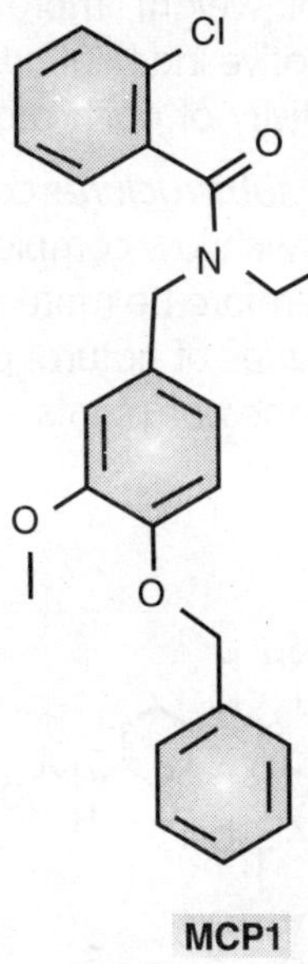

MCP1

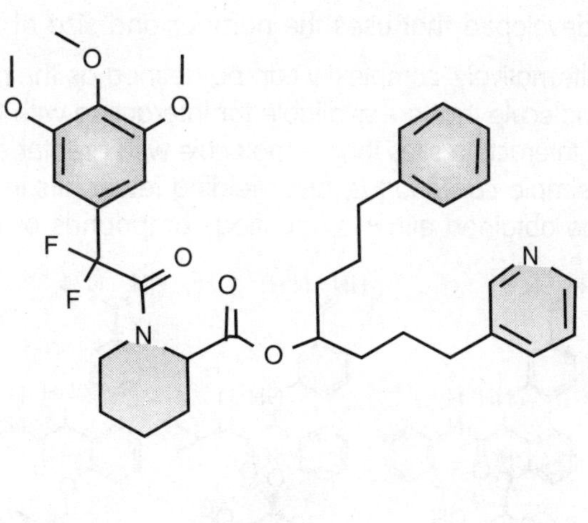

FKB-001

In a recent example, biostructural information was used to classify kinases according to the similarity of their small-molecule ligands, and develop a fast virtual screening method-structural interaction fingerprint (SIFt)-that allowed rapid identification of novel and selective transforming growth *factor-phenotype*, to cellular response, to animal *pharmacology*, to efficacy in the clinic.β kinase inhibitors.

This method, based on experimental data, combines for the first time protein sequence *homology* and the chemical structure information of the ligand into one truly *chemo-genomic* method. In another example of combinatorial parallel chemistry, we have recently used the Ugi three-component reactions (Ugi 3-CR) to construct a library of 16,840 protease inhibitors.

It has been demonstrated previously that the Ugi-3CR reaction provides a useful chemical scaffold for the design of serine protease inhibitors: N-substituted 2-substituted-glycine N-aryl/

alkyl-amides have been identified that are potent factor Xa, factor VIIa, or thrombin inhibitors. The three variable substituents of this scaffold, provided by the *amine, aldehyde,* and *isonitrile* starting materials, span a favorable *pyramidal pharmacophoric* scaffold that can fill the S1, S2, and S3 pockets of the respective protease.

This library was screened against five proteases (factor Xa, *trypsin, uro-kinase, tryptase,* and *chymotrypsin*) at different concentrations to create an *exhaustive* structure-activity relationship (SAR) data set. Using this SAR, both selective and potent *inhibitors* could be identified against all five proteases, such as the four *prototypic* factor Xa inhibitors shown in Figure elsewhere in this chaptere.

Products in Nature

By intuition, natural products are thought to exhibit a higher degree of complexity than traditional *synthetic drugs*. This hypothesis was verified by analyzing the chemical core structures of natural products. In addition, natural products tend to have a higher molecular weight, more oxygen atoms, and fewer nitrogen atoms than synthetic drug molecules. A quantitative index of *complexity* was developed that uses the number and size of rings and the connectivity of each atom.

Alternatively, complexity can be defined as the number of *domains* or *substructures* contained in a molecule that are available for interacting with the target. A molecule with low complexity has fewer interaction sites than a molecule with greater *complexity*, that can therefore be more selective than simple *compounds*, also yielding fewer hits in primary screens. *Libraries* of natural products can be obtained either as purified compounds or as extracts from *microbes* or plants.

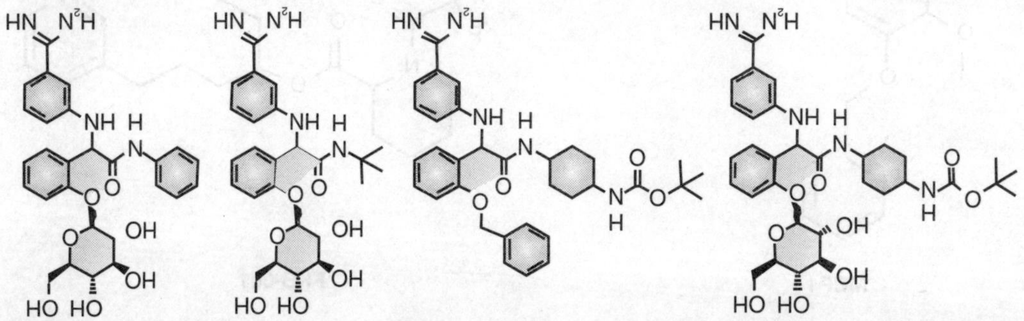

Although most natural products do not follow the rules of five, finding natural-product drug candidates has been a successful enterprise in the past-almost half of the small-molecule drugs on the market are either directly natural products or chemically derived from those.

However, the total synthesis of these complex structures poses a serious challenge to organic chemistry, resulting often in the inability to follow up first hits quickly by *resynthesis* and *derivatization*. Thus, the synthesis of complex natural products using efficient strategies, often *accompanied* by the development of novel reagents and reactions, was considered the "*royal*" discipline of organic chemists in the past.

Today, we observe that those natural -productoriented organic chemists are broadly using the developed know-how to bring together the pharmacophores of structurally diverse natural products with methods of *combinatorial* chemistry.

6-epi-Dysidiolide

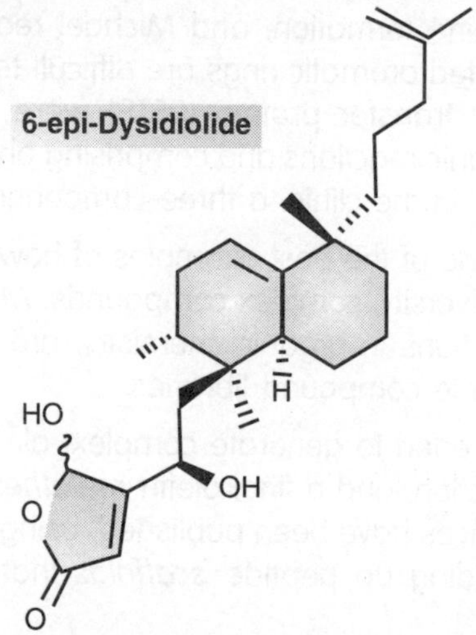

As a representative example, the rather complex natural-product dysidiolide was synthesized by methods amenable to *combinatorial* synthesis. Using a *cycloaddition*-based approach to the dysidiolide core, a *solid-phase* synthesis of *epi-dysidiolide* and analogs thereof was developed.

The total synthesis proceeds in >10 steps on solid phase and includes various transformations, including an asymmetric Diels-Alder reaction, oxidation with singlet oxygen, and olefin metathesis. This synthesis sequence is among the most advanced and demanding solid-phase syntheses developed so far for chemical genomics experiments. It demonstrates that the total synthesis of complex natural products in multi-step sequences on solid phase is feasible.

Moreover, the biological investigation of dysidiolide analogs has yielded selective Cdc25 inhibitors with high potency in *enzymatic* and *cellular assays.*

HISTORICAL SYNTHESIS

Many natural products have been directly developed in the past as drugs—without the possibility for a significant synthesis of potentially better analogs. This observation has inspired a new strategy of *synthesizing* natural-product-like compounds using *combinatorial,* diversity-oriented synthesis (DOS). Thereby, the generation of molecular *complexity* by reaction sequences that are amenable to *combinatorial* chemistry has become an important theme of recent research.

Optimized reactions have been utilized to create compound libraries that are comparable in terms of structural diversity with natural products. In one example, the alkaloid *martinellic* acid has been the target for transition metal-catalyzed hetero-Diels-Alder reactions. Surprisingly, a reaction was found that yielded the desired scaffold in one step by a novel three-component *coupling.*

In another example, a hetero-Diels-Alder scaffold was used to synthesize *peptidomimetics,* aiming to arrive at *conformationally* and proteolytically stable products derived from biologically active peptides. In a different approach, one may synthesize molecular scaffolds that have not been found in nature. These templates are often also called new *chemotypes,* expressing the hope of finding interesting and unexpected biological activities.

For example, *cycloadditions*, imine formation, and Michael reactions are used to generate such novel backbones. Poly-substituted aromatic rings are difficult to synthesize-in a recent work, inhibitors of the cholesterine ester transfer protein (CETP) were synthesized, using a highly sophisticated assembly of basic organic reactions and comprising phenotype, to cellular response, to animal pharmacology, to efficacy in the clinic. a three-component reaction.

This work certainly represents one of the best examples of how combinatorial chemistry has merged with the synthesis of high-diversity, complex compounds. *Multi-component* reactions, first *misinterpreted* as rather rare exceptions in organic chemistry, are now commonly regarded as useful tools to generate highly diverse compound libraries.

Only four synthetic steps are needed to generate complex, almost baroque structures using an Ugi-type MCR, a Diels-Alder reaction, and a final olefin *metathesis*. Other examples of similar diversity-generating reaction sequences have been published, using oxidative bond formations or *macrocyclization* strategies by building up peptide *scaffolds* that are than cyclized via *olefin metathesis*.

In most recent developments, it has been possible to devise synthetic strategies whereby a variety of chemical scaffolds can be obtained in one scheme. Whereas in former examples of *combinatorial* synthesis a large number of compounds were synthesized that all shared the same chemical *backbone*, this new technology is opening new ways toward a diversity that can be used for chemical genomics in a designed strategy.

CONCLUDING REMARKS

The success of merging genomics and chemistry will depend on finding small molecules that can interact specifically with a given protein target molecule. This goal might be especially difficult to achieve when dealing with proteins that are similar in their sequence, three-dimensional structure, or function, as are kinases. One way, as outlined in this chapter, is to improve chemical methods to arrive at suitable small molecules.

However, the concept of chemical genomics might also inspire and call for completely new *technologies* that integrate chemistry and genomics even better. As an illustration, we would like to cite the work of Shokat. To allow for the fast selection of suitable molecules, the three-dimensional structure of the v-Sic kinase was changed through site-directed *mutagenesis*, creating a new binding pocket in the vicinity of the active site while retaining its *catalytic* properties.

It was then shown that an ATP analog, created through introducing a side chain by chemical means into ATP, resulted in a highly specific substrate that is converted only by the mutated v-Src kinase and not by other members of this large enzyme family. It was thereby possible to study the *pharmacological* role of v-Src kinase without affecting other kinases. This method is the first experimental merger of *genomics* and *small-molecule* chemistry, allowing the creation of tools for chemical *genomics studies*.

Other ideas along these lines may include variants of dynamic *combinatorial* libraries or click chemistry, where the target protein assists the synthesis of the small molecule by selecting a binding molecule out of a large pool of possible *molecules*. The examples described here in any case illustrate a new quality of *investigating* the *interaction* of small molecules within complex biological systems. How can one use the chemical-genomics answers obtained to design better drugs, more rapidly and more *efficiently*?

Typically, the biological responses towards a chemical entity are complex and *multi-dimensional*, such as gene-expression patterns, *protein binding*, *phenotype reversals*, and *phenotype* inductions. *Multi-dimensional* optimization algorithms will be required to translate such biological feedback into the creation of better drugs.

In low-dimensional optimization spaces, traditional structure-activity relationships work well. Higher-dimensional search spaces, as provided by chemical genomics, need *heuristic optimization* procedures, as *exemplified* by *neuronal networks* or *genetic algorithms*.

Genetic *algorithms* utilize *biological* genomics information in combination with a recently introduced, DNA-like description of small synthetic molecules, to provide an opportunity for optimizing small molecules more efficiently using nature's evolutionary principles.

Ultimately, one could *envisage* a co-evolutionary drug-discovery process that would enable the discovery of novel drug targets by using small molecules, which are then improved by respective biological feedback loops into highly potent and selective new drug candidates.

Chapter 10

Diseases Due to Chromosomal Inversion

A number of findings revealed that chromosome inversions are more frequent than deduced from classical *cytogenetic* studies. Indeed, some *paracentric cryptic* inversions have been found to be flanked by segmental duplications, either causing a Mendelian disease owing to the *interruption* of specific genes at *inversion breakpoints* or being present in the normal population as a *polymorphism*.

In the latter case, in the *heterozygous* state they predispose to further unbalanced rearrangements such as inv dup rearrangements or simple deletions and duplications. The importance of this *susceptibility* factor has been well clarified with respect to some genomic disorders involving chromosome 8p and it is now emerging as a possible model that may explain the genetic basis of other recurrent chromosome rearrangements.

BALANCE REARRANGEMENTS

Chromosome inversions are the most common rearrangement *differentiating* humans and the great ape species at the *karyotypic* level. Inversions in which a breakpoint is in heterochromatic regions (1qh, 9qh, 16qh, and Yq) are relatively frequent and are regarded as variants. The most common inversion not involving centromeric *heterochromatin* is the inv(2)(p11q13) that is also considered a benign variant.

Other polymorphic inversions are also *frequent* and include inv(5)(p13q13) and inv(10)(p11q21.2). Apart from these cases, inversions are approx 10 times more rare than the other balanced rearrangements (Robertsonian *translocations*: 1 in 1000; reciprocal translocations: 1 in 625;) having a frequency ranging from approx 0.012 to 0.07% for the *pericentric* inversions and approx 0.01 to 0.05% for the paracentric inversions.

In contrast to translocations for which the most frequent cause of *ascertainment* is the presence of reproductive difficulties, the majority of inversions are ascertained at *prenatal* diagnosis or because of an abnormal phenotype. The risk of unbalanced offspring from an *inversion* carrier, putting together the data of pericentric and the paracentric inversions, is approx 1%, much lower than that of reciprocal translocation carriers (2.7% in families ascertained through a balanced *proband* and 19.2% among the families ascertained through an unbalanced proband).

Accordingly, the reproductive fitness of inversion carriers is 0.926 ± 0.085, higher than that of reciprocal translocation carriers (0.70 ± 0.048;). Reproductive fitness in Robertsonian translocation carriers varies between 0.768 ± 0.056 for the D/D carriers and 0.921 ± 0.123 for the D/G carriers.

HETEROZYGOUS INVERSIONS

Studies on inversion *heterozygotes* in man and in other species have reported crossover suppression in the inverted region and increased *recombination* elsewhere on the same chromosome. When recombination occurs within the inverted region between the normal and the inverted chromosomes two *complementary* recombinant chromosomes arise. Each of them is duplicated for the distal region of one arm and deleted for the opposite one.

In the case of *paracentric* inversions the two recombinants are *dicentric* and *acentric*. The size of the inversion seems to be the main factor *influencing* the type of synapsis between the inverted and the normal *chromosome*. For a more complete discussion on this topic, *see* ref. elsewhere in this chapter.

PARACENTRIC INVERSIONS

In the last 15 years, a series of *submicroscopic paracentric* inversions have been discovered while studying specific genes at a genomic level. Some of these inversions lead to the breakage of a dosage sensitive gene and, therefore, cause a *Mendelian* disease, other inversions break within either an extragenic region or disrupt nondosage sensitive genes and are, thus, considered benign.

Most of these cryptic inversions are mediated by large (usually >10 kb), highly *homologous* low-copy repeat (LCR) structures (also called segmental duplications or duplicons) that can act as recombination substrates in *nonallelic homologous* recombination (NAHR). There are clear evidences that some of the benign inversions are not mere *neutral polymorphisms* but may predispose to other rearrangements.

The story began in 1993 when Lakich et al. discovered that nearly half of the patients with severe *haemophilia* A had a breakage of the factor VIII gene caused by an inversion mediated essentially by *intrachromatid* recombination between DNA sequences in the A gene in intron 22 and one or other of two inverted copies of this sequence *(int22h-2, int22h-3)* located, respectively, 500 and 600 kb more *telomeric*.

A few years later a similar mechanism was shown to be *responsible* for 13%> of Hunter disease patients. In this case the inversion occurs because of abnormal recombination between the *IDS* gene and its *pseudogene* located 90 kb away, resulting in the *disruption* of the gene. Other recurrent cryptic inversions mediated by homologous LCRs have been reported in normal *individuals* and, at least some of them, are *regarded* as *benign*.

The inversion at the Xq28 *emerin/filamin* region was found while trying to *elucidate* the *discrepancies* observed between the genetic and physical map distances. This inversion, present in the *hemizynate* state in 33% of females and in the *hemizygous* state in 19% of males, is mediated by two 99% sequence identical segmental duplications flanking the emerin and *filamin genes*.

Although inversion carriers are completely normal, some of the emerin deletions associated with *Emery-Dreifuss* muscular *dystrophy* have been *hypothesized* to be the result of inversion-mediated rearrangements. Moreover, it is not impossible that some of the dicentric X chromosomes pter-qter::qter-pter are formed as a *consequence* of unequal *crossovers* between the same LCRs that mediate the inversions.

Another recurrent benign cryptic inversion has been detected at the region surrounding the NPHP1 locus at 2q 13. Saunier et al. discovered that homozygous deletion of 290 kb responsible for *nephronophthisis* 1 is mediated by two copies of segmental duplications with the same *orientation*.

These LCRs are *surrounded* and partially embedded into two homologous segmental *duplications* having opposite orientation and mediating an approx 500-kb inversion present in 1.3% of the population in the *homozygous* state. No negative effect has been *demonstrated* in association with this inversion. It is obvious that *embryos* with any dicentric chromosome 2pter-2q13::2q13-2pter would not be viable.

The paradigmatic situation is that of the Y chromosome. This chromosome has an abundance of LCR (*amplicons*), which render this chromosome susceptible to a multitude of rearrangements that, when involving the long arm, are often the cause of *spermatogenic* failure. It has been assumed that the *polymorphisms* observed at loci on the Y chromosome and mtDNA are selectively neutral and, therefore, existing patterns of molecular variation could be used to deduce the histories of populations in terms of drift, population movements, and cultural practices.

However, Jobling et al. demonstrated that the 3-Mb Yp inversion present in the male population is the preferential background for the *PRKX/PRKY* translocation underlying most XX males and some XY females. This is a clear demonstration that this Y inversion "*polymorphism*" is not neutral. It seems likely that other cryptic inversions are responsible for the complex series of deletions and duplications mediated by several types of LCRs associated with the AZFc locus.

RECURRENT TRANSLOCATION

The finding that a Y chromosome inversion was the basis for a recurrent translocation was at first considered *peculiar* to the sex chromosomes. Further relevance of these data emerged following our studies reporting that some recurrent chromosome rearrangements at 8p occur as a consequence of an 8p submicroscopic paracentric inversion present in the parent transmitting the *abnormal chromosome*.

At that time, we already knew that the recurrent inv dup(8p) rearrangement, an inverted duplication of the chromosome 8 short arm associated with deletion of the very distal 8p (8p23.2-pter), was not the primary product of an abnormal recombination but instead was produced by the breakage of a dicentric chromosome 8qter-8p23.1::8p23.1-8qter leading to the formation of an inv dup(8p) and, though formally not demonstrated, to a chromosome 8 deleted for part of its short arm (8p-).

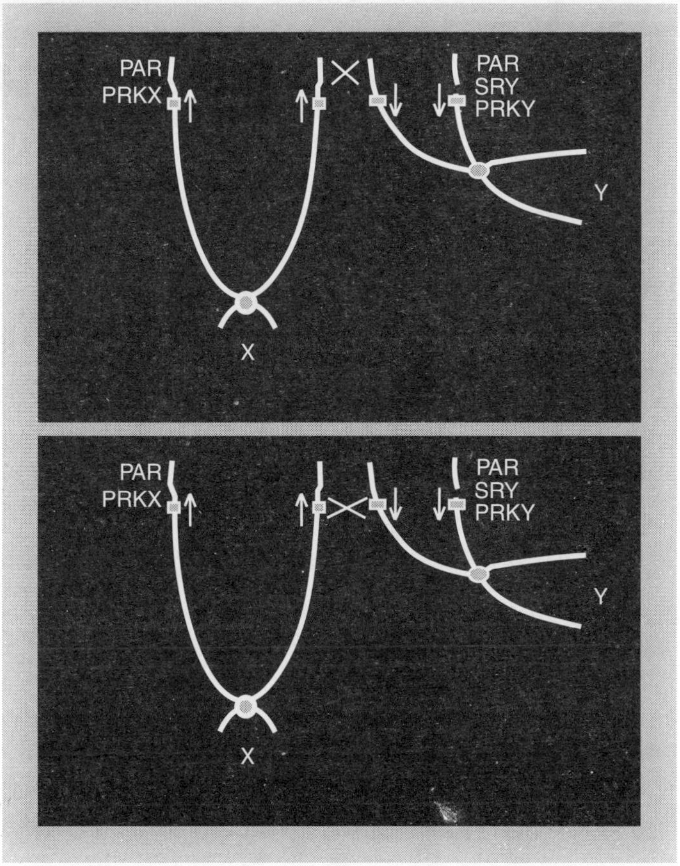

Figure 10.1: Mechanism of the Xp/Yp translocation at the basis of XX *males and XY females.* (Top) *Recombination between the Xp and Yp chromosome normally occurs at the pseudo-autosomal region 1 (PAR).* PRKX *gene lies on the X chromosome approx 1 Mb centromeric to PAR. In* sub*jects with inverted PRKX and PRKY acquire the same orientation (Bottom), thus, allowing an ectopic recombination. The result* is *the transposition* of SR *Yon the Xp chromosome and the absence* of SR Yon *the Y chromosome.*

In the dicentric chromosome the two duplicated regions are separated by a small single copy region. The fact that the original inv dup (8p) was a dicentric chromosome led us to hypothesize that the reciprocal analphoid chromosome might be present in some subjects who acquired of a neocentromere.

From a literature review we were able to identify two of such cases where the analphoid chromosome was a supernumerary marker that was mosaic with a normal cell line. Molecular definition of this marker (+der[8p]) demonstrated that it was exactly the reciprocal of the dicentric chromosome 8 and that both rearrangements were mediated by a pair of olfactory receptor (OR)-gene clusters mapping to 8p23.1.

The refined distance between the two clusters (the single-copy region) is approx 3.5 Mb. We reasoned that, according to the classical *cytogenetics,* the production of two *reciprocal* rearrangements, one dicentric and the other acentric, could only result following a crossover within

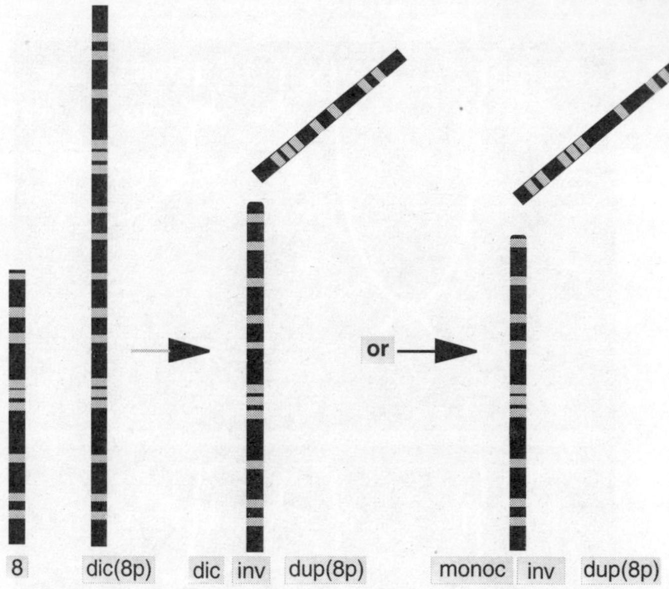

Figure 10.2: The dicentric chromosome resulting from nonallelic homologous recombination can undergo :I breakage at the level of the second centromere or a more proximal breakage, leading to an inv dup(8)s and a del(8p).

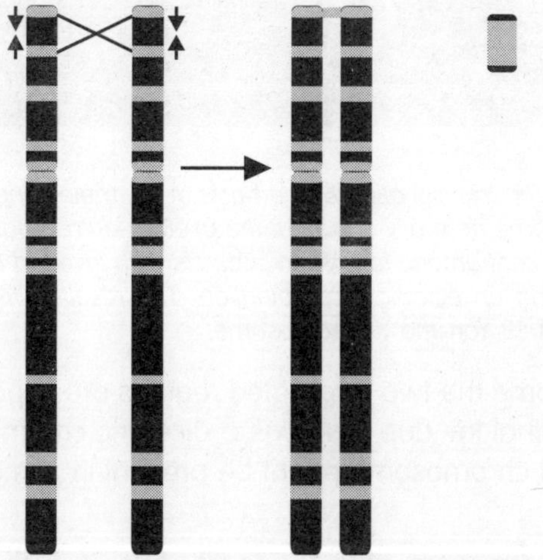

Figure 10.3: Cryptic paracentric inversion between the two pairs of olfactory receptor (OR) gene cluster mapping to 8p23.1.

a *paracentric* inversion. We, thus, studied by fluorescent *in situ hybridization* the apparently normal chromosomes 8 in the parent-of-origin of several inv dup(8p)s (thus far 16 cases) and of one case of +der(8p) and discovered that a *heterozygous* paracentric inversion with the same breakpoints of the dicentric and the acentric chromosomes was present in all of them.

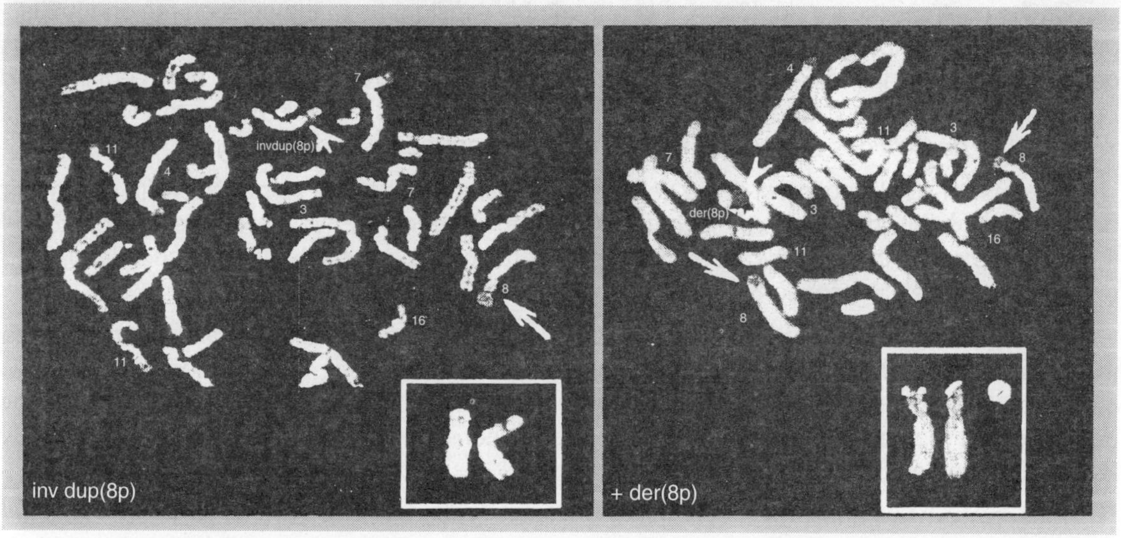

Figure 10.4: The inv dup(8p) and the analphoid supernumerary chromosome +der(8p) are the two reciprocal products of the same abnormal recombination. Both fluorescent in situ *hybridization images show the results obtained with the bacterial artificial chromosome (BAC) clone RPI 1-287P18 specific to the distal 8p olfactory receptor genes cluster.*

The presence of the inversion also explained a highly unlikely crossover pattern at 8p23 observed in members of some Centre d'Etude du Polymorphisme Humain families that showed apparent triple recombination events in a very short region. Studies on normal populations revealed that the 8p23 inversion is present in the *heterozygous* state in 26% of the European population and in 39% of the Japanese population.

Thus, the inversion represents a genomic polymorphism that, in the heterozygous state, renders *misalignment* and abnormal recombination more likely, just as it occurs in *cytogenetically* identifiable inversions. In other words, the inversion predisposes the individual to a susceptibility to the formation of what were considered *de novo* chromosome rearrangements (inv dup[8p] and the + der[8p]).

ANOMALOUS CHROMOSOME

An unexpected finding is that both subjects with the inv dup(8p) or with +der(8) are single cases in their families, although the parent who has transmitted the anomalous chromosome is a carrier of the inversion and, therefore, one would have anticipated more than one unbalanced child. However, if we consider the *analphoid* chromosome, the occurrence of a neocentromere is a rare event.

Thus, most of the zygotes containing a +der(8p) will lose it very soon after acquiring a normal chromosome complement. As to the dicentric chromosome, we had assumed that it did undergo a breakage at the second meiotic division thus leading to a gamete having the inv dup(8p) and a gamete deleted for a portion of 8p. Floridia et al. demonstrated that the size of the inv dup(8p) may differ according to the size of the 8p duplication which may involve even the centromere (from 8p22 to 8pcen) or may be much smaller (from 8p21.2 to 8p22) .

On the contrary, the inv dup(8p) size is remarkably constant as to the deletion region (8p23.2-pter) and the single copy region at 8p23.1 flanked by the two clusters of OR genes. These findings demonstrated that the dicentric *breakage* may occur in different positions between the two *centromeres* leading to a dicentric inv dup(8p) and a reciprocal acentric 8q, or to inv dup (8p)s with smaller *duplications* and reciprocal 8p- with different degrees of deletion.

It seems very likely that most of the deleted chromosomes 8 are not *compatible* with *embryonic* development resulting in *premature termination* of the *pregnancy*. Moreover, we demonstrated that in some cases the dicentric does not break at meiosis II but is inherited as such leading to an almost completely trisomic 8 zygote.

Because of its instability, the dicentric may undergo breakage generating different cell lines in the embryo. Should the *inactivation* of one centromere occur very early, thus stabilizing the dicentric 8, the resulting embryo will be *trisomic* for an almost entire chromosome 8. It has been clearly demonstrated that survival of trisomy 8 is possible when the *aneuploid* cell line arises relatively late in development.

Thus, embryos with *trisomy* 8 owing to the presence of a dicentric chromosome are expected to be prematurely aborted. Because no evidence of increased *spontaneous* abortions has been found in mothers of inv dup(8p) subjects (manuscript in preparation) we postulate that the trisomy 8 fetuses result in preclinical abortions. In conclusion, as Madan and Sutherland et al. already observed with paracentric inversion carriers, also subjects carrying the 8p23 cryptic inversion have a very low risk for unbalanced progeny.

HETEROZYGOUS INVERSION

Osborne et al. observed a heterozygous inversion of the Williams-Beuren syndrome region in 4 of 12 parents transmitting the disease-related chromosome. Their data had been fully confirmed by Bayes et al. who found that one-third of the transmitting progenitors were *heterozygous* for an inversion between the *centromeric* and the *telomeric* segmental duplications at 7q11.23, which are in opposite orientation.

The mechanism by which the inversion generates an interstitial deletion is probably owing to the fact that the normal and the inverted *chromosome synapse* for all their length and "*balloon out*" at the inversion region, thus allowing NAHR between the segmental duplications lying in opposite orientation.

We also demonstrated that the t(4;8)(p16;p23) recurrent translocation, reported in several cases either in the balanced or unbalanced form, is mediated by two pairs of *homologous* OR-gene clusters located at 4p16 and 8p23, respectively. In five *de novo* cases of unbalanced and balanced *translocations*, all of maternal origin, we could *demonstrate* a double *heterozygous* inversion between the two pairs of OR-gene clusters at 4p 16 and 8p23 in all the mothers.

It seems likely that the two pairs of *homologous* chromosomes cannot synapse in their distal short arm regions involved in the inversion and this allows the occurrence of a crossingover between homologous regions located into *nonhomologous* chromosomes. 4p16 and 8p23 *heterozygous* inversion was detected in 12.5% (4/40) and 26% (13/50) of control subjects, respectively, whereas 2.5% (1/40) were scored as double heterozygous.

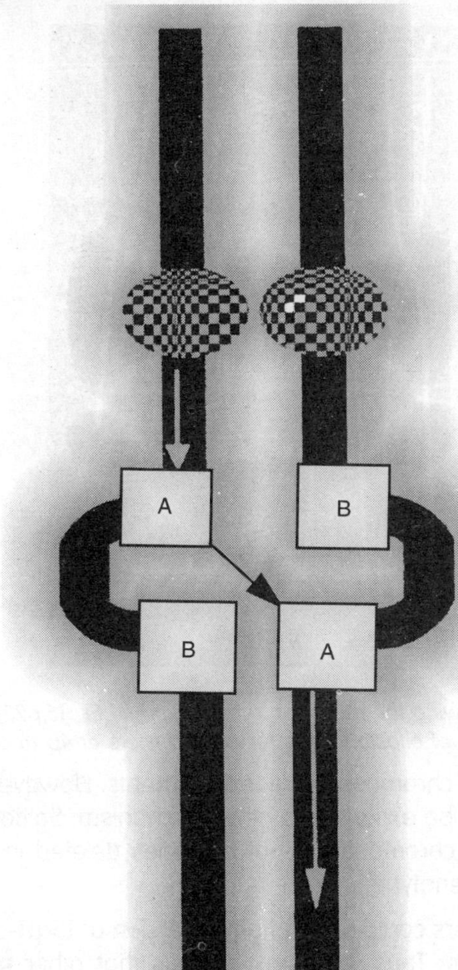

Figure 10.5: Showing the paracentric inversion leading to a deleted recombinant chromosome.

Heterozygous cryptic inversions as the basis of unbalanced rearrangements have also been found in mothers having a 15g11-q13 deleted *Angelman syndrome* (AS) child. The inversion was detected in the mothers of four out six AS cases with the breakpoint 2-3 (BP2/3) 15g1 1-q13 deletion, but not in seven mothers of AS because of paternal uniparental disomy 15.

The BP2-BP3 chromosome 15q 11-q13 inversion was detected in 4 of 44 subjects (9%) of the general population. From these data it emerges that the heterozygous inversion between homologous segmental duplications is an important mechanism causing abnormal synapsis and NAHR and, thus, that these types of inversions constitute an important factor for susceptibility to

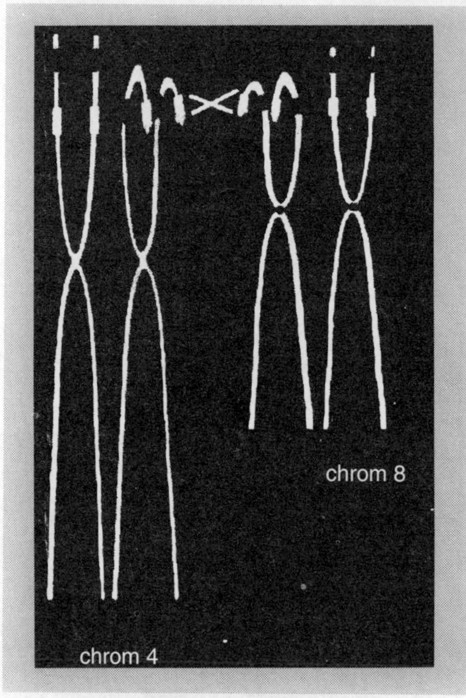

Figure 10.6: Showing the recurrent translocation t(4;8)(p16;p23) is mediated by two pairs of olfactory receptor gene clusters at 4p 16 and 8p23.

the occurrence of unbalanced chromosome rearrangements. However, it is also clear that not all recurrent rearrangements can be explained by this mechanism: Saitta et al. did not find evidence of inversion of 22q11.2 in the chromosome that becomes deleted in subjects with a DiGeorge/ velocardiofacial syndrome phenotype.

Similarly, several researchers could not find any inversion at 15q11-q13 in the fathers of deletion Prader-Willi syndrome subjects. Thus, it seems obvious that other mechanisms beyond to the inversion may cause synapsis displacement between homologous segmental duplications, in turn, leading to NAHR.

Recent studies demonstrated the existence of large segments of the genome, ranging in size from 100 kb to 2 Mb, varying several folds in copy number in the human population and *denominated* large-scale copy number *variations* (LCVs).

A higher than expected association between LCVs and known segmental duplications has been noted both by Iafrate et al. and Sebat et al. It has been *demonstrated* that the instability of some genomic regions is owing to the presence of *segmental duplications*. This suggests that LCVs and genomic rearrangements might have a common *mechanistic basis*.

CONCLUDING REMARKS

Inversions and more specifically cryptic inversions flanked by inversely oriented segmental duplications could be much more frequent than previously estimated. Segmental duplications represent approx 5% of our genome. A systematic study of the genomic regions flanked by segmental duplications might reveal other inversion genomic *polymorphisms*.

Ours and other studies revealed that cryptic inversions flanked by LCRs are an important but not the only *susceptibility* factor for the occurrence of unbalanced constitutional chromosome rearrangements. After years of darkness on the molecular causes of structural chromosome abnormalities and on the possible susceptibility factors behind them, it seems that the tangle begins to be unravelled.

Segmental duplications are the cause of *genomic* instability and heterozygous carriers of cryptic inversions are at enhanced risk to form gametes with chromosome *imbalances*. Investigations on the presence of heterozygous variants with respect to LCVs in the parents *transmitting* structural chromosome abnormalities will likely determine whether these variants represent another risk factor.

Chapter 11

Whole Genome

As I write this chapter, there are 409 published complete genomes, with at least 1,662 additional genomes in the process of being sequenced (Genomes Online Database; www.genomesonline.org). It would be no surprise that, when this book becomes available, the number of sequenced genomes will have almost doubled.

Assigning of functions to the proteins encoded in these genomes constitutes a major *challenge* of the genome era. Ideally, one would like to have the function of every protein characterized through experimental analysis. However, despite recent developments in high-throughput techniques, the experimental characterization of proteins lags far behind the availability of new sequences; therefore, the *annotation* of genomes relies mostly on *computational* methods.

The most common and straightforward approach for *computationally* assigning a function to a protein sequence consists in the detection of homologous proteins with a known function to subsequently transfer its functional annotation. This methodology, known as *homology-based function prediction,* is based on the fact that proteins that have a certain level of sequence similarity have similar structures and, therefore, are likely to perform related functions.

However, this established approach has several important limitations, mainly because there is no simple relationship between sequence similarity and function. For instance, it has been shown that enzyme classification numbers tend to be conserved completely only among proteins with more than 80% sequence identity.

Below this level, changes in substrate specificity or enzyme activity might lead to incorrect annotations. Moreover, this technique is of little help when there are no *experimentally* characterized *homologues.* The combined effect of these drawbacks leads to a situation in which there is no clear functional prediction for about 40% of the genes in most *genomes.*

Complementary to *homology-based* methods, alternative approaches for the computational characterization of proteins have been proposed. In particular, the so-called genomic context methods exploit relations between genes on the genome, such as gene fusions, *chromosomal* proximity, their distribution across species, and the like.

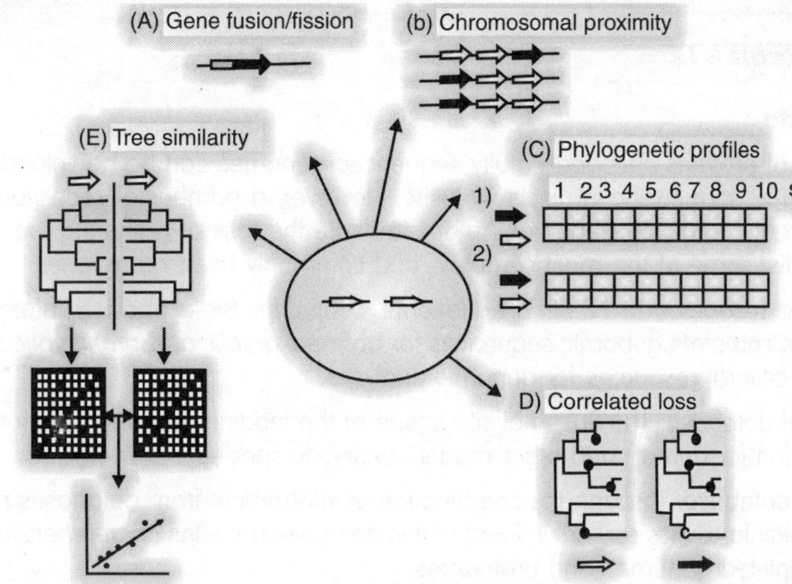

Figure 11.1: Overview of the comparative genomic-based methods described in the chapter. A functional link between a pair of genes (center) can be inferred from: (A) gene fusion or fission, both genes form a single gene in other genomes; (B) chromosomal proximity, both genes are close to each other in a significant number of genomes; (C) phylogenetic profiling, both genes have similar distributions across diverse organisms; (D) correlation of gene loss events, the history of losses of both families is similar; (E) mirror tree, similarity of the phylogenetic trees of both protein families can be measured by the correlation coefficient of the distance matrices from both trees

The kind of information provided by such techniques is qualitatively different from that of the classical, homologybased one: While the detection of characterized homologues of a sequence gives us information on its molecular function, a genome context analysis is useful to by using FASTA or some species of BLAST.

The aim of this chapter is to provide the essential theoretical background and the necessary practical tools to allow the reader to implement comparative genomic-based function prediction approaches to his or her own data. The *protocol* assumes a set of sequences coding for proteins of unknown function as the starting point.

This set may consist of just a single protein sequence, a large number of them coming from a fully sequenced genome, or any number in between.

By comparing this set of sequences with other sequences encoded in complete genomes, we search for evolutionary conserved traits that might give us hints about their functional role. Working in a UNIX environment, such as GNU/Linux (www.gnu.org), and having some skill in a computer scripting language, such as python (www. python.org), greatly facilitates the comparative genomics operations needed for applying the techniques described here.

Alternatively, no computer literacy is needed to use integrated servers such as STRING , in which by pasting the sequence of interest we can obtain *precomputed* results from comparative genomics analyses. This strategy, however, is limited to cases in which the number of proteins to *analyze* is reduced, since little *automation* is allowed.

REQUIREMENTS

Sequencing Data

Repertoires of proteins encoded in fully sequenced genomes can be downloaded from their respective genome sequencing projects' website. However, a number of *repositories* that store data from different completed genomes greatly facilitate the task of gathering the sequences of interest. Here,1 list some of the most complete and commonly used *repositories*:

1. Genome resources at NCBI. The National Center for Biotechnology Information (NCBI) provides complete genomic sequences for both *eukaryotic* and *prokaryotic* organisms as well as several resources for genomic analyses.
2. Ensembl database: The Ensembl site is one of the leading sources of genome sequence annotation for human and other model eukaryotic species
3. Integr8 database: Through the combination of information from databases maintained at major bioinformatics centers in Europe, this database provides *comprehensive* information on completed genomes and proteomes.
4. GOLD: The Genomes Online Database (GOLD) provides the largest available and most detailed *monitoring* of genome sequencing projects, both completed and ongoing. Therefore GOLD is the site at which to check for what *genomes* are, or soon will be, available to plan a *comparative-genomics* project.

These databases can be searched online through their Web interfaces or, alternatively, stored locally. In the latter case, the use and repetition of automated searches is favored.

Genomic Diversity

Sequence comparison is inherent to *comparative genomics*, since it allows the *establishment* of *evolutionary* relationships among the protein sequences encoded in complete genomes. A large number of algorithms to search for similar sequences in a database exist and are thoroughly described elsewhere. Here, I provide just some indication about two of the most commonly used tools:

1. BLAST. The Basic Local Alignment Search Tool (BLAST) finds regions of local similarity between sequences. In addition to a BLAST score, the program provides an expect value (e-value) that indicates the statistical significance of the hit by taking into account the size of the database and the length and *composition* of the hit.

   To put it simple, it gives a measure of the chance of observing a hit of that score or higher just by chance. Therefore, the lower the e-value, the higher is the statistical significance of the hit. The use of filters to mask regions of "*low complexity*" is recommended.
2. Psi-Blast. The Position-Specific Iterated (PSI)-BLAST is a very sensitive implementation of the BLAST program and, therefore, appropriate for finding distantly related proteins. PSI-BLAST starts with a standard BLAST search and use the hits with e-values lower than a given threshold to derive a position-specific scoring matrix (PSSM).

This PSSM takes into account, specifically, the type of residues found at a given position in the multiple sequence alignment of the hits. The next iteration continues by using the PSSM from the previous search and updating it with the new hits. The process is repeated until convergence (no new sequences are found) or a specified number of rounds is completed.

For genome-context function prediction, we use either BLAST or Psi-Blast to define homologous groups of proteins, to *subsequently* build groups of *orthologues*, or perform a *phylogenetic* analysis.

In doing so, and to avoid hits restricted to a specific protein domain, it is useful to request not only that the homologous proteins share a region of similarity under a given e-value threshold but that this region covers a significant fraction of the full length of the sequences considered. BLAST and Psi-Blast can be used through the NCBI website (www.ncbi.nlm.nih. gov/blast) or can be downloaded and installed locally (www.ncbi.nlm.nih.gov/ BLAST/download.shtml).

Evolutionary Reconstruction

Phylogenetic reconstruction is needed for a number of genomic-context function prediction approaches, in particular for the establishment of *phylogeny*-based *orthology* relationships and the detection of *coevolving* protein families.

The process of phylogenetic reconstruction involves two main steps that use specific programs: the alignment of the homologous proteins considered and the *phylogenetic* inference itself.

1. Multiple sequence alignment. An alignment of protein sequences basically aims at placing homologous residues of different proteins on top of each other. Programs such as Clustal W [5] and MUSCLE implement fast, efficient multiple sequence algorithms and are easy to use. They basically require as an input a multiple sequence file in the appropriate format (usually FASTA or PHYLIP).

   A manual curation of the alignment or the removal of unreliably aligned regions of the alignment (e.g., regions with many gaps) is recommended before the phylogenetic analysis. However, this is not always possible when working at a large scale.

2. Phylogenetic reconstruction. Phylogenetic reconstruction aims at explaining the differences observed in a protein alignment by means of the evolutionary distances among the sequences involved, given *a subyacent* evolutionary model. Several strategies exist for *Phylogenetic* reconstruction, but in this chapter we will focus on just two of them, *neighbour* joining (NJ) and maximum likelihood (ML). Neighbor joining is a fast heuristic algorithm that estimates a "*minimal evolution*" tree, from a matrix of pairwise distances of the sequences involved. Since it is very fast and fairly accurate, it usually is the method of choice for large-scale analyses. Clustal W is a fast, easy to use program that can perform both the multiple sequence alignment and the *neighbour* joining analysis.

Maximum likelihood (ML) methods try to find the tree with the maximum likelihood to have produced the *variation* observed in the *alignment*. To produce results in reasonable time, all practical ML methods rely on certain *heuristics*. I recommend PhyML, which has gained popularity because of its speed and ease of use.

Installation of Data

The local installation of data and software allows the researcher to perform tailored comparative genome analyses, that is, to use the preferred algorithms and programs on the set of genomes of choice. However, this is not always simple, and some researchers might prefer to use standard *analyses* on standard datasets that, in return, have faster results and user-friendly *interfaces*.

Fortunately, several platforms on the Internet allow access to precomputed comparative genomics analyses or enable doing such *analyses* online. The choice is huge, and here I focus on three complementary ones: a general comparative genomics server (NCBI), a server that integrates *phylogenetic* tools (Phylemon), and a genome-context-based function prediction server (STRING). Most of the techniques described here can be performed through these servers, although, as I pointed out before, the analysis cannot be applied to large datasets.

1. NCBI. In its section "*Genomic Biology*", the NCBI offers a number of tools for comparative genomics. These include, organism-specific BLAST searches, sequence repositories, and *predefined* clusters of orthologous groups.
2. Phylemon. Phylemon is a recently developed Web server that integrates a complete suite of phylogenetic analysis tools. Applications range from sequence alignment to very sophisticated *phylogenetic* analyses and evolutionary tests. A useful property of the server is that all *applications* can be easily *interconnected* in a user-friendly pipeline.
3. STRING. The Search Tool for the Retrieval of Interacting Genes/Proteins (STRING) is a database that integrates known and predicted interactions between proteins of fully sequenced genomes.

   By establishing orthology relationships among genes in different species, the predicted interactions are transferred across organisms. STRING implements the most important types of genomic context, and therefore, it can be used directly to predict the function of a query protein.

PROCEDURES

Identifying Different Genomes

A basic requirement of comparative genomics is to identify which genes are common in different genomes. This ideally is done by establishing orthology, rather than just homology, relationships. The detection of orthology is not straightforward when comparing distantly related genomes, and several operational definitions do exist.

The most widely used method for automatically establishing orthology relationships is based on the detection of best reciprocal hits (BRHs), that is, pairs of sequences from different species that, reciprocally, are the best hit of each other in a sequence search [9]. Extensions of the BRH approach include the definition of "tri-angular" BRH relationships across a minimum of three species, such as COG.

More recently, some methods approximate the classical, phylogenybased approach for detecting orthology. Such methods generally rely on the detection of duplication and speciation events by comparing the gene tree with the species tree [11]. Here, we selected some of the most widely used techniques to define orthology relationships, which range from less (BRH) to more (phylogeny based) time consuming; therefore, the choice of the methods depends on the size and characteristics of the set to analyze.

Alternatively, one can use predefined orthologous sets, if these are available for the species of interests. Websites like COG, KOG, Ensembl, or multiparanoid [15]) provide such predefined orthologous sets.

Tips to Apply

1. Perform a BLAST search for every protein encoded in genome A against a data base consisting of all proteins encoded in genome B.
2. Select the hit with the highest score for each protein.
3. Perform the reciprocal search (a BLAST search for every protein in genome B against the database of proteins in genome A) and select the best hit for each protein.
4. Compare the lists of best hits from both searches and define BRHs as those pairs of proteins from different genomes that, reciprocally, are the best hits of each other.

COG-like Method

1. Proceed as in Sect. elsewhere in this chaptere with all pairs of genomes involved.
2. Detect "triangles" of BRHs; for example, if a protein pA from genome A reciprocally is a best hit of protein pB in genome B and of protein pC in genome C, and similarly pB and pC also are BRHs, then pA, pB and pC constitute a cluster of orthologues.
3. Merge triangles of BRHs into a single COG if they share one side.
4. Repeat step 3 until no more COGs are fused.

Evolutionary Inference

1. Retrieve homologues in the genomes of interest by using BLAST or Psi-Blast.
2. Perform a multiple sequence alignment.
3. Remove gap-rich regions in the alignment (e.g., those columns having a gap in more that 20% of the sequences).
4. Perform a phylogenetic reconstruction based on that alignment.
5. Compare the resulting tree with the phylogeny of the species involved and map duplication and speciation events. Do this automaticaly by using the RAP algorithm implemented by Dufayard and colleagues (http://pbil.univ-lyon l .fr/ software/RAP/RAP htm) [16].
6. Orthologues then are defined as genes from different species that derive from a common ancestor only through events of speciation.

Fission and Fusion Gene Technique

A gene fusion (or fission) event is revealed when two or more proteins encoded by separate genes are found to be encoded by a single gene in a different species. This genomic event constitutes the most direct type of genomic context association, and it usually is indicative of a very close functional relationship, often involving physical interactions. At the molecular level, the fusion of two proteins can enhance their functional interaction, such as by facilitating the channeling of a substrate.

The discovery of functional relationships between fused proteins on a small scale has been intuitively exploited in the past. The most widely known example perhaps is the fusion of the alpha and beta subunits of tryptophan synthase in yeast and other fungi. An extension of such strategy at a genomic scale was proposed in the late 1990s.

Based on the fact that a significant fraction of the observed fusions involved proteins known to interact functionally, the authors proposed the use of this *technique* on a large-scale to predict other *interactions*. In agreement with the aforementioned substrate *channeling* effect, most of the observed fusion events involve metabolic enzymes, although not *necessarily* from *consecutive* steps in a given pathway.

Although gene fusion is a rare event, the number of proteins for which a fusion can be found increases *exponentially* with the number of genomes compared. For instance, when comparing three bacterial genomes, Enright et a]. found 88 gene *fusions* events, while just two years later and using 30 bacterial genomes, 10,073 fusions were detected.

Molecular Detection

1. Perform an all-to-all BLAST comparison of the genomes involved. For each protein, store the best hit in each genome and information regarding the region of similarity e.g., residue x to residue y in the query protein aligns with residues a to b in its best hit.
2. Fusion (or fission) events are evidenced by those proteins that are the best hit of two proteins in another genome and their regions of similarity with the query protein do not overlap.

Inspecting Gene Order

Comparisons of bacterial genomes have shown that there generally is a lack of gene order conservation, even among closely related species. This indicates that recombination events rapidly shuffle the genomes during evolution. Despite this overall trend, some gene clusters, mainly in prokaryotes, are conserved in terms of their relative order over large evolutionary distances. A closer inspection of such genes reveals that they usually constitute operons and are functionally related.

The use of gene order conservation to predict functional interactions was first proposed in 1998 by two independent studies that measured, respectively, the conservation of genes in runs (gene sets separated by less than 300bp) and the conservation of pairs of neighboring genes. In eukaryotes, the chromosomal proximity of the genes rarely is indicative of functional association, although several remarkable examples do exist.

In any case, one can use chromosomal proximity to predict the function of eukaryotic proteins if they have orthologues in prokaryotic species, assuming that the function has been conserved. Next, I describe the use of chromosomal proximity using the two aforementioned strategies.

Molecular Conservation of Genes in Runs

1. Establish orthology relationships among the genes encoded in the genomes involved. Discard very closely related species (e.g., belonging to the same genus), since gene order might be too conserved in these cases.

2. For each genome, define runs as sets of genes localized in the same strand of the chromosome and separated by less than 300 nucleotides.
3. A cooccurrence event between two genes is defined as these genes being part of the same run in a given genome.
4. The chromosomal proximity score between two genes is defined as the number of cooccurrence events observed divided by the total number of genomes analyzed that contain at least one of the genes. Scores higher than 0.3 usually are considered statistically significant.

Cytological Proximity Gene Neighborhood

Proceed as in Sect. elsewhere in this chapter, but define a cooccurrence event between two genes when these are localized next to each other in the chromosome.

Evolutionary Profiling

The evolutionary profiling techniques, compare distributions across complete genomes of proteins to subsequently predict functional links between those proteins that exhibit similar distributions. The technique is based on the observation that proteins having a similar species distribution often are found to be involved in the same biological process.

In addition, the finding of proteins with complementary distributions might result from a nonorthologous gene displacement and thus suggests that both protein families perform a similar function. The distribution of a gene across different organisms can be formalized by means of a phylogenetic profile, a binary series that codes the presence or absence of a protein in a set of genomes, then detects proteins with similar profiles.

Distances between profiles can be computed by simply counting the differences (Hamming distance) or using more sophisticated scores, like mutual information or Pearson correlation coefficient.

Isolating Genes with Similar Profiles

1. Derive orthology relationships among the genes encoded in the genomes of interest by using any of the approaches described already.
2. Code the information of presence or absence in the form of a phylogenetic profile, indicating absence as a 0 and presence as 1. For example, the profile 00010010000111, indicates that the gene is absent from all genomes except those coded in the positions 4, 7, 12, 13, and 14.
3. Compare profiles using any of these distance measures:
   a. Hamming distance. The Hamming distance between two strings of equal length is the number of positions with different numbers. It measures the number of substitutions required to transform one profile into the other. The hamming distance between the profiles 001000111001 and 00010001 1001 is 3.
   b. Mutual information. Mutual information is a measure of the independence of two distributions of discrete variables. Mathematically, this measure is equivalent to the log-odds ratio of the expected cooccurrence of pairs of genes, based on their individual

frequencies, to the oserved cooccurrence. Thus, the mutual information between the profiles *i* and *j* (M_{ij}) is

$$M_{ij} = \sum_{ij}\left(P_{i,j} / P_i P_j\right)$$

where P_{ij} is the frequency of observed cooccurrence events (both genes are in the same genome) and P_i and P_j are the individual frequencies of each gene.

Observing Complementary Phylogenetic Profiles

Proceed as in Sect. elsewhere in this chapter but search for genes having dissimilar phylogenetic profiles (i.e., greater Hamming distances). Perfect complementarity is found between the profiles of two genes if the Hamming distance equals the length of the profile.

Profiling Technique

A variant of the aforementioned phylogenetic profiling technique incorporates information on the evolutionary relationship of the species to identify genes that have been lost concomitantly during evolution.

This method originally was used for the prediction of genes involved in pathogenicity by searching for those genes present in a pathogenic organisms but absent in closely related nonpathogenic species or strains. More recently, the method was extended to the detection of genes that have been concomitantly lost along the phylogeny of a group of species.

A prerequisite of the method is that the differences in the observed phylogenetic profiles are the result of differential loss across the lineages rather than differential gain of genes. Another requirement is that the evolutionary relationships among the species considered are known.

For the detection of similar geneloss patterns,

1. Proceed as in the previous section to derive phylogenetic profiles of the genes considered.
2. Use the species phylogeny and a parsimony method to infer the presence or absence of a gene on every ancestral node. The presence of a gene is inferred when any of the

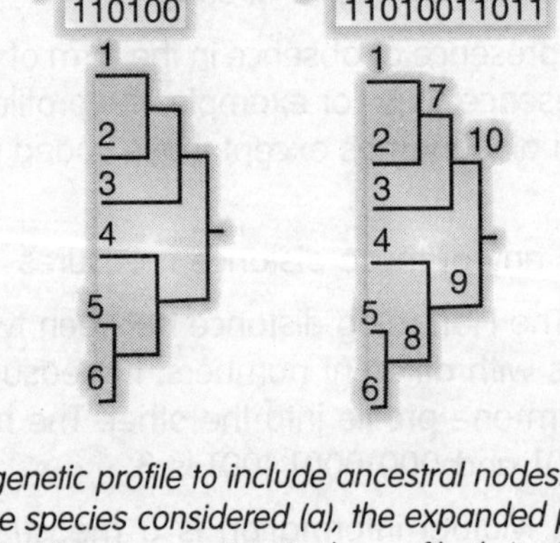

Figure 11.2: How to expand a phylogenetic profile to include ancestral nodes. Given a phylogenetic profile and the evolutionary relationships among the species considered (a), the expanded profile can be derived by assigning to every ancestral node (numbered in bold) a new position in the profile that will describe the inferred presence (1) or absence (0) of that gene in the ancestor (digits in bold).

descendants have that gene. In this manner, the original phylogenetic profile is expanded to a new, bigger one that includes the presence or absence of the genes in the ancestral nodes.

3. Hamming distances and mutual information can be applied to these extended profiles. However, Gabaldon and Huynen found that the following coevolution score provided the best results: concordances of the type 1,1 (both genes are present) minus discordances of the types 0,1 and 1,0 (only one of the genes is present). Note that species or ancestral nodes in which neither of the two genes is present are not taken into account in the score.

Basis of Coevolution

The coevolution of two proteins can be followed more closely by using the evolutionary information contained in the protein alignments of their respective families. For specific cases of interacting protein families, such as insulin and its receptors, it has been shown that their phylogenetic trees are more similar to each other than expected, based on the species divergence.

This suggested the existence of correlated evolution resulting from similar evolutionary constraints. Pazos and Valencia used this property to search for interacting partners within the *Escherichia coli* proteome by measuring the correlation between the distance matrices used to build the phylogenetic trees, a technique they call *mirror tree*.

1. Perform a phylogenetic reconstruction of each protein family, using the neighbor joining approach.
2. For each pair of trees, compute the correlation coefficient of the distance matri ces used to make the tree.
3. Proteins with high correlation coefficients in their distance matrices have similar phylogenetic trees and therefore show coevolution.

POINTS TO REMEMBER

1. When the techniques presented here are applied to sequences coming from genomes for which no complete sequence is available, the prediction of orthologues might be compromised, since the "true" orthologue can be missing in the incomplete genome.
2. The complete set of proteins encoded in a single genome is also known as the *proteome,* although this term can be applied to the complement of proteins expressed in the cell during a given period or under specific conditions.
3. Since the Internet is a dynamic entity, the websites and links provided in the chapter might be relocated or disappear. Usually the authors will indicate the new address. If this is not the case, one can try searching with Google or a similar search tool or writing to the authors of the site of interest.

   It is important to check for the date of the last update of the site, since it might get obsolete. In these cases and those in which the site has apparently disappeared, look for equivalent sites.
4. Low-complexity sequences are stretches of residues with a very limited variability in terms

of amino acid composition. For example, the protein sequence QQQKKQQQQQQLLK GQQQQ has low complexity. Regions of low complexity are problematic because they can produce artifactual hits with high scores. If the low-complexity filter is used, BLAST will substitute the low complexity regions by a string of Xs and not use this regions in the computation of the alignment scores.

5. Different programs produce or require different formats, and one often needs to convert the results from one program into the format required for the next program in the pipeline. The program readseq from EBI is a very useful tool that allows the conversion across many different formats.
6. For this purpose, a multiple sequence alignment viewer such as jalview and the program Gblocks for automatically selecting conserved blocks in the alignment can be very useful.
7. It is recommended to use the option "correct for multiple substitutions" to provide better estimates of the distances. This correction takes into account that, when fairly distant sequences are compared, the observed differences are a minimal estimate of the real amount of mutations, since a single residue may have mutated several times.
8. Several models are available for performing maximum likelihood analysis, some differences in the length of the branches and, eventually, the topology of the tree can be expected when using different models. A general model, such as JTT, will work fairly well for most cases, but if we want to use the proper model, we can run PhyML with several models and use the one producing the best likelihood. Regarding the use of a gamma-distribution, choosing a fourclasses gamma distribution with the alpha parameter and the proportion of invariable sites being estimated from the data usually gives good results.
9. The concept of orthology is central to comparative genomics. Orthologous genes are a special kind of homologous genes derived from an ancestral sequence through speciation only. Paralogues, in contrast, are homologous genes that derive from an ancestral sequence through events of gene duplication. Since orthology has a phylogenetic definition, the ideal way of determining orthology is by reconstructing the phylogeny of a given family and subsequently mapping on the tree the events of speciation and duplication to establish the orthology relationships.
10. Be aware of the so-called promiscuous domains, like those involved in signal transduction, that appear in very diverse functional and genomic contexts. Fusion events involving such domains usually are not informative.
11. This quantity may be altered, for example, when working with eukaryotic genomes where the distance between the genes is greater.
12. Yet another modification of the method allows for genes to be gained at certain lineages. The decision whether a single gene gain is preferred over multiple gene losses must be set in the form of a gain vs. loss threshold.
13. If the species tree is not fully resolved, the method still can be applied by representing the uncertainty with multiple furcations in the tree.

Chapter 12

Genome Sequencing

Computational biology provides insights into the nature of genomes and organisms, and provides tools for understanding how an organism's characters or phenotypes are determined by its genome sequence. In prior chapters, we presented a number of computational methods addressing a variety of specific biological questions. In this concluding chapter, we indicate in more detail how these tools can be employed in the context of complete genomes.

Computational analysis of genome sequence data has transformed the approach to answering biological questions because now they can be formulated in the context of all genes operating as a coordinated system. This more integrated approach complements the reductionist approach of traditional molecular biology.

Comparisons can be performed within genomes and between genornes. *Within-genome* comparisons focus on the genome of organism X and scan it from beginning to end, analyzing variations in base composition, *k*-tuple frequency, gene density, and numbers and kinds of transposable elements, and identifying any duplicated regions.

Between-genome comparisons employ closely related organisms (e.g., to help identify conserved genes, gene organizations, and control elements) or more distant organisms (to identify genes that are restricted to particular clades of a phylogenetic tree). Such data can help trace evolutionary trajectories of organisms.

What properties can be used to describe genomes? We start with compositional measures stated in terms of k-word content. As indicated in Chapter elsewhere in this book, even simple measures such as this can prove very informative. Second, we examine the fraction of the genome represented by transposable elements. These elements may or may not represent a large proportion of the genome.

Although they do not code for organism-specific proteins, the sequences of these elements can be used as tracers for defining the evolutionary histories of groups of like organisms. Third, we examine how sequence organization in chromosomes reflects duplication within genomes or evolutionary relationships between genornes.

We have already touched on this question in Chapter elsewhere in this book. Fourth, we provide a brief introduction to the identification and characterization of genes. Finally, we indicate how the predicted proteome can be functionally annotated and how proteomes of organisms can be compared.

Statistical Measurements

Summaries and statistical descriptions of genomes are essential for comprehending genome content. The page that you are reading now can accommodate approximately 3500 characters. Approximately 500 to 1000 such pages would be required to print out the sequence of an "average" microbial genome, and about a million pages would be required to print out the sequence of the human genome.

The human mind cannot grasp such quantities of data in this form. In this section, we illustrate what can be learned from simple *k*-word statistics, especially when these statistics are combined with other data. Even as simple a number as genome size, when combined with gene number, imposes important constraints on genome composition.

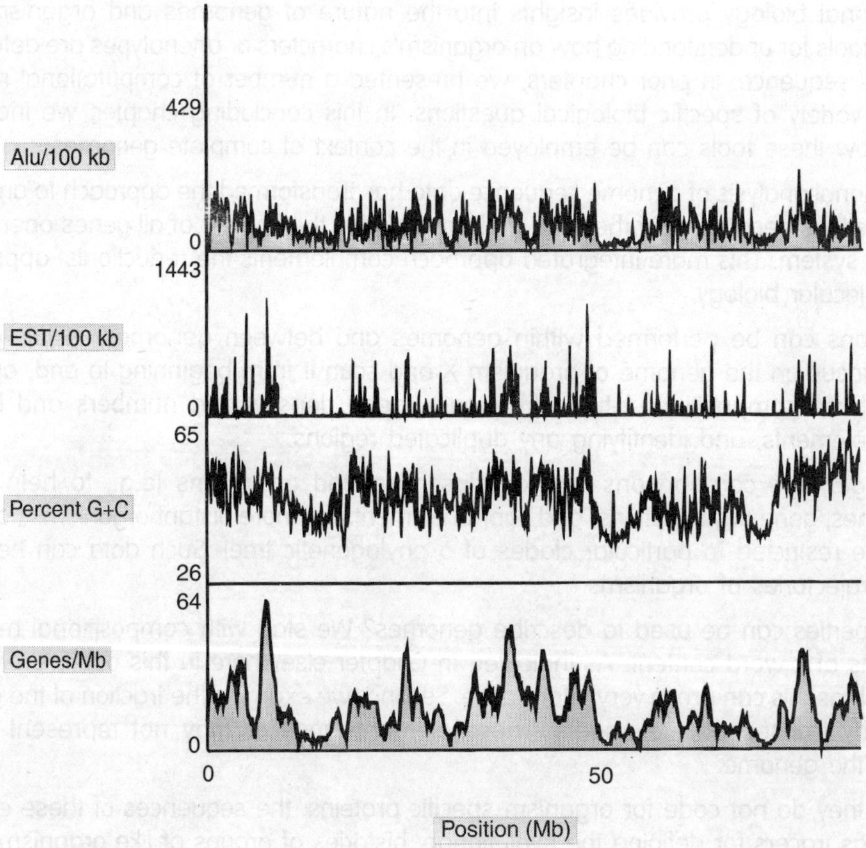

Figure 12.1: Composition of human chromosome 17 (Hsa17). Various properties have been summarized for 100 kb intervals along the length of the chromosome.

This is discussed in detailed already. k-word compositions of genomes measured as a function of position along the chromosome are not uniform. Even simple compositional measures can be very informative. The simplest measure is the base composition (corresponding to $k = 1$). Figure 12.1 shows distributions of *GC* content, transposable elements, and genes along human chromosome 17.

Distributions of these features along chromosomes provide a statistical analog of a genetic map, and they are useful for understanding chromosome structure and function. The third panel from the top shows the %G+C as a function of position.

The three regions where this quantity is the lowest correspond to regions where genes are particularly sparse. In the human genome, gene-rich regions typically have a higher %G+C than regions that are gene-poor.

Another statistic based on k = 1, and particularly useful for prokaryotic genomes, is the *GC* skew, defined in Chapter 2 as *(#G - #C)/(#G + #C)*. This statistic is computed *for* sliding windows of length w along the genome. Prokaryotic chromosomes are usually circular and usually have a single replication origin and a terminus located approximately 180° away from the origin.

This means that in one half of the chromosome, the genomic DNA strand as written in the sequence file corresponds to DNA produced by *leading strand* DNA synthesis, while the other half of that strand is produced by *lagging strand* DNA synthesis. In many bacteria, the leading strand exhibits an excess of *G* relative to C, or a positive *GC* skew. The *GC* skew has been used to infer the origin of replication in sequenced genomes.

An example is shown in Figure 12.2 for *Borrelia burgdorferi,* which has a linear rather than circular chromosome (lower panel). The predicted replication origin would be near the 450 kb position based upon the pattern of GC skew.

Genome regions having unusual shifts in one or more statistics may have particular biological interest. For example, *Yersinia pestis,* the causative or ganism for plague, or "Black Death," is a potential biowarfare agent. *Yersinia pestis* is closely related to another *Yersinia species, Y. pseudo*

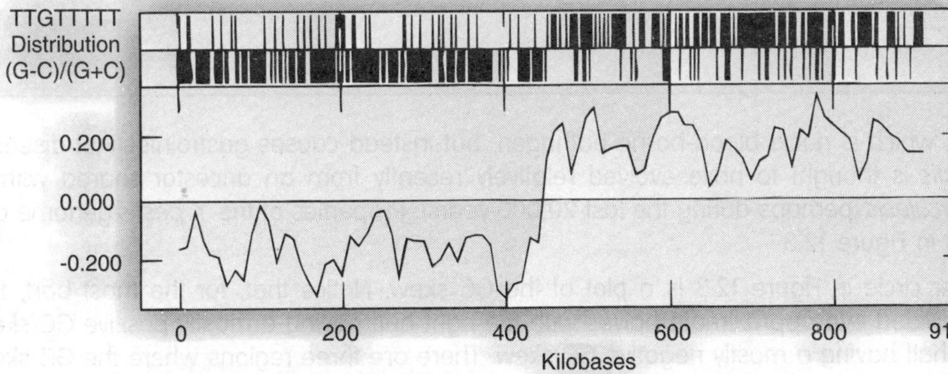

Figure 12.2: GC skew for the Borrelia burgdorferi *linear chromosome (lower panel) and distribution of eight-letter word* TTGTTTTT *on "top" and "bottom" strands (upper two panels).* GC *skew measures the relative excess of* G *(compared with C) on a particular strand.*

Table 12.1: Statistics describing genomes of common and model organisms. Human mitochondrial DNA is included as an example of an organellar genome.

Number of Organism	Genome Genes	Size (Mb)	Chromosomes
(Human mtDNA	37	0.016	~$10^3 - 10^4$)
Mycoplasma genetalium (bacterium)	517	0.58	1
Escherichia coli (bacterium)	4,288	4.64	1
Saccharomyces cerevisiae (baker's yeast)	6,000	12.05	16
Caenorhabditis elegans (nematode worm)	18,400	97	5+X
Drosophila melanogaster (fruit fly)	13,600	180	4
Arabidopsis thaliana (dicotyledonous plant)	25,500	125	5
Fugu rubripes *(Pacific* puffer *fish)*	31,000	365	22
Homo sapiens (mammal)	25,000	3,080	23
Allium cepa (onion)	NA	15,000	8

tuberculosis, which is not a blood-borne pathogen, but instead causes gastrointestinal disease. *Yersinia pestis* is thought to have evolved relatively recently from an ancestor shared with *Y. pseudotuberculosis* (perhaps during the last 20,000 years). Properties of the *Y. pestis* genome are summarized in Figure 12.3.

The inner circle in Figure 12.3 is a plot of the GC skew. Notice that, for the most part, the genome is divided into approximate halves with the right half having a mostly positive GC skew and the left half having a mostly negative GC skew. There are three regions where the GC skew reverses sign for comparatively short portions of the genome, suggesting relatively recent inversion events.

Genes involved in *Y. pestis* pathogenicity or adaptation are marked in dark blue on the two outer circles of Figure 12.3. It is known that genes conferring pathogenicity traits are often located in clusters, called pathogenicity islands, and that these may be acquired by horizontal transfer (from conjugative plasmids, bacteriophages, or other gene transfer mechanisms).

DNA from such sources will not, in general, exhibit the same base composition or other statistical properties as the genome that receives it. Notice that the pathogenicity genes at the 2.15 Mb position on the genome (between 5 and 6 o'clock) are clustered on one strand and that a peak of elevated %G+C appears at the same position (second circle from the inside).

This type of pattern could result if a block of pathogenicity genes had been acquired from another organism. Genomes can also be described by dinucleotide measures (k = 2; Sections 2.5 and 2.6). For example, probability distributions for dinucleotides can be organism-specific (Karlin et al., 1998).

In human DNA, 5'-CG-3' (also known as CpG, where p stands for the phosphate residue) occurs with about 20% of the frequency anticipated for iid bases having the base composition of human DNA (IHGSC, 2001). However, there are regions of the genome where their frequency is closer to the predicted value. These regions are called CpG islands.

In the human genome, there arc approximately 29,000 CpG islands (a number similar to the predicted gene number), and most of them are less than 1800 by long. There is a correlation between the density of CpG islands and the gene density, which is expected from prior experiments showing association between CpG islands and the 5' ends of vertebrate genes.

Unique and Repeated Sequences

Studies of eukaryotic DNA by reassociation kinetics in the 1960s and 1970s showed that eukaryotic genomes contain different sequence components that vary by copy number. Some genomic sequences appear once or a few times *(unique sequence),* while others appear many times *(repeated sequences).* One class of repeated sequences is the telomeric repeats (the short repeated sequences found at the ends of chromosomes), which are important for chromosome stability.

Another class of repeated DNA consists of multiple copies of one or more types of *transposable elements.* These can represent a substantial fraction of some eukaryotic genomes. For example, about 45% of the human genome is attributable to transposable elements. In contrast, the percentages of transposable element DNA in *Arabidopsis, Caenorhabditis elegans,* and *Drosophila* genomes are 10.5%, 6.5%, and 3.1%, respectively (IHGSC, 2001).

For efficient computational analysis of genomes, it is necessary to know the number of transposable element classes, the number of elements in each class, and how the elements are distributed along the genome. These multiple copies can complicate DNA sequence assembly in whole-genome shotgun approaches. Highly repetitive sequences are usually "masked" (omitted) before applying exon prediction tools, or before making interspecies comparisons.

The types and copy numbers of transposable elements within the human genome are shown in Figure 12.4. There are generally two categories of each type of transposon. One category encodes functions required for autonomous transposition, and the other category lacks one or more of these functions (i.e., elements of this category are defective).

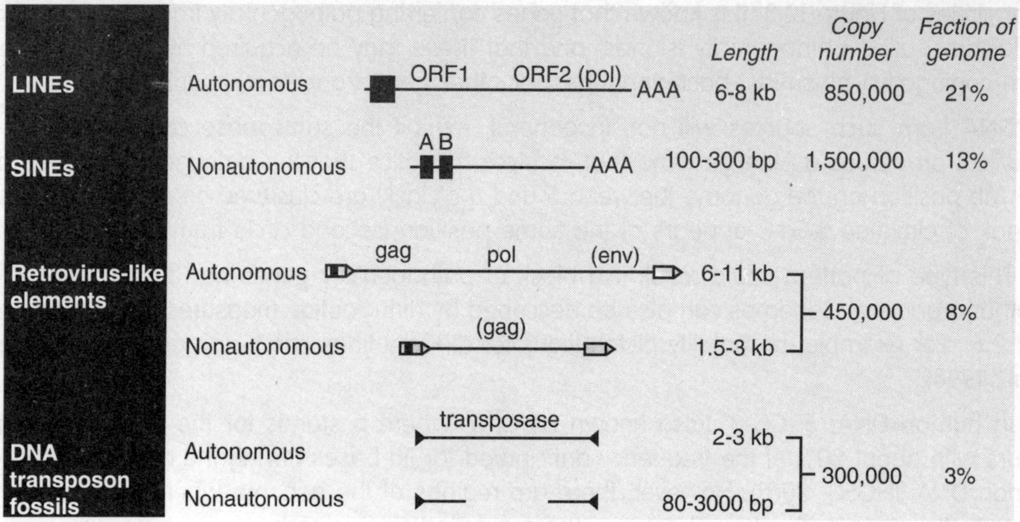

Figure 12.3: Types and numbers of transposable elements found in the human genome.

Transposition of defective elements requires gene products supplied in *trans* from related elements. One type of transposon is the LTR transposon, where LTR stands for long terminal repeat, a characteristic of retroviruses. Its nonautonomous counterparts possess the LTRs, but they lack the reverse transcriptase.

Another general type of element is the LINE (long interspersed nuclear element), which also uses reverse transcription to transpose but does not require or possess LTRs. Its nonautonomous counterparts are the SINEs (short interspersed nuclear elements.) The third type of element consists of DNA transposons (whose mechanism does not require reverse transcription) and their defective partners, which lack the transposase encoded by the autonomous counterpart.

Also shown in Figure 12.4 are the numbers of each type of element. Over a million copies of Alu elements (a type of SINE) are found within the human genome. They comprise about 10% of the human genome, and they appear on average once every three kb. They do not contribute to the protein encoded by those genes into which they are inserted.

Since the average human gene extends over 27 kb, we would expect (on average) to find nine such elements in the noncoding regions of an average gene, provided Alu elements are targeted uniformly throughout the genome. (It should be obvious why Alu elements are absent from coding sequences.) In fact, targeting preferences to different parts of the genome are not the same (Alu elements seem to target AT-rich DNA more frequently), but the observed distribution of these elements is complicated by post-transpositional losses (deletion).

There are more than 500,000 copies of LINE L1 elements in the human genome, suggesting that they appear on average once every 6 kb. Thus, we would also expect to find several L1 elements in regions the size of an average gene. L1 elements, like Alu sequences, also seem to preferentially target AT-rich DNA.

Most transposable elements in genomes are defective. If all were active, the mutational "load" associated with the large numbers of transposable elements found in many eukaryotes would probably be incompatible with species survival. Because most genomic transposable elements have been mutationally inactivated during the course of evolution, they are not under selection for function, and therefore they continue to accumulate additional mutations.

If we align instances of any particular type of transposon, we can define a consensus sequence for that element (presumably corresponding to a functional element having no mutations). We can then distribute all elements into "bins" having differing levels of sequence divergence from this consensus. The degree of sequence divergence from the consensus is a measure of how long ago the elements in each bin were transpositionally inserted: higher levels of divergence correspond to more remote times.

An example of this is shown in Figure 12.5. Bins of each element type having large proportions of elements correspond to periods when transposition of that element was particularly active. It is evident that most Alu transpositional events were relatively recent in humans but that LINE L2 elements were primarily active in the remote past and are no longer transposing at a significant rate.

The distribution of transposable element sequences as a function of sequence divergence provides a measure of the evolutionary trajectory of the organism. Even recently transposed elements can provide evolutionary information over shorter timescales. For example, Alu insertions can be used to analyze phylogenetic relationships among great apes (Salem et al., 2003).

Genome and Chromatin Material

Even before the advent of genome sequencing, it was clear that DNA within a chromosome could be divided into distinguishable regions. The first division of DNA into types is its assignment to either *euchromatin* or *heterochromatin.* Recall that chromatin is composed of DNA with the bound histones and other chromosomal proteins.

Euchromatin is more actively transcribed, less condensed during interphase, and exhibits cytological staining properties different from heterochromatin, which remains condensed during interphase and is relatively inactive transcriptionally. Heterochromatic regions are commonly found near the centromeres of chromosomes. Recently, heterochromatin has been defined operationally as that portion of the genome that cannot be readily cloned in high-capacity cloning vectors such as BACs (Adams et al., 2000).

This latter property, thought to relate to the relatively large number of repeated sequences in heterochromatin, has slowed the completion of genomic sequencing in some organisms. Organisms can differ significantly in the amounts of heterochromatin that they contain. For example, heterochromatin represents about 6.5% of the human genome but approximately 33% of the *Drosophila melanogaster* genome. Draft sequences at the time of initial publication may exclude large proportions of the heterochromatic regions.

In the case of *Drosophila,* this amounted to about one-third of the total genome (Adams et al, 2000). Genetic mapping studies allow comparisons within and between genomes at a map resolution determined by marker or gene densities. Map comparisons can reveal genome segments that have been duplicated within genomes or segments whose genetic organization has been conserved between genomes.

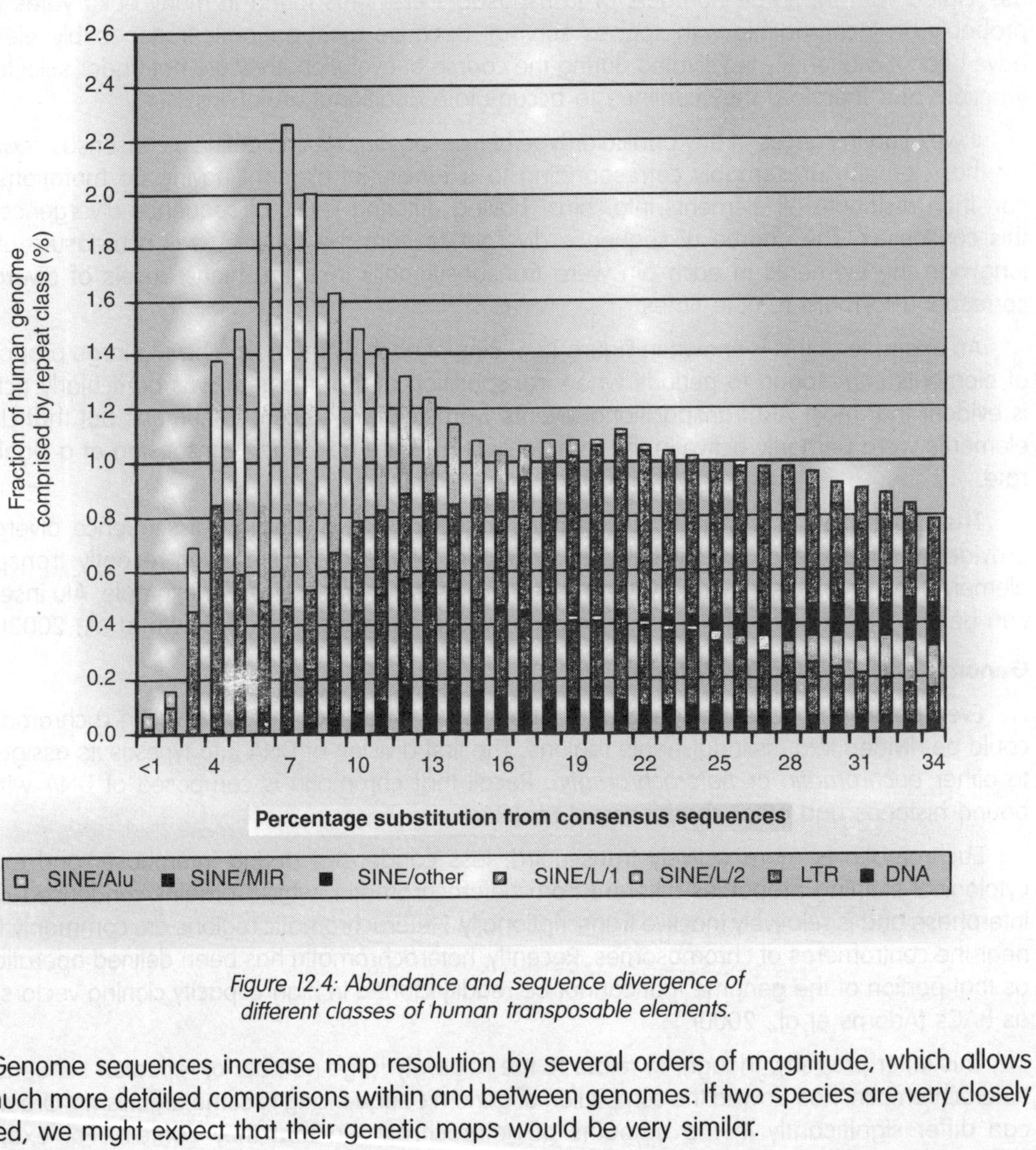

Figure 12.4: Abundance and sequence divergence of different classes of human transposable elements.

Genome sequences increase map resolution by several orders of magnitude, which allows for much more detailed comparisons within and between genomes. If two species are very closely related, we might expect that their genetic maps would be very similar.

In other words, the gene orders and relative locations may be similar, at least over short distances. We already saw an example of this, where contents of a mouse chromosome and human chromosomes were compared. We also saw in that chapter how alterations in gene orders could serve as measures of the evolutionary distance between organisms.

Conserved Segment

As indicated in Section elsewhere in this chapter, two or more genes are said to be *syntenic* if they reside on the same chromosome. A set of genes $g_1, \ldots, g_n$ that is syntenic in organism A

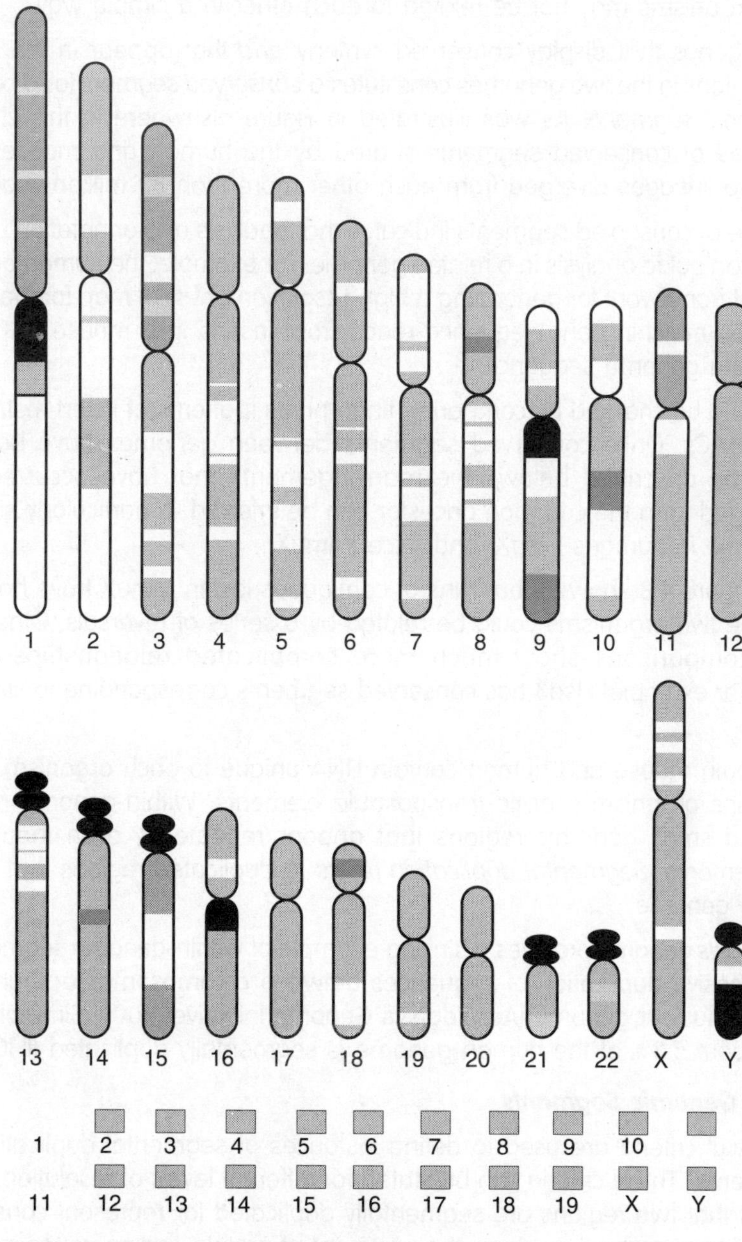

Figure 12.5: Regions containing conserved segments in human and mouse chromosomes. Ideograms of human chromosomes are shown at the top. Color coding indicating the identity of mouse chromosomes (bottom) shows which regions of the human chromosomes are similar to the respective mouse chromosomes. Note that nearly all of the small human chromosome Hsa20 corresponds to a portion of the large mouse chromosome Mmu2. Portions corresponding to Mmu2 are also found on Hsa2, Hsa9, Hsa10, Hsa11, and Hsa15.

and also syntenic in organism B represents a *conserved synteny.* Note that the chromosomes in the respective organisms may not be related to each other in a simple way.

A group of genes that display conserved synteny *and* that appear in the same order and relative map positions in the two genomes constitutes a *conserved segment* (also called a *conserved linkage* or *syntenic segment*). As was illustrated in Figure elsewhere in this chapter, there are multiple instances of conserved segments shared by the human and mouse genomes, even though these two lineages diverged from each other more than 83 million years ago.

The existence of conserved segments indicates that analysis and annotation of such segments in one genome can guide analysis in a related genome. For example, the human genome sequence provided a useful framework for generating a high-resolution physical map for the mouse genome. This was done by matching end-sequence reads from inserts in a mouse BAC library with the assembled human genome sequence.

The tiling could be checked by comparing fingerprints (patterns of insert restriction fragments) of overlapping BACs. Once conserved segments between genomes have been identified by approaches to be described below,. the rearrangements that have occurred since the two organisms diverged from the common ancestor can be inferred. A particularly simple example is the X chromosome in humans (HsaX) and mice (MmuX).

The plot in Figure 4.3 showed that runs of contiguous loci in MmuX have homologs in HsaX. These runs in the two organisms could be related by a series of reversals. Other mouse-human chromosome comparisons show much more complicated relationships involving many chromosomes. For example, Hsa3 has conserved segments corresponding to six different mouse chromosomes.

Of course, both mouse and human contain DNA unique to each organism. Moreover, each organism contains organism-specific transposable elements. Within-genome comparisons can reveal large and small genomic regions that appear repeatedly, even though they are not transposable elements. *Segmental duplication* refers to duplicated regions that encompass only a fraction of the genome.

The *Arabidopsis* genome provides a striking example of within-genome segmental duplication. The obvious extensive duplication of sequences between chromosomes accounts for about 60-80% of the entire nuclear genome (Arabidopsis Genome Initiative, 2000; Simillion et al., 2002). In contrast, only about 5.3% of the human genome is segmentally duplicated (IHGSC, 2004).

Recognition of Genomic Segments

What statistical criteria are used to define instances of segmental duplication or conserved sequence segments? These criteria can be stated for different levels of resolution. At the first level, we might assert that two regions are segmentally duplicated (or represent conserved segments in two different organisms) if they share the same set of protein-coding genes in the same order. Such sets of genes are called *collinear gene clusters.*

Mural et al. (2002) defined collinear gene clusters as clusters having at least three genes meeting this criterion. At higher resolution, we can perform sequence alignments of genomic segments with other segments in the same or a different genome. A segmental duplication (or conserved segment) of segment X would be identified if other segments are found having lengths

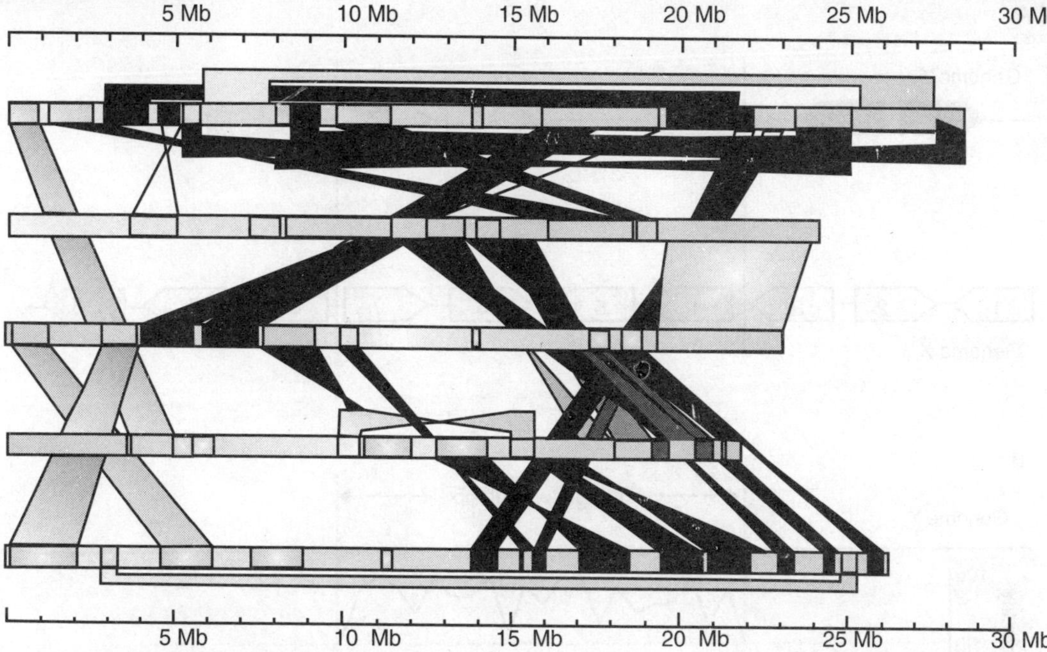

Figure 12.6: Segmental duplications within the Arabidopsis thaliana *genome. Black boxes denote the centromeres in each of the five chromosomes. Color-coded connecting ribbons indicate similar duplicated regions between and within chromosomes. Ribbon edges that cross indicate inversion of gene order in one segment compared with the other. Reprinted.. with permission, from The Arabidopsis Genome Initiative (2000)* Nature 408:796-815. *Copyright 2000 Nature Publishing Group.*

and levels of sequence identity above arbitrary thresholds. (Clearly, transposable element sequences should be ignored before the comparison is performed.)

There are two different implementations of this second approach. We could set a low sequence identity threshold, but demand alignment over larger regions, as was done in one of the analyses of the *Arabidopsis* genome. In that case, segments longer than 1000 by and having more than 50% sequence identity were sought. Or we could use a set of smaller "anchor" sequences between regions being compared.

These are relatively short (e.g., 200 bp) DNA segments (not necessarily within genes) having higher levels of sequence conservation (e.g., more than 85% sequence identity). Regions corresponding to segmental duplication or conserved segments would be recognized as "runs" of corresponding anchor sequences in the same order in the two regions being compared.

A third way of identifying segmental duplication (applicable for comparisons within but not between genomes) relies on statistics associated with sequence reads generated during whole-genome shotgun assembly. This is illustrated in Figure 12.9. Recall that whole-genome shotgun assembly employs reads from the ends of large- and small-insert clones.

These reads are relatively short, usually around 500-800 bp. The average coverage (read "depth") is sufficiently large (typically greater than 5) to ensure collection of sequence data from almost all of the genome. Now consider a DNA segment Z that became duplicated sometime in

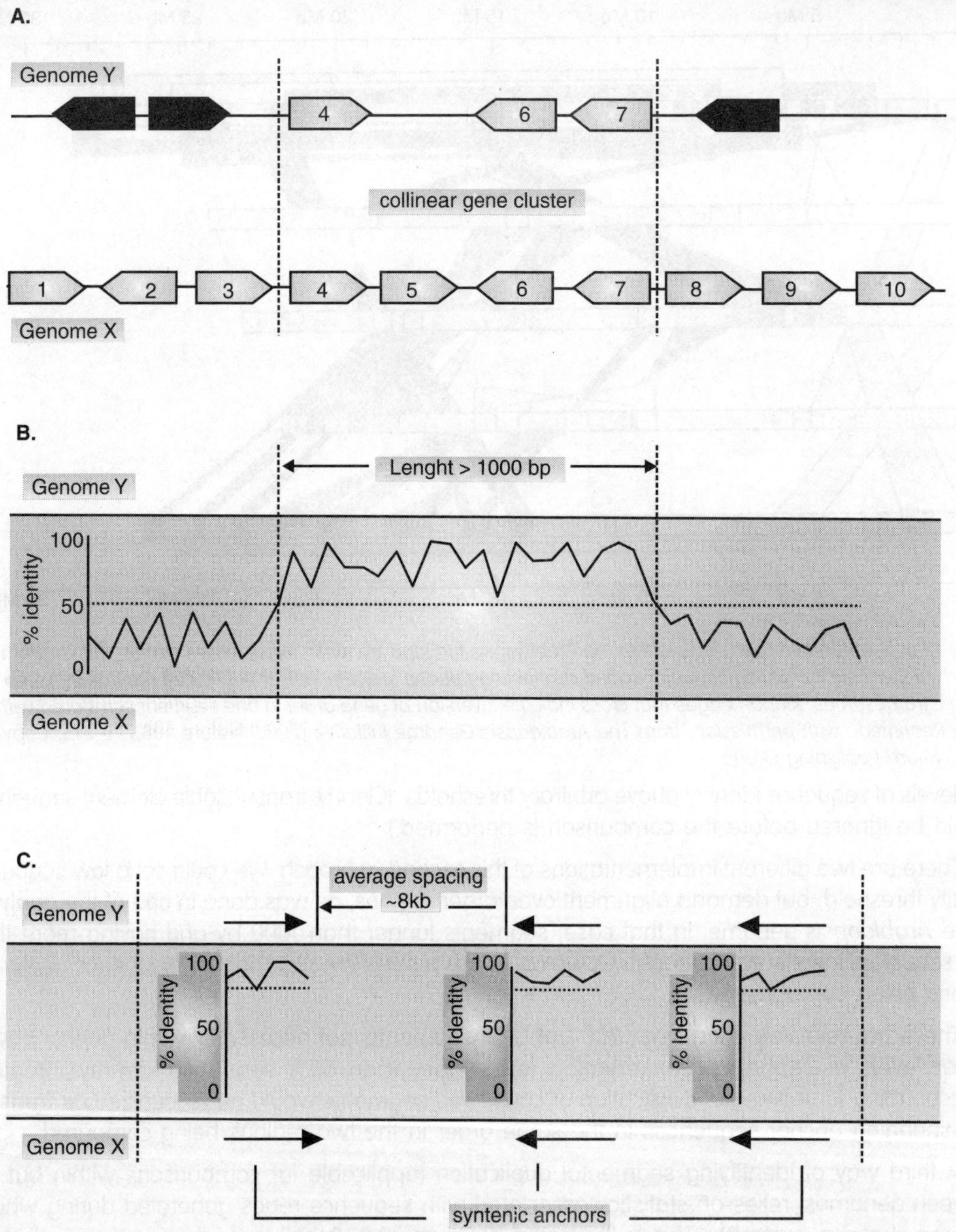

Figure 12.7: Criteria for identifying conserved sequence segments in two genomes, X and Y. Conserved segments are regions between the vertical dashed lines. Panel A: Conserved segment defined by collinear gene clusters. Shaded genes are unrelated to unshaded ones. Panel B: Conserved segment identified by alignment with low identity threshold (dotted horizontal line) and minimum alignment length. Panel C: Use of anchor sequences having high identity thresholds and identified as reciprocal "best hits" in genomes X and Y.

the past. Since that time, both copies, Z1 and Z2, have independently accumulated mutations, so that the reads pooled from these two regions will have lesser amounts of sequence identity than would be the case for unique sequences.

The method starts with the assembled genomic sequence and then aligns with it all reads that exceed a specified threshold of sequence identity (e.g., 95%). Because sequence reads from *both* duplicated regions can align within *each* region, there is an increase in the coverage (read depth) within the duplicated region (twofold in the illustration).

Moreover, the percentage identity of reads covering these regions is lower than for unique sequence regions. Regions of the genome having increased read depth and decreased levels of sequence identity may be identified as segmental duplications, as shown in Figure 12.9.

Phylogeny of Genome

As indicated earlier, one way that genomes evolve is by duplication of genome segments with subsequent genome rearrangement and deletion. In 1970, Susumu Ohno proposed an alternative model. He suggested that vertebrate genomes have evolved by *whole-genome duplications* of ancestral genomes. Such processes have occurred in higher plants (angiosperms): extant variants having two genome equivalents (tetraploids) and three genome equivalents (hexaploids) are well-known.

For *Arabidopsis* and baker's yeast *(Saccharomyces cerevisiae),* evolution of contemporary genomes as a result of one or more earlier whole-genome duplication events is now well-supported. These two alternative mechanisms for genome evolution-independent segmental duplication with rearrangement or whole-genome duplication-have been epitomized as the "Slow Shuffle" or the "Big Bang," respectively (Smith et al., 1999).

Whole-genome duplication can provide raw material for evolution because it produces paralogs of every gene in the genome. One member of each paralogous pair is available to supply the necessary ancestral gene function, while the other may be mutationally inactivated, deleted, or may evolve new functions (usually related to the original function). The presence of numerous duplicated genes within a genome is one indicator of an earlier, whole-genome duplication.

However, over time, deletions, translocations, and inversions may erase some of the evidence of the original duplication event, making it hard to detect from within-genome sequence comparison. Although the baker's yeast genome is the result of a whole-genome duplication (see below), its genome is only 12% larger than the unduplicated genome of a comparable yeast species, *Kluyveromyces waltii,* and the number of predicted genes is only 10% greater.

There are several ways to infer whole-genome duplication events that have led to a contemporary genome. An obvious potential indicator is the presence of *duplicate genes.* However, these can also arise from independent duplication events, and if a sufficient proportion of paralogs has been lost, it may be difficult to distinguish the results of whole-genome duplication from the consequences of many smaller independent duplications.

Sometimes sequence comparisons allow estimates of the time elapsed since the paralogs were formed, and many paralog pairs dated to the same historical time interval may indicate whole-genome duplication rather than a set of independent segmental duplication events. *Patterns*

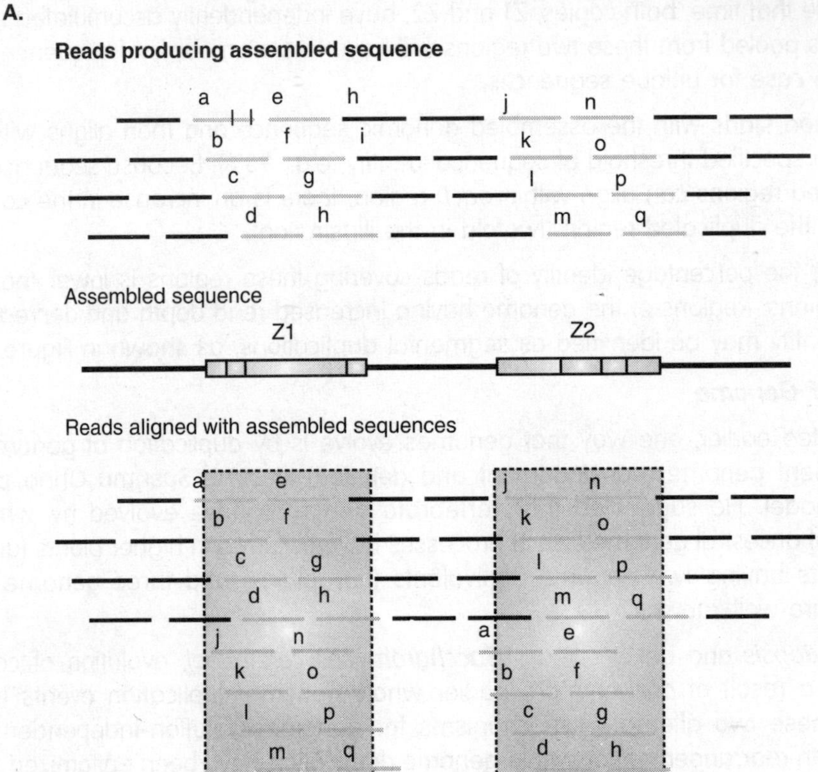

Figuer 12.8: Identifying duplicated regions from the number of reads generated during whole-genome shotgun sequencing. Panel A: Effects of segmental duplication on the number of reads and degree of sequence conservation between duplicated regions. Panel B: Illustration of how read "depth" and dissimilarity increase within duplicated regions in a portion of the human genome.

of collinear gene clusters or anchor sequences, described above, may also be indicators of whole-genome duplication.

Collinear gene clusters can be used to detect whole-genome duplications even if the fraction of paralogous genes is small. We illustrate this for two different methods using either between-genome or within-genome comparisons. *Between-genome comparisons* for identifying genome duplications leading to species D requires a closely-related species A whose genome is unduplicated and not extensively rearranged.

A simplified illustration of this process is shown in Figure 12.10. Actually, a particular organism whose genome has been duplicated might lie at any stage in the overall path illustrated, and the comparison shown in Figure 12.10 E will apply to syntenic segments rather than the whole genome because of translocations and inversions (not illustrated) that also occur.

The processes of mutation and deletion are also concurrent rather than strictly sequential, as shown in this simplified diagram. We can simulate the overall process by using standard playing cards. Select one suit (◇ for example), and shuffle cards of that suit to represent the gene order of an unduplicated genome. Then lay out ♠ and ♥ in two adjacent horizontal rows, each with cards in the same order as the ◇ cards.

The ♠ and ♥ together represent the duplicated genome. To simulate the result when only one member of each paralog remains after a series of deletion events following the initial duplication, sequentially "delete" either a ♠ or a ♥ of each card type, going from left to right. The suit to be deleted for each card type is chosen randomly by a coin toss. One realization of this procedure is shown below:

◇	6 7 5 4 Q K A 2 3 9 10 J 8	(ancestral order)
♠	5 4 A 2 9 J	(descendant after
♥	6 7 Q K 3 10 8	duplication and deletion)

There is no obvious way to determine from ♠ and ♥ cards alone that they resulted from a duplication event since each "gene" is a single copy. But observe the result of alignment with the unduplicated Q genome:

```
♠ – – 5 4 – – A 2 – 9 –  J –
      | |     | |   |    |
◇ 6 7 5 4 Q K A 2 3 9 10 J 8
  | |     | |     |   |     |
♥ 6 7 – – Q K – – 3 – 10 – 8
```

The "genes" shared by ♠ and ◇ are in corresponding gene order, as is also true for genes shared by ♥ and ◇ Furthermore, when aligned with the unduplicated ◇ genome, ♥ and ♠ genes are *interleaving (i.e.,* ♠ appears where ♥ is missing, and vice versa).

This illustrates two results expected from genome duplication using the model specified in Figure elsewhere in this chapter: conserved gene order, and interleaving of genes within corresponding duplicate segments.

These observations are unlikely by the alternative (slow shuffle) model. (Try aligning the original ◇ sequence with ♣ that have been shuffled from the same starting sequence and then cut into

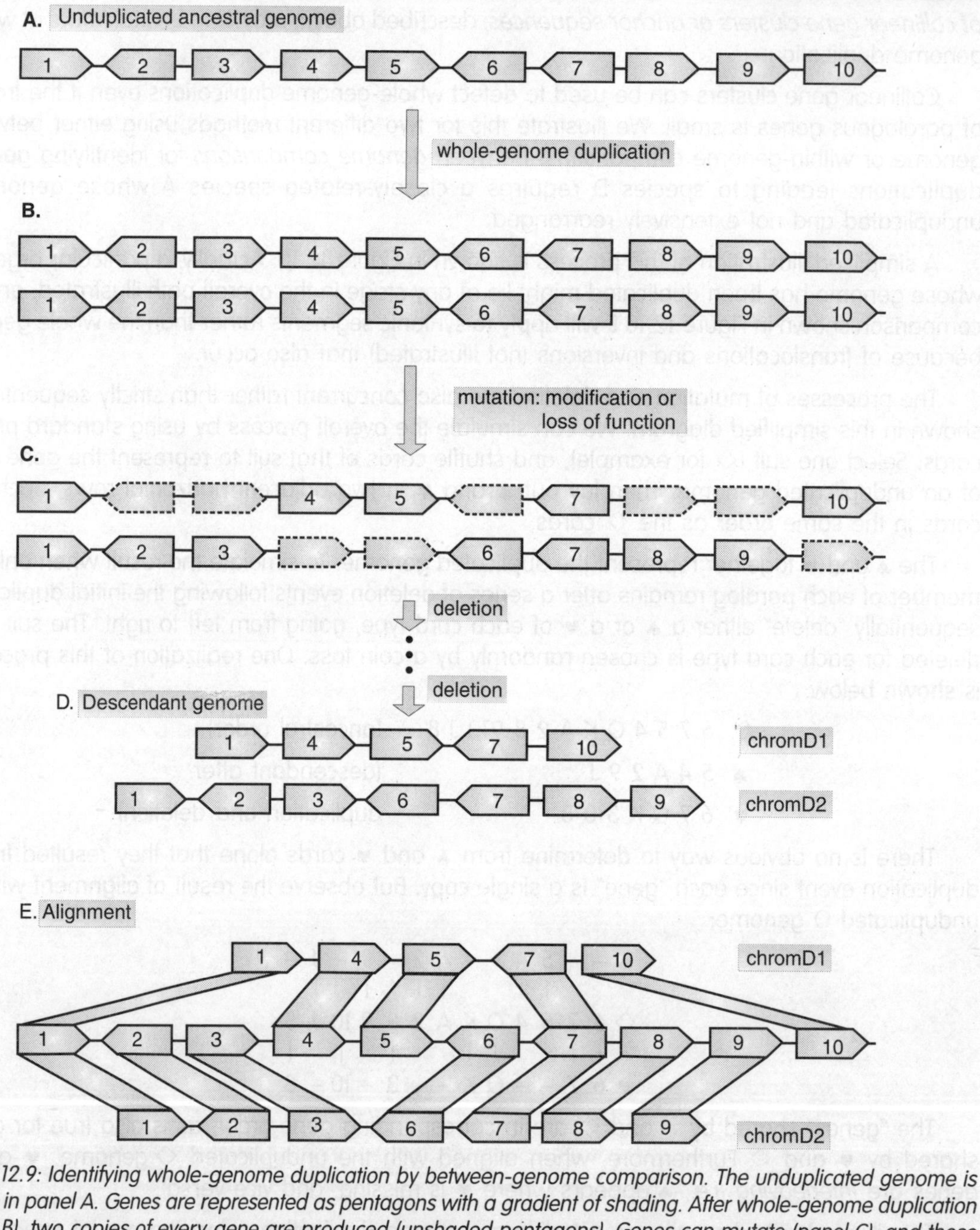

Figure 12.9: Identifying whole-genome duplication by between-genome comparison. The unduplicated genome is shown in panel A. Genes are represented as pentagons with a gradient of shading. After whole-genome duplication (panel B), two copies of every gene are produced (unshaded pentagons). Genes can mutate (panel C), and these mutations may destroy functions (indicated by broken bounding lines) or alter them (shaded genes labeled with primes). Deletions may occur (panel D), subject to the constraint that essential genes be retained. Alignment of chromosomes D1 and D2 with the unduplicated genome of another descendant of the shared common ancestor (panel E) supports the whole-genome duplication because nonparalogous genes in the two chromosomes appear interleaved and in the same order as genes in the unduplicated chromosome.

two rows of six and seven cards.) Whole genome duplications in yeast have been recognized from the conserved gene order or interleaving of genes in duplicate regions.

This approach verified that the *Saccharomyces cerevisiae* genome is a product of whole-genome duplication, as evidenced by comparison with the genomes of yeasts *Kluyveromyces waltii* or *Ashbya gossypii* . The duplicate mapping and interleaving of genes near the centromeres of *S. cerevisiae* chromosomes Scer12 and Scer10 relative to *K. waltii* chromosome Kwa15 is shown in Figure 4.11. Note that for *S. cerevisiae* there are 16 chromosomes, compared with 8 from *K. waltii.* This is consistent with a whole-genome duplication event, which will double the number of centromeres.

The *S. cerevisiae* gene set includes about 88% of the *K. waltii* genome and contains about 457 paralogous gene pairs (8% of the *S. cerevisiae* gene set) (Kellis et al., 2004). Because of genome rearrangements and other effects in the independent lineages leading to *K. waltii* and *S. cerevisiae* from their common ancestor, gene orders and orientations are not conserved over entire chromosomes, but rather are limited to conserved segments spanning 27 genes on average.

The second method, *within-genome comparison,* has been used to identify multiple genome duplications even when substantial genome rearrangements have occurred. The method is illustrated for collinear gene clusters hose homologous genes have been detected by using BLAST. Two whole-genome duplications are depicted. Although this should produce four collinear gene clusters, cluster C has been lost as a result of subsequent deletion events.

Thus, the multiplication level (number of duplicate copies) for this genome region is three rather than four. Shared homologous genes define a collinear gene cluster contained within A′ and B′ and another cluster contained within B′ and D′. A′ and D′ do not share any homologous genes, but *transitive homology* can be inferred: since region A′ is homologous to B′ and region B′ is homologous to D′, we conclude that A′ is homologous to D′. This type of analysis gives a better assessment of multiplication levels. This method has been used to determine the number of whole-genome duplications in the history of *Arabidopsis* (Simillion et al., 2002).

Multiplication levels for direct homologies of collinear gene clusters measured along the genome mostly ranged from two to four. When transitive homologies were included, many regions with

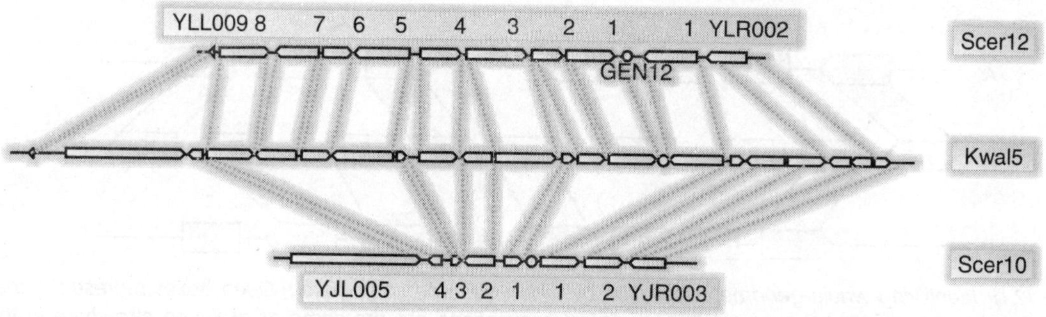

Figure 12.10: Example of duplicate mapping of centromeric regions from S. cerevisiae *chromosomes Scer12 and Scer10 to the same region near the centromere of K. waltii chromosome Kwa15. Although neither S. cerevisiae chromosome has all of the genes found in this region of Kwa15, the gene orders are the same as in Kwa15, and the genes of* S. cerevisiae *interleave to tile almost the entire portion of Kwa15 shown.*

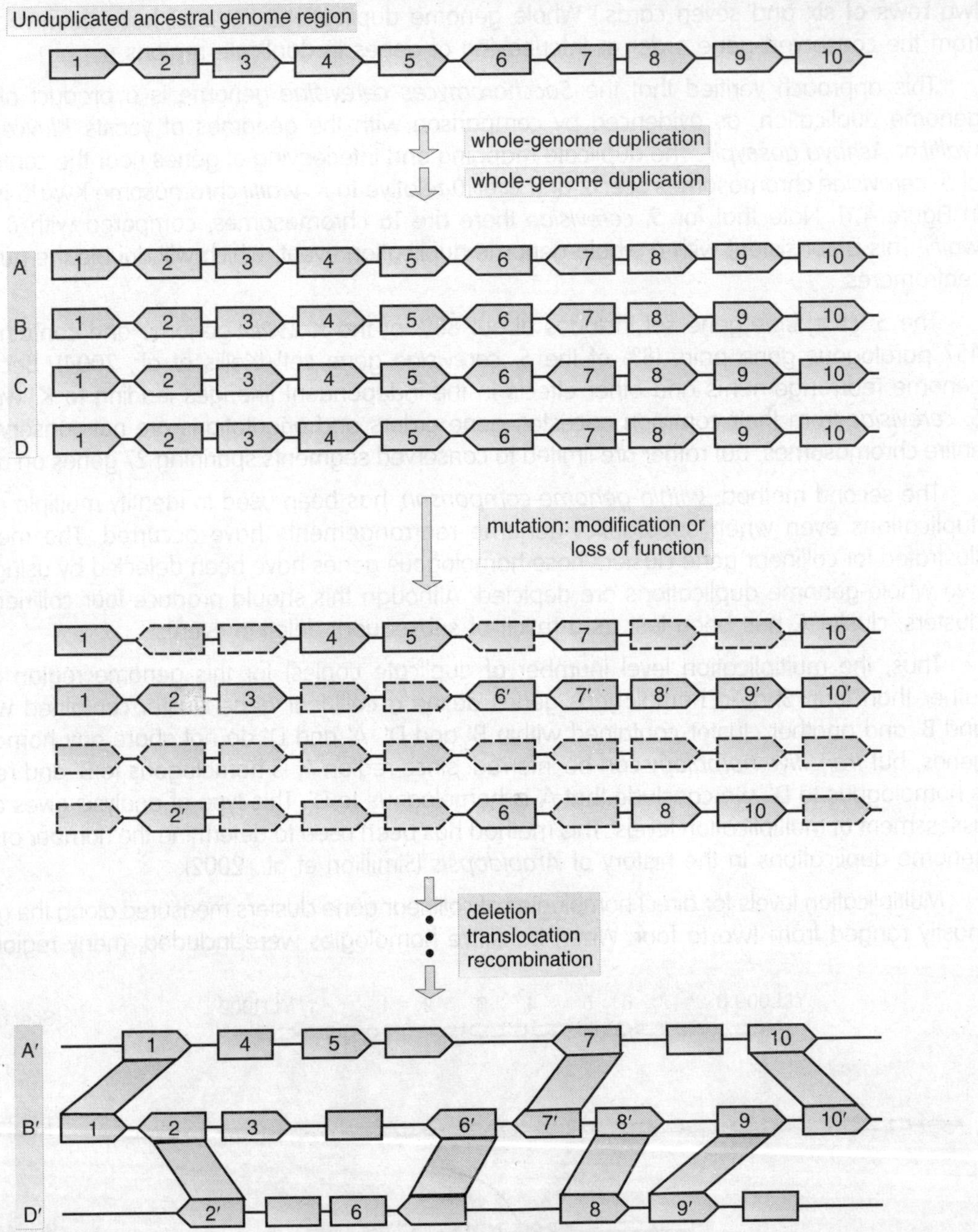

Figure 12.11: Identifying whole-genome duplication by within-genome comparison. Open boxes represent genes not involved in this duplication event. Other graphical conventions are the same as in Figure elsewhere in this chaptere. A segment of the unduplicated genome is shown at the top. Two rounds of whole-genome duplication produce four copies A, B, C, and D. After subsequent deletion, mutation, translocation or other events, collinear gene clusters within (A′, B′) and (B′,D′) remain, indicating a multiplication level of three for this region. Homology of D′ with A′ is inferred even though they share no homologous genes.

multiplication levels of five to eight were observed. This suggested that three whole-genome duplication events had occurred ($5 > 2^2$ and $8 = 2^3$).

The fraction of synonymous substitutions per synonymous site, K_s, was used to estimate dates of 75, 160, and 220 million years ago for these duplication events. Other methods for identifying and dating genome duplication events are available (Van de Peer, 2004). Some of these methods employ clusters of homologous genes that appear together at different chromosomal locations, but not necessarily in the same order.

The statistical significance of these hoinologous clusters can be estimated by randomly shuffling the gene neap and identifying any homologous clusters that appear. For each pair of homologs, additional homologs closer than a specified distance (measured either by window size or by the number of unduplicated genes that separate them) constitute randomly-generated gene clusters. The probability of observing these gene clusters is approximated after 1000 such simulations.

Analysis of the human genome by using clusters of paralogous genes, or paralogons, indicates that at least one whole-genome duplication has occurred in the chordate lineage (McLysaght et al., 2002). The estimated date for this duplication is 350-650 million years ago. As additional complete genome sequences from throughout the phylogenetic tree become available, it will be possible to refine hypotheses about genome evolution by whole-genome duplication.

Coding and Non-Coding Sequences

Two major tasks to accomplish after assembly of a complete genome sequence are identifying gene sequences and assigning gene functions. Functional assignment is also discussed elsewhere in this chapter. Here we focus on identifying genes and their associated regulatory sequences. Three types of information employed in gene finding are signals (e.g., splice sites), sequence content (e.g., k-word frequencies), and similarities with cDNAs or ESTs (Stormo, 2000).

The first two types of information depend upon the local sequence context, while the latter type employs comparison with database entries. The different categories of genome sequence are shown schematically in Figure 12.13.

Once transposable elements and other repeated sequences have been identified and suitably masked, the remaining single-copy genomic sequence can be further characterized as either *coding sequence* or *noncoding sequence*.

Information represented in coding sequences will eventually appear in gene products either as RNA (e.g., tRNA) or proteins. Coding regions represent only a small proportion of the genomes of complex eukaryotes such as plants and animals. The percentage of the DNA that codes for genes indicates the magnitude of the gene identification problem. The yeast *S. cerevisiae* genome, for example, contains 70% coding sequences, which means that the signal-to-noise ratio is high.

The human genome, in contrast, contains only about 1.2% protein coding sequences, and just 5% of each protein-coding gene actually codes for amino acid residues (the signalto-noise ratio is low). Such genome properties will affect which experimental and computational approaches are appropriate.

The classical experimental approach to gene finding is to generate mutations in an organism, characterize the resulting phenotypes, genetically map the locations of the mutations in the genome, and then identify the altered gene products biochemically. This traditional method is arduous and

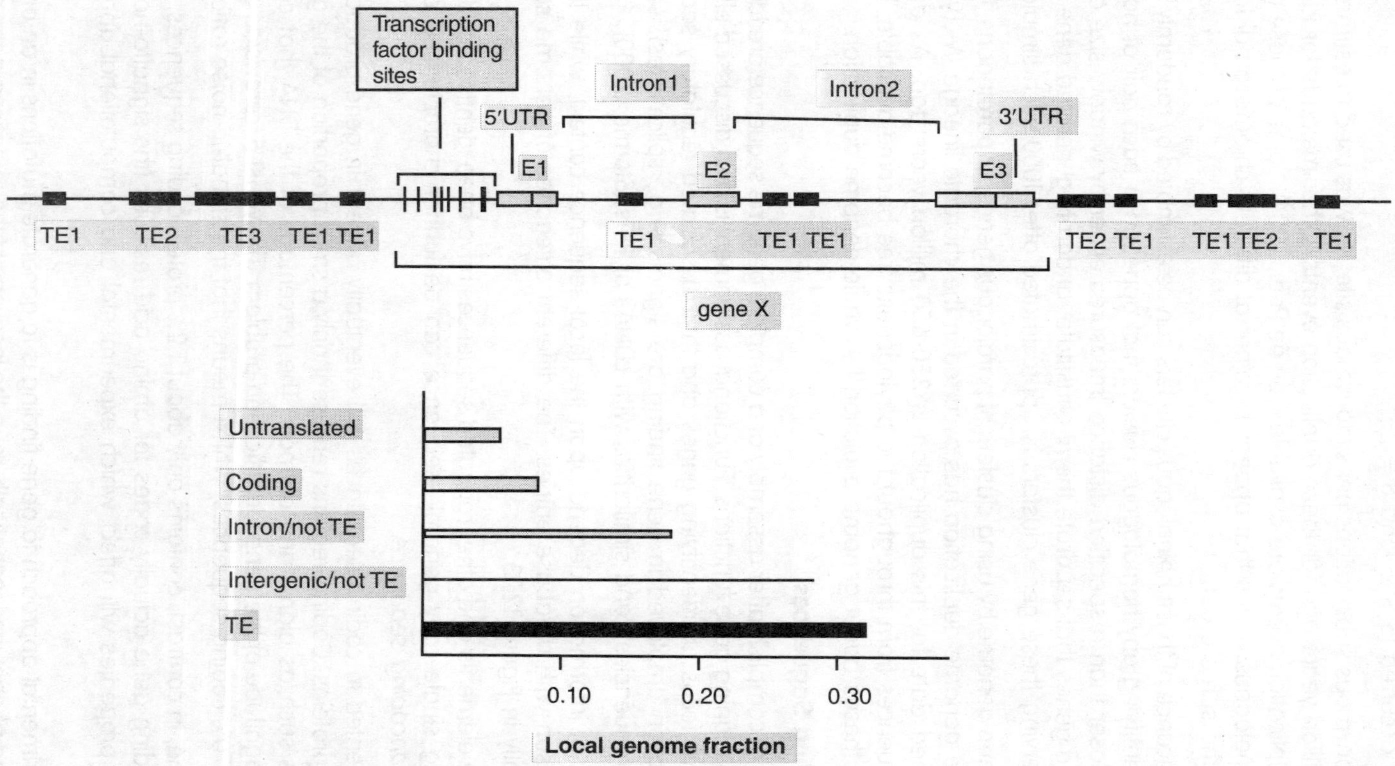

Figure 12.12: Types of sequences found in eukaryotic genomes. Coding sequences (open boxes within exons El, E2, and E3) may represent only a small fraction of eukaryotic genomes. Noncoding sequences may be intergenic (thin lines), intronic sequences within genes (thick lines), or untranslated portions of exons (grey boxes). Intersecting vertical lines represent transcription factor binding sites (thin lines) or the core promoter (thick line). Transposable elements of various types (filled boxes) may represent a few percent or the majority of a genomic sequence. The fractional amount of each sequence type in this genome region is shown at the bottom.

Table 12.2: Measures of coding sequence content for eukaryotic model organisms.

Genome	No. Genes	% Coding[a]	Genomic DNA per gene (bp)	Gene size (average bp)	mRNA size[b] (average nt)	Exon size (average bp)	Exons/gene (average)
Saccharomyces cerevisiae	5,500	70	2060	1450	1450	1450	1[c]
Caenorhabditis elegans	18,400	27	5100	2700	1400	240	6
Drosophila melanogaster	13,600	20	8550	3250	1770	425	4
Arabidopsis thaliana	26,400	26.3	4870	1970	1280	164	5.2
Homo sapiens	25,000	1.2	137,000	27,000	1340	145	9-10

[a] % coding computed relative to assembled sequence, which may not include heterochromatin.
[b] mRNA size is for the processed transcript, with introns removed.
[c] An estimated 240 yeast genes contain one or more introns. *96%* of the yeast genes have no introns.

Data: S. *cerevisiae* (Goffeau et al., 1996)
C. elegans (C. elegans Sequencing Consortium, 1998)
D. rnelanogaster (Adams et al., 2000)
A. thaliana *http://mips.gsf.de/proj/thal/db/tables/table* s_gen_frame.html
H. sapiens (International Human Genome Sequencing Consortium, *2001; 2004)*
Exon/intron statistics for last four organisms above are also available in: Deutsch and Long, *1999.*

time-consuming. Genetic engineering methods for systematically targeting particular genes for mutation (i.e., gene "knockout" methods) are available and practical for some organisms. For example, 96% of the open reading frames of baker's yeast *(S. cerevisiae)* were deleted one at a time, and effects upon cell morphology and growth were measured under a variety of conditions (Giaever et al., 2002).

However, knockout technologies may not be available for all organisms, or in the case of humans are not applicable for ethical reasons. For such cases, and in general as a preliminary, it is usual and convenient to employ computational methods for initial characterization of sequenced genomes.

Identifying Eukaryotic Genome

Computational gene prediction in a genomic sequence is an important and complex problem, particularly for eukaryotic genomes. Complete characterization of a gene includes identifying the promoter, all transcription factor binding sites, all introns and exons, the transcriptional start site and termination sites, and sites involved in pre-mRNA processing.

These tasks are all computationally nontrivial in their own right, so the complete identification of genes based only upon their local sequence context is rarely achieved for eukaryotes. Identifying eukaryotic promoter sequences is an important (and difficult) problem. Promoter sequences are recognized by general transcription factors, such as TATA-binding protein; TBP. Not counting regulatory regions, promoters often span a region 200 by or more upstream of the transcriptional start site at +1. Critical promoter elements are the TATA box, the CAAT box, and the GC box.

The TATA box and CAAT box (when present) occur about 25 by and 100 by upstream (in the 5′ direction) of the transcriptional start site, respectively, but these locations can vary. The GC box may be found about 200 by upstream of the transcriptional start site. There are "TATA-less" promoters, which may contain multiple copies of the GC box at a variety of positions upstream of the transcriptional start site.

Positional weight matrices for the TATA box and the start site are shown in Table 12.3. A major task for eukaryotic gene prediction is identifying exons and introns from sequence data. Introns contain at least three potential signals recognized by the splicing apparatus for processing mRNA: the 5′ donor site, the 3′ acceptor site, and the branch site.

The splicing occurs on an RNA substrate, but gene finding is usually performed on DNA, so we write this as the DNA version. The simplest description of an exon-intron relationship is:

5′ exon GT ... branch site ... AG 3′ exon

where 5′ and 3′ refer to the exons immediately flanking a particular intron (not necessarily the exons at the extreme 5' and 3' ends of genes), and the vertical lines separate exons from the intron, which lies between the vertical lines. Conserved dinucleotides GT and AG obviously can occur very frequently by chance, so additional positions are used to define the splice junctions.

These extended sequences provide a better representation (patterns less likely to occur by chance), but as we might expect of consensus methods, these patterns fail to represent completely the range of allowable splice sites. For example, a small proportion of introns contain GC at the 5′ splice junction.

Table 12.3: Positional weight matrices (profiles) for a variety of features associated with human genes. Positions are labeled by various biological conventions. Entries are probabilities. Intron profiles are for low %G+C loci.

Branch profile

	-5	-4	-3	-2	-1	0	+1
A	0.25	0.25	0.00	0.00	0.39	1.00	0.18
C	0.19	0.22	0.60	0.02	0.24	0.00	0.33
G	0.15	0.17	0.00	0.00	0.32	0.00	0.03
T	0.41	0.36	0.40	0.98	0.05	0.00	0.46

5' Splice site profile (low G+C):

	-3	-2	-1	0	+1	+2	+3	+4	+5
A	0.38	0.62	0.12	0.00	0.00	0.71	0.73	0.11	0.21
C	0.31	0.10	0.04	0.00	0.00	0.02	0.06	0.06	0.10
G	0.18	0.12	0.77	1.00	0.00	0.24	0.08	0.75	0.14
T	0.13	0.16	0.07	0.00	1.00	0.03	0.13	0.08	0.55

3' Splice site profile (low G+C):

	-15	-14	-13	-12	-11	-10	-9	-8	-7	-6	-5	-4	-3	-2	-1	0	+1
A	0.15	0.14	0.13	0.11	0.10	0.10	0.11	0.12	0.13	0.11	0.10	0.26	0.07	1.00	0.00	0.26	0.24
C	0.24	0.21	0.20	0.22	0.21	0.22	0.25	0.28	0.28	0.25	0.22	0.25	0.55	0.00	0.00	0.11	0.15
G	0.10	0.12	0.10	0.09	0.10	0.09	0.10	0.10	0.08	0.05	0.05	0.15	0.01	0.00	1.00	0.50	0.20
T	0.51	0.53	0.57	0.58	0.59	0.59	0.54	0.50	0.51	0.59	0.63	0.33	0.37	0.00	0.00	0.13	0.41

TATA box profile

	-3	-2	-1	0	+1	+2	+3	+4	+5	+6	+7	+8	+9	+10	+11
A	0.12	0.06	0.69	0.02	0.70	0.57	0.69	0.42	0.24	0.13	0.20	0.18	0.19	0.18	0.15
C	0.39	0.22	0.04	0.15	0.06	0.06	0..09	0.05	0.18	0.35	0.36	0.31	0.32	0.30	0.30
G	0.35	0.15	0.07	0.11	0.12	0.05	0.16	0.26	0.46	0.41	0.34	0.34	0.31	0.35	0.36
T	0.14	0.57	0.20	0.72	0.12	0.32	0.07	0.27	0.12	0.12	0.11	0.17	0.17	0.18	0.19

Start site profile

	-9	-8	-7	-6	-5	-4	-3	-2	-1	0	+1	+2	+3
A	0.21	0.15	0.21	0.21	0.18	0.24	0.57	0.31	0.15	1.00	0.00	0.00	0.26
C	0.28	0.38	0.38	0.24	0.35	0.49	0.05	0.43	0.53	0.00	0.00	0.00	0.20
G	0.31	0.27	0.22	0.42	0.27	0.22	0.36	0.17	0.27	0.00	0.00	1.00	0.41
T	0.19	0.20	0.19	0.13	0.20	0.05	0.02	0.10	0.04	0.00	1.00	0.00	0.10

Positional weight matrices (expressed as probabilities of occurrence of each base at each position) provide a more complete description for the 5′ and 3′ splice sites and the branch site (Zhang, 1998). Another set of matrices has been compiled for high %G+C genes (Zhang, 1998), and the donor and acceptor sequence preferences for the minority of introns having GC at the 5′ splice site have also been determined (Thanaraj and Clark, 2001). With such matrices and the background base frequencies q_a, scoring matrices having $\log_2(p_{ai}/q_a)$ as their elements can be prepared to aid in exon prediction. For eukaryotic genomic sequences, computational gene prediction tools such as GeneMark, Genie, or Genscan can be employed. Some of these are ab initio gene finders, meaning that they generate a prediction based upon the statistics and signals associated with the input string of the DNA sequence being searched.

These methods have associated false-positive and false-negative error rates that affect our ability to identify genes in any organism. Their accuracy (measured as the average of the sensitivity and specificity of exon prediction) can range from around 45% to 75%, depending upon which tool is employed (Rogic et al., 2001).

With metazoan genes, which typically have several exons per gene, it is clear that the likelihood of correctly predicting an entire gene (all exons) is low (30%-40% correct predictions for H M M Gene and Genie applied to a portion of the *Drosophila* genome; Reese et al., 2000).

Statistical Study of Exon and Intron

The sizes of exons and numbers of introns affect gene identification. Average exon and intron statistics for several genomes are presented in Table 12.2. These numbers indicate how complicated genome analysis will be for each organism. For example, only 4% of yeast genes contain introns; therefore, large open reading frames in yeast will be strongly correlated with actual genes, and gene finding will be comparatively easy.

In contrast, human protein coding genes contain nine to ten introns on average, and correct gene prediction will require correct identification of about 18 to 20 exonintron or intron-exon boundaries for each gene. Exon size distribution data can be useful when an-alyzing genes. Exon size distributions for three different eukaryotes are shown in Figure 12.14.

These distributions are very broad and skewed right (see also Deutsch and Long, 1999). Lengths of candidate exons predicted from 5′ and 3′ splice signals (see below) can be tested against the exon length distribution for that organism to determine whether the lengths are improbably short or long.

Computational Approach

A simple computational approach for gene finding is a sequence similarity search using alignment applications such as BLASTX to compare translations of genomic sequence in all six reading frames against a protein database. This is particularly appropriate if most genes are not interrupted by introns. If the organism's genes have many introns, the translated string used for searching may not accurately correspond to an exon, which will reduce the sensitivity of the comparison.

A similar approach is to compare genomic sequences directly to entries in EST (expressed sequence tag) databases. ESTs contain short sequences derived from cDNAs and are frequently determined by singlepass, high-throughput sequencing of thousands of cDNAs derived from particular tissues and particular physiological conditions. ESTs usually contain more sequencing errors than does finished genomic DNA.

Also, they frequently only represent the 3′ end of the mRNA—not the full-length transcript. Estimates of gene numbers may be confounded by fragments of a single gene that are attributed to separate transcripts and by the failure to recognize pseudogenes.

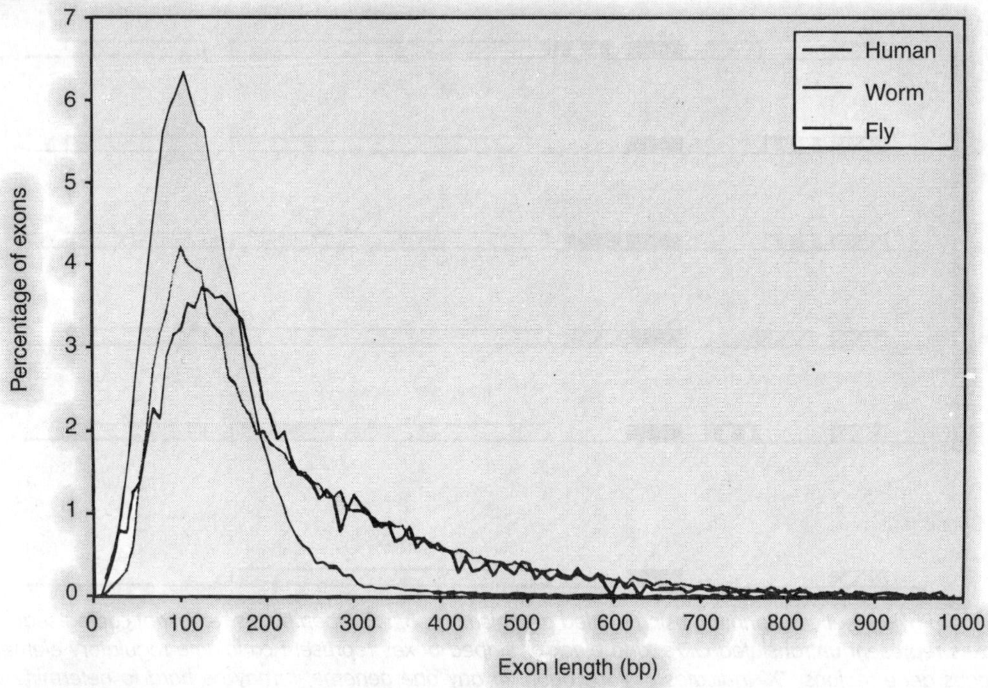

Figure 12.13: Distributions of exon lengths in H. sapiens, C. elegans, and D. melanogaster. Only a small proportion of human exons exceed 300 by in length, indicating that such a threshold might be used as an additional probabilistic criterion for exon scoring.

An additional problem for both prokaryotic and eukaryotic genomes is the absence of significant database matches for substantial fractions of predicted genes (30-40% for early sequencing projects).

Comparative genomics provides an additional approach to gene finding (coding sequences *and* regulatory sequences) that employs several related genomes to reduce the signal-to-noise ratio. After applying gene-finding methods to the genomic sequence of any single organism, we usually do not know whether all exons have been correctly identified, nor do we know what transcription factors regulate the expression of that gene.

Now suppose that the corresponding genomic sequences from other related organisms are known. We perform multiple alignment on their corresponding regions. Since the gene and its regulatory sequences have been evolutionarily selected for function, we expect to find a higher level of sequence conservation within the coding and regulatory sequences than in the introns or "spacer" regions between transcription factor binding sites.

This concept is diagrammed in Figuere 12.15. After multiple sequence alignment, correct translational start sites, stop sites, and exon boundaries can be inferred from the "consensus" assignment for the whole set. Regulatory sequences are particularly difficult to identify from a single genome because the motifs may be short and may display considerable sequence degeneracy at positions within the sequence pattern.

If we were to use PWMs to search for regulatory sequences, there would typically be a

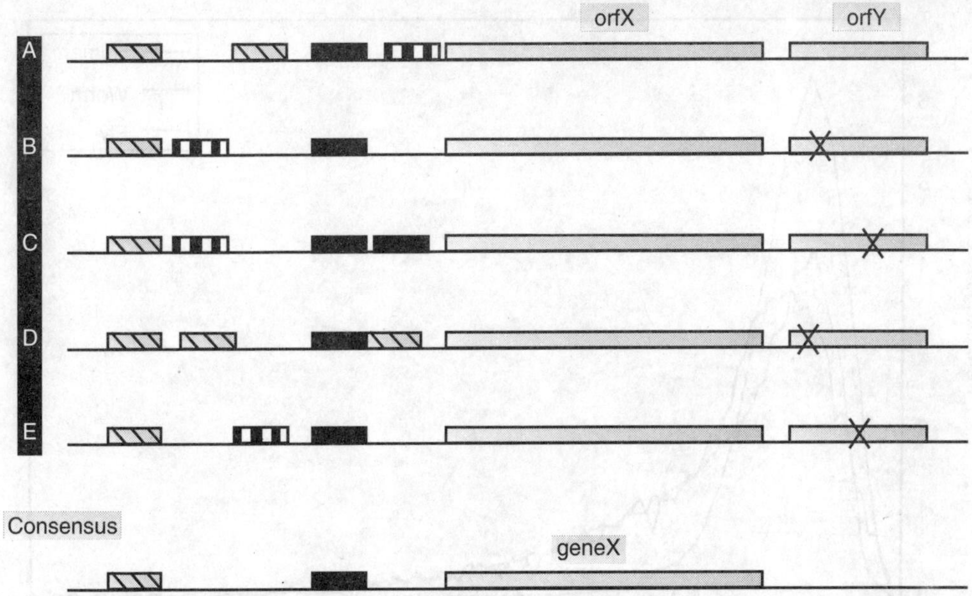

Figure 12.14: Phylogenetic footprinting using related organisms A, B,...,E. Open boxes represent coding sequences, grey boxes represent untranslated ORFs, and black or striped boxes represent candidate regulatory elements in orthologous gene regions. "X" indicates a stop codon. In any one genome, it may be hard to determine which regulatory elements are functional, but conserved elements shared by the whole set most likely represent the actual functional elements. Similar logic can be used to identify functional ORFs. Although orfY appears in A, its absence in the other strains (note stop codons) suggests that it may not be functional.

substantial false positive error frequency. But with aligned genomic sequences, the regulatory sequences become more apparent because of their conserved sequences and positions relative to the coding sequences. Identifying regulatory sequences by comparing sequences from corresponding genomic regions of related organisms is called *phylogenetic footprinting.*

The power of such methods is illustrated by the comparative study of *S. cerevisiae* and three close relatives (Cliften et al., 2003; Kellis et al., 2003). As we mentioned earlier, yeast gene identification is relatively easy, because of the 6062 ORFs listed in the Saccharomyces Genome Database, only 240 were indicated to have one or more introns. Even with this simple genome, the comparative genomic approach showed that about 500 of the approximately 6000 previously annotated "genes" were probably not genes at all.

The phylogenetic footprinting approach identified 40 new regulatory motifs in addition to many of the 55 regulatory motifs that had previously been identified (Kellis et al., 2003). With the availability of additional genome sequences throughout the phylogenetic tree, this comparative approach will be a powerful addition to other methods of computational gene analysis.

Marking and Naming Genes

The size, number of genes, and fraction transcribed are important properties distinguishing genomes. Genome sizes and estimates of numbers of genes already give us significant information. We start by noting that there is no clear relationship between the amount of DNA and the genetic complexity of an organism. For example, *Arabidopsis* has more genes than *Drosophila* but less

DNA. Next, note the number of genes contained in the *Mycoplasma genitalium* genome. This number (about 500) gives us an upper bound for the minimum number of genes required to operate an autonomous, free-living organism.

Experimental gene knockout studies on this organism suggest that the minimal genome for autonomous existence is in the range of 250-350 genes. If we compare the numbers of genes of *M. genitalium* and *E. coli,* we see that *E. coli* has about 8 times the number of genes found in *M. genitalium*.

This may mean that *E. coli* has accumulated genes that are useful for growth under a broader range of environmental conditions. Genes for the utilization of particular sugars such as lactose and xylose exemplify this phenomenon (genes in the lactose operon are not significantly expressed in the absence of lactose). If we compare the numbers of genes of *S. cerevisiae* (a eukaryote) and *E. coli* (a prokaryote), we find numbers that are similar, suggesting that the eukaryotic lifestyle is not just a matter of having more genes but rather having particular sets of gene types.

For example, there are 1105 yeast genes that are essential for growth in rich medium (Giaever et al., 2002) compared with an estimated minimal genome of 250-350 for *Mycoplasma.* Of course, numbers of genes are insufficient to represent the complex organization of eukaryotic regulatory networks.

Comparison between metazoan genomes is instructive. For example, *C. elegans* (a nematode "worm") is a very simple organism that as an adult hermaphrodite consists of 959 somatic cells. (This is the actual *cell number,* not the number of *cell types.)* Its genome is predicted to contain about 18,400 genes. The adult *D. melanogaster is* a much more complex organism, both *morphologically* and behaviorally, yet it is predicted to have 13,600 genes.

As indicated above, we conclude that it is not just the number of genes that determines the complexity of an organism but rather the types of genes or the complexity of the interactions between genes and gene products that matters. A comparison of the human genome with that of *Arabidopsis thaliana is* also informative.

Both are multicellular eukaryotes, and both contain roughly the same number of genes. We tend to view humans from the standpoint of behavioral complexity, while we say that plants just "vegetate." However, plants have additional biochemical complexity that humans do not possess (e.g., human cells cannot manufacture all of the essential amino acids, nor can they engage in photosynthesis), and thus the similarity in number of genes is perhaps understandable.

Note, however, that the human genome is about 25 times larger than the *Arabidopsis* genome, even though the numbers of genes are comparable. This size difference could be explained by a number of hypotheses, including the suggestion that more of the human genome is devoted to gene control, or alternatively the more trivial explanation that nonfunctional DNA has accumulated to a greater extent in the human genome.

We saw earlier that the human genome contains a higher proportion of transposable elements than does the genome of *Arabidopsis.*

Cataloging Genes

We now consider how to annotate and catalog predicted genes. Our goal is to assign one or more functions to each predicted polypeptide. Having annotated genes to the extent possible, the next job is to catalog them and to make a list of all different types of genes by functional category. This indicates the repertoire of functional pathways available to the organism.

Molecular Functions

We must specify what is meant by "function." Recognizing a need for a common descriptive vocabulary and classification scheme for the various concepts subsumed by the term "function," the Gene Ontology Consortium has defined three families of hierarchical descriptors corresponding to molecular functions, biological processes, and cellular components (Gene Ontology Consortium, 2004).

In terms of *molecular function,* a protein might, for example, be a permease, a phosphatase, or a DNA-dependent ATPase. Although they are precise designations, these descriptions do not describe the *processes* in which the proteins participate. For example, the permease might be involved in utilization of a sugar, the phosphatase might be part of a signaling pathway, and the DNA-dependent ATPase might be a helicase involved in DNA replication.

In terms of *cellular components* to which the proteins belong, the permease might be associated with the plasma membrane, and the helicase (DNA-dependent ATPase) might be associated with the replication fork. Clearly, a complete description of the function of any gene product will include (but not necessarily be limited to) a specification of its molecular function, biological process, and cellular location.

Ordinarily, protein functions cannot be deduced from first principles using the predicted polypeptide sequences. Instead, functions are predicted based upon patterns of gene expression from microarray data or by sequence similarity to other proteins whose functions have been characterized biochemically. Assume that we are starting with the collection of predicted polypeptide sequences for a particular genome.

Each one of these polypeptides can be used as a query sequence to search appropriate databases. The hits (database matches) that display significant similarity with the query sequence are retrieved and listed. If the similarity is sufficiently high and if the functions of some of the hits are known, then it may be reasonable to attribute some of their functional annotations to the query sequence.

As we indicated in our discussion of alignment, sequence similarity means the degree of agreement of the character states at positions of one sequence aligned with another. (Remember that similarity over short segments of two strings need not imply an evolutionary relationship between two proteins: the agreement could have occurred by chance.)

We indicated in Section elsewhere in this chaptere that genes or proteins that have descended from a common ancestor are said to be *homologs,* which are expected to show sequence similarity, provided that the evolutionary distance separating them is not too great. Recall that if gene homologs g_B and g_C (descended from gene g_A) are observed in organisms B and C, respectively, then g_B and g_C are said to be *orthologs.*

The functions of orthologs are often the same as the function of the ancestral gene g_A, so for annotation of sequences, orthologous relationships are particularly helpful (i.e., if the function of g_B is known but that of g_C is not, the function of g_C is likely to be the same if g_B and g_C are orthologs). But suppose that after the lineage of organism B split from the C lineage, duplications of g_B occurred, leading to production of g_{B1} and g_{B2}.

Genes g_{B1} and g_{B2}, products of intragenomic duplication, are said to be *paralogs*. One of these (g_{B1}, for example) may retain the same function as g_C, but g_{B2} may have acquired a different function as a result of mutation and selection. This is possible because the other copy, g_{B1}, is still available to supply the original function. Gene g_C (function unknown) might show sequence similarity to both paralogs g_{B1} and g_{B2}.

In this case, the match with g_{B1} would provide a correct functional assignment, but the match with g_{B2} would not. Identifying orthologs is an important activity in comparative genomics because experimentally determined functions of genes in tractable model organisms can be associated with their orthologs found in newly examined organisms. For example, ribosomes in *E. coli* have been shown experimentally to contain 52 different protein species.

Identification and characterization of these proteins required many years of genetic and biochemical research. Because ribosomal functions are highly conserved, we expect orthologs of most of these 52 proteins to appear in most bacteria. Orthologous relationships make it unnecessary to repeat all of the extensive "wet lab" work to identify and characterize ribosomal proteins for each new eubacterium whose genome is sequenced.

As increasing numbers of complete genome sequences become available, more and more orthologous relationships can be established. A consistent way of defining and cataloging these relationships is through the use of *clusters of orthologous genes* (or COGs; Tatusov et al., 1997). Originally defined for prokaryotes (for which many more complete genome sequences are available), they have been extended to include orthologous groups within eukaryotic genomes as well. These groups are called KOGs. COGs are produced as *follows:*

1. Start with a data set containing predicted proteins encoded in a set of complete genomes (43 prokaryotic genomes in a recent release).
2. Perform an "all-against-all" series of BLASTP comparisons using the predicted protein sequences. This means that each protein sequence in an organism is used as a query sequence for a BLASTP comparison against all other protein sequences in all of the organisms (including the one from which it came).
3. Multiple "hits" from within-organism comparisons are indicative of paralogs, which are treated as a single group-not as multiple instances.
4. Identify best hits for each gene in each organism with comparable genesin other organisms.
5. Define a COG as the set of mutually consistent best hits found between genomes in a minimum of three organisms. Genes correspond to nodes, and the mutual relationships define the edges of a triangular graph defining the minimal COG.
6. Expand the graph of any COG by merging graphs with COGs that share edges in common with it.

This process is illustrated in Figure 12.16. Because genome D may be evolutionarily remote from A, the similarity between a particular protein in D and its ortholog in A may not be recognized, whereas its similarity to the corresponding genes in B and C is significant. The last step allows orthology between the genes in D and A to be identified, because of their relationships to orthologs in B and C.

Each COG corresponds to a particular function represented in a collection of genomes. There are currently 3300 COGs. Some functions may appear in all genomes, whereas others have a more restricted distribution. For example, small ribosomal subunit protein S3 appears in all 43 reference genomes, but ribonuclease I appears in only three. COGs have been organized into the functional categories listed in Table elsewhere in this chaptere.

Functional Assignments

In the previous section, we described the assignment of a gene's function based upon *orthology,* which was recognized on the basis of sequence similarity between gene products from two different organisms. This type of functional assignment includes all three aspects of function molecular activity, biochemical process, and cellular localization-provided those- annotations are available for one member of the orthologous pair. However, there are proteins that participate in the same functional process but are not orthologs. (For example, DNA polymerases and helicases both may be involved in DNA replication).

How can we recognize function at the level of process or cellular location if the participating proteins are not orthologs or if there are no suitable annotated orthologs available? As we mentioned earlier, this situation is not uncommon: 30% or more of the genes found in a newly sequenced organism may have *no* database matches.

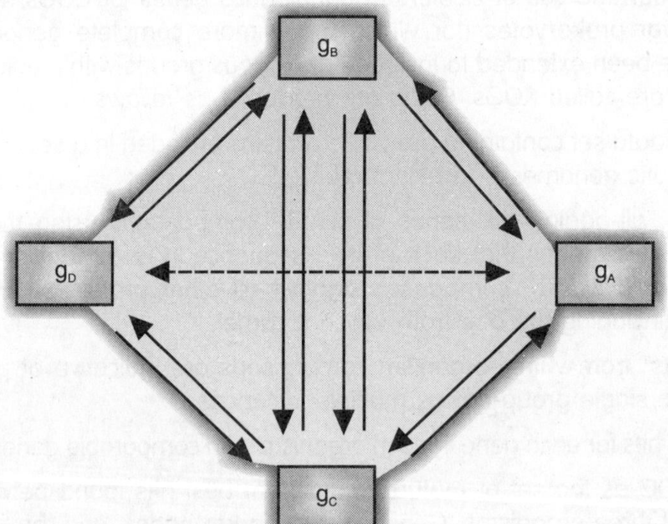

Figure 12.15: Definition of COGs. Orthologs of gene g in organisms A, B, and C can be recognized because the predicted polypeptides are each other's best "hit" for every genome pair (indicated by arrows). The minimal COG consists of this type of pairwise relationship for a particular gene represented by orthologs in three genomes (shaded boxes). Another COG including genomes B, C, and D might also be recognized. Even though g_D and g_A may have diverged from each other enough to make their orthologies hard to detect (dashed line), the common side shared by these two COGs is sufficient to join them, thus placing these orthologs from all four organisms into relationship.

At least three genomic methods for identifying shared processes are particularly successful for prokaryotes and also are applicable to eukaryotes. These methods rely on patterns of gene distribution in three different contexts: distribution of genes within genomes (*gene neighbors*), association of independent polypeptides in one organism with *gene fusions* in another, and distribution of genes among a reference collection of organisms (*phylogenetic profiles*).

Table 12.4: Summary of clusters of orthologous genes (COGs) by functional categories.

Code	Number of COGs	Number of Domains	Description
Information storage and processing			
J	217	6449	Translation, ribosomal structure, and biogenesis
K	132	5438	Transcription
L	184	5337	DNA replication, recombination, and repair
Cellular processes			
D	32	842	Cell division and chromosome partitioning
O	110	3165	Post-translational modification, protein turnover, chaperones
M	155	4079	Cell envelope biogenesis, outer membrane
N	133	3110	Cell motility and secretion
P	160	5112	Inorganic ion transport and metabolism
T	97	3627	Signal transduction mechanisms
Metabolism			
C	224	5594	Energy production and conversion
G	171	5262	Carbohydrate transport and metabolism
E	233	8383	Amino acid transport and metabolism
F	85	2364	Nucleotide transport and metabolism
H	154	4057	Coenzyme metabolism
I	75	2609	Lipid metabolism
Q	62	2754	Secondary metabolite biosynthesis, transport, and catabolism
Poorly characterized			
R	449	11948	General function prediction only
S	750	6416	Function unknown

These methods can be applied individually, together, or in combination with experimental approaches, such as expression profiling, to identify *networks* of genes (not necessarily orthologs) that participate in a shared functional process. The *gene neighbors* approach relies on the observation that genes having related functions tend to be more closely linked genetically than genes having unrelated functions.

This association in map positions is particularly strong for prokaryotes, whose genes are commonly organized into operons. (Even the eukaryote *C. elegans* has 25% of its genes within operons.) The close proximity of related genes allows the sharing of promoter elements for coordinated gene expression. Suppose that in genome *i* we observe that gene *A* (for which the functional process is known) is located next to gene B (for which the functional process is unknown).

Moreover, suppose that these genes also appear in close proximity in genomes *j, k,l,* ... . The conserved neighboring relationships across a range of organisms suggests that these two genes may share in the same functional process (Dandekar et al., 1998; Overbeek et al., 1999).

The second method depends upon the observation that proteins often contain several functional domains that fold in a structurally independent manner. For example, genes *trpG* and *trpD,* which are involved in the biosynthesis of tryptophan, encode glutamidotransferase and anthranilate phosphoribosyl transferase activities, respectively. The products of these two genes appear on separate polypeptide chains in *Serratia marcesens.*

However, these activities appear together on a single polypeptide chain (a product of the *trpD* gene} in *E. colt'.* Similarly, in *E. coli,* the DNA polymerase III holoenzyme contains ten subunits. One of these is the a subunit, which is encoded by *dnaE* and displays polymerase activity, and another is the e subunit, encoded by *dnaQ* and having 3'-5' exonuclease activity. In *Bacillus subtilis,* the replication polymerase PolIIIa* contains the domains for polymerase and 3'-5' exonuclease fused on a single polypeptide chain.

Notice that in both of these examples, each pair of polypeptides (or domains) functions in the same process (a result of experimental observation). Fusion of these domains into a single polypeptide chain is predicted to facilitate the process on biochemical grounds. If we reverse the logic, we can predict that if *two polypeptides A_i and B_i occur separately in organism i but each is similar to different portions of a larger polypeptide in organism j (i.e., gene g_j appears to be a fusion of regions encoding polypeptides resembling A_i and B_i), then A_i and B_i probably participate in the same biological process.*

Note that genes *Ai* and B_i are not similar to each other, so their functional assignment does not depend upon orthology. In addition, appearance of these domains in a single polypeptide chain, where a degree of interaction will occur, suggests that these domains also physically interact when they occur on separate polypeptide chains.

Tests of this method using independently annotated genes showed that 32%-68% of genes associated by domain fusion shared keywords in their annotations (for yeast and *E. coli,* respectively), while pairs of genes selected at random shared keywords in 14%-15% of their annotations.

The third method, identification of phylogenetic profiles, employs statistical tools similar to those that we used for classification and clustering. The procedure is, in a sense, the "transpose" of what we were doing before. Before, we used matrices that looked like the one below to summarize the properties of OTUs

	characters					
OTU	i	ii	iii	iv	. . .	n
1	0	1	1	0	. . .	0
2	1	0	1	1	. . .	0
3	0	1	1	0	. . .	0
4	0	1	1	0	. . .	1
.	.	.	.	.	. . .	.
.	.	.	.	.	. . .	.
m	1	0	0	0	. . .	0

Instead of using anatomical characters, we could use the presence (1) or absence (0) of particular proteins p_1, P_2, ... , Art in m completely sequenced genomes (the genomes *must* be completely sequenced) as our characters. At least some of the *pi* should be annotated functionally in some model organism.

OTU	p_1	p_2	P_3	p_4	. . .	p_n
1	0	1	1	0	. . .	0
2	1	0	1	1	. . .	0
3	0	1	1	0	. . .	0
4	0	1	1	0	. . .	1
m	1	0	1	1	. . .	0

Now transpose the matrix to represent the distribution of each protein across taxa. Each row represents the phylogenetic profile of a particular protein. Remember, some of these proteins pi will have been functionally annotated in one or more genomes.

Proteins	OTU_1	OTU_2	OTU_3	OTU_4	...	OTU_m
P_1	0	1	0	0	...	1
P_2	1	0	1	1	...	0
P_3	1	1	1	1	...	1
P_4	0	1	0	0	...	1
p_n	0	0	0	1	...	0

Boldface entries highlight the vectors of two proteins having similar profiles.

The key idea for using such data is that genes encoding proteins participating in the same functional pathway tend to be coordinately present or absent in a collection of genomes. The phylogenetic profile of a protein is the *n,* dimensional vector representing its presence (1) or absence

(0) in each of *n,* genomes (OTUs). Proteins having similar profiles are inferred to be functionally linked. In the illustration above, proteins p_i and p4 are expected to be functionally related, and if the functional process is known from annotation of one of them, that process will be attributed to the other. Or, if a new protein $p;_L$ with no database matches has the profile

	OTU_1	OTU_2	OTU_3	OTU_4	...	OTU_m
p_x	0	1	0	0	...	1

then it would be inferred to be functionally related to p_1 and p_4. This approach has been evaluated for vectors generated from 16 genomes by examining the number of pairs of proteins matched by their phylogenetic profiles that also share keywords from prior independent annotations. This number was compared with the number of keyword matches for gene pairs selected at random and found to be substantially higher.

Systematic Placing

Our attempts to conceptualize the complex biological world raise questions such as, "What is an archaebacterium?", "What is a eukaryote?", or "What is an animal?" Traditional systematics has employed large numbers of phenotypic characters to answer these questions, guided by an evolutionary perspective. With the availability of whole-genome sequences, it is informative to ask these questions again from the perspective of gene content.

After all, the genes and the networks of interactions between genes and gene products give rise to those phenotypes that we associate with each organism. Within-genome analysis gives a picture of what biochemical or physiological functions the cells of a particular organism can perform. Between-genome comparisons reveal gene categories that are or are not shared by taxa (e.g., What gene categories are shared by vertebrates?

What gene categories are found in animals but not fungi?). Comparison of gene contents among organisms requires a unified conceptual framework for categorizing genes. For example, genes might be grouped into the functional categories defined by COGs. Some of the higher-level categories for each of the three conceptual definitions of function proposed by the Gene Ontology Consortium are shown in Table 12.5.

To see how these definitions apply, consider just three categories of genes from *H. sapiens* and *Arabidopsis thaliana* using molecular function to define categories. In the "defense and immunity" class, there are over a thousand human genes, while *Arabidopsis* has only about 200 such genes. In the "intracellular signaling" category, humans and *Arabidopsis* have similar numbers of genes (about 1800 and 1700, respectively).

For "general metabolism," *Arabidopsis* has over 1000 more genes than humans (about 4600 for *Arabidopsis* compared with 3300 for humans). The comparisons for the first and last functional categories are as expected from the known biology of plants and animals (e.g., plants don't produce antibodies, and humans don't photosynthesize or produce cellulose).

An alternative way of comparing proteome content across organisms is by the inventory of domains found among the proteins encoded by their genomes. Many proteins can be described in terms of their constituent motifs, profiles, and domains. A protein motif is a short polypeptide sequence pattern associated with a particular function. An example is the ATP/GTP-binding site

Table 12.5: Examples of gene ontology high-level categories (Gene Ontology Consortium, 2004).

Biological process	Cellular component	Molecular function
Behavior	Chromosome	Antioxidant activity
Cell communication	Cilium	Apoptosis regulator activity
Cell recognition	Cytoplasm	Binding
Cell-cell signaling	Cytoplasmic chromosome	Cell adhesion molecule activity
Host-pathogen interaction	Cytoplasmic vesicle	Chaperone activity
Endogenous stimulus response	Cytoskeleton	Chaperone regulator activity
External stimulus response	Cytosol	Defense/immunity protein activity
Signal transduction	Endoplasmic reticulum	Catalytic activity
Cell growth/maintenance	Endosome	Enzyme regulator activity
Cell cycle	Golgi apparatus	Motor activity
Cell growth	Lipid particle	Protein stabilization activity
Cell organization/biogenesis	Microtubule organizing center	Protein tagging activity
Cell proliferation	Mitochondrion	Signal transducer activity
Chemi-mechanical coupling	Peroxisome	Nutrient reservoir activity
Cell homeostasis	Plastid	Structural molecule activity
Metabolism	Ribosome	Transcription regulator activity
Response to stress	Vacuole	Translation regulator
Transport	Nucleus	Transporter activity
Death	Nuclear chromosome	Triplet codon/amino acid adapter
Development	Nuclear membrane	
Cell differentiation	Nucleolus	
Embryonic development	Nucleoplasm	
Morphogenesis	Plasma membrane	
	Thylakoid	

motif A (amino acid sequence (A/G)-x-x-x-x-G-K-(S/T), where x is any amino acid residue). A domain is a coherent structural polypeptide unit that is associated with a particular function. A domain may include one or more motifs. A *profile* is a probabilistic description of a protein sequence pattern, perhaps in terms of a positional weight matrix. Table 12.6 provides a comparison of the domain content of humans, pufferfish, *C. elegans,* and fruitflies, listing only some of the more common domains.

The numbers of different types of domains encoded provide a measure of the genetic capabilities of the organisms. For example, note the lower number of homeobox domains in *C. elegans* compared with the vertebrates. Identifying genes shared by animals but absent in other eukaryotes is a partial answer to the question, " What is an animal?" Sets of proteins shared by two or more organisms are determined by using alignment scores obtained by using FASTA or some species of BLAST.

This type of comparison can be performed in a pairwise manner (e.g., what fraction of *Drosophila* genes are also found in *C. elegans,* and vice versa). There are n(n–1) /2 possible pairwise comparisons between n organisms, and for each of these, there are two choices for the

Table 12.6: . Comparison of animal proteome complexities based upon selected domain content. Domains are listed in decreasing order of frequency among human proteins. Values correspond to the approximate number of instances of each domain in each organism.

InterPro Accession	Name	Number of each domain type in: *H. sapiens*	*F. rubripes*	*C. elegans*	*D. melanogaster*
000822	Zn-finger, C21-12 type	800	500	200	350
000276	Rhodopsin GPCR	750	500	400	100
003006	Immunoglobulin/MHC	750	500	50	150
000719	Eukaryotic protein kinase	550	650	400	200
003593	ATPase	400	350	250	250
001909	KRAB box	300	0	0	0
001680	G-protein beta WD-40 repeat	300	300	150	200
001841	Zn-finger, RING	300	300	150	100
001849	Plekstrin-like domain	250	300	50	50
000504	RNA-binding RNP-1	250	200	100	100
002110	Ankyrin	250	250	100	100
001356	Homeobox	250	300	100	100
001611	Leucine-rich repeat	250	250	50	100
000561	EGF-like domain	250	250	50	50
002048	Calcium-binding EF-hand	250	250	50	100
001452	SH3 domain	200	250	50	50

reference genome. For example, in a *Drosophila/C. elegans* comparison, we could state that 7.8% of the *Drosophila* genes are found in *C. elegans* or that 5.9% of *C. elegans* genes are found in *Drosophila*.

The number of shared proteins depends upon two factors, one statistical and one conceptual. The conceptual issue is whether comparisons are to be made at the level of proteins or from their functional domains. If comparisons of complete genes are made, then the result may be confounded by gene fusions. (We have noted this phenomenon in the previous section.) The statistical issue is the cutoff score chosen as the criterion for orthology.

The score might be an E-value from an all-versus-all BLAST search. Are two genes or-thologs when $E < 10^{-100}$? When $E < 10^{-50}$? When $E < 10^{-10}$? Obviously, with $E < 10^{-10}$, more orthologs will be identified than if $E < 10^{-100}$ were chosen. For example, the number of *Drosophila* genes shared by the fly, worm, and yeast was determined (Rubin et al., *2000)*. This is one approximate measure of the gene set shared by eukaryotes.

At a cutoff of $E < 10^{-100}$, there are 435 fly genes shared with the worm and yeast. Pairwise comparison of yeast with the worm and the worm with yeast yielded 330 and 370 shared genes, respectively, at a similar cutoff. These results would suggest that the "core proteome" distinguishing eukaryotes is approximately 300-400 genes.

When shared genes at a cutoff of $E < 10^{-50}$ (about 1000 genes) were divided into functional categories, the number of genes in each category was found to be about the same in both the worm and yeast (e.g., approximately equal numbers of genes involved in nucleic acid metabolism in both organisms and approximately equal numbers of genes involved in signal transduction in both organisms).

This would be expected for those core functions shared by eukaryotic cells. Another between-genome study compared the human proteome to proteomes of the fly, worm, and yeast (IHGSC, 2001). Reciprocal best hits instead of numerical cutoff scores were used to define orthologous groups of genes that were shared among all four organisms. (These groups included paralogs.) There were 1308 such groups.

When orthologs only were considered (one gene in each of the four organisms), this number was reduced to 564 proteins, which would correspond to the eukaryotic "core proteome" defined by this clustering criterion (which is different from the cutoff criterion described in the previous paragraph). This number (564) is nevertheless of the same order of magnitude as the size of the core proteome estimated from genes shared by the fly, worm, and yeast.

Over half of the proteins in this group (where functional assignments could be made) were associated with "housekeeping" functions (e.g., transcription, translation, DNA replication, etc.). Rather than searching for genes shared by a group of organisms, we can also look for genes that distinguish one group from another. For example, genes shared among archaebacteria but not with eubacteria or eukaryotes have been sought (Graham et al., 2000).

In this study, an all-versus-all comparison was made among genes in *Methanococcus jannaschii, Methano bacterium thermoautotrophicum, Archaeblobus fulgidus,* and *Pyrococcus* species. Genes having $E < 10^{-20}$ were clustered (including orthologs and paralogs). Graham et al. (2000) identified 351 clusters containing no matches to eubacterial or eukaryotic genes. These represented 1149 protein-coding genes. Of these, 81% were hypothetical proteins (i.e., having no functional annotation).

Of the 351 clusters, 20% were represented in each of the four genomes. The archaealspecific genes constitute a genome signature for this domain of life, and when functions become known, these genes may help define what an archaebacterium is at the biochemical level.

CONCLUDING REMARKS

Genome sequences and massively parallel gene expression data added to prior biochemical and genetic knowledge create a more comprehensive way of thinking about the biology of organisms. Now it is possible to examine organisms at a number of different levels of generalization, ranging from the very particular (Does the organism encode anthranilate synthase?) to the very general (How are human developmental functions regulated and expressed?).

Given the breadth of genome sequencing efforts still in progress, it is premature to designate the present time as the "Post-Genome Era," but it is clear that questions concerning proteomics and interaction networks are becoming more prominent. This means that genes and gene products soon will be understood from an integrated perspective.

Any student of genomics (undergraduate, graduate, post-doctoral, or professor) needs to develop skills to think at all levels of generalization and to be equipped with the *computational* skills to do so. The reward for learning these skills is the ability to think and work in one of the most revolutionary scientific areas of our time. The result will be an appreciation and understanding of how biology "works" to produce *living* organisms from an inorganic world.

Chapter 13

Protein Homeostasis

Proteolysis, or the processing and *degradation* of proteins, has emerged as one of the most widely studied processes in biology today. Long viewed as a dead-end process, of importance only for the removal of obsolete peptides and proteins, proteolytic events are now associated with numerous biological events apart from the processing and degradation of *polypeptides*, ranging from protein maturation, signal transduction, cell-cycle control, and controlled cell death, to the generation and *presentation* of *antigenic* peptides in the adaptive *immune* system.

Realizing that proteolysis plays an important role in many bio-chemical processes, it has become apparent that proteolytic events, and the *enzymatic* activities involved, are *tightly* controlled. This control can be exerted at the *transcriptional* level, but also on the *maturation* state and *localization* of *proteolytic* activities. A case in point is the caspase family of proteolytic activities.

Normally dormant, these cysteine proteases are triggered by a variety of signals and, upon activation, are involved in inducing apoptosis through a series of tightly *controlled* pathways. The processing and presentation of antigenic polypeptides can be divided into two distinct pathways (for a *comprehensive* review on both pathways,). In the class I pathway, endogenous *proteins* are targeted for *proteasomal* degradation through *polyubiquitination.*

Upon docking of the *poly-ubiquitinated proteasome* substrates to the 19S cap, the protein is unfolded and transferred into the inner 20S core, which harbors the six individual catalytic subunits, two copies of each of three distinct activities named hereafter as β1, β2, and β5. These *endopeptidase* activities residing in the 20S core cleave polypeptide substrates to produce *oligopeptide* fragments of typically 3 to 20 amino acid residues in length.

Further *processing* and *degradation* is effectuated by a series of *aminopeptidase* activities, which act upon the N-terminal site of the *proteasome* products and cleave off a few *amino acid residues.* In this way, the vast majority of cytosolic and nuclear proteins are processed and degraded completely to produce amino acids for ensuing *incorporation* in newly *synthesized* proteins.

A small number of proteasome products, or their *aminopeptidase-trimmed counterparts,* escape to the *endoplasmic reticulum* with the help of the transporter associated with antigen presentation (TAP). Within the *endoplasmic reticulum* (ER) lumen, these may be further processed by ER-specific *aminopeptidases* to a final length of eight or nine amino acid residues and are finally loaded onto major *histocompatibility* complex (MHC) class I complexes that are assembled in the ER.

Presentation of the loaded MHC class I peptide complexes allows the immune system to sample the nature of the cytosolic and nuclear protein content, which may contain *exogenous* proteins such as viral proteins. Of note is the presence in *immunocompetent* cells of a second protea-some particle, dubbed the *immunoproteasome* (as opposed to the constitutive proteasome expressed uniformly throughout the organism), which closely resembles the constitutive proteasome but now contains three distinct *proteolytic activities* named β1i, β2i, and β5i.

The class II pathway is responsible for the presentation to the immune system of extracellular proteins. Professional antigen-presenting cells (APCs) are charged with this task and do so by means of endocytic uptake of *extracellular* material and ensuing stepwise endolysosomal proteolysis through numerous proteolytic activities such as *cysteine* and *aspartic* acid proteases of the cathepsin family and *asparagine endopeptidase* (AEP).

In a fashion rather resembling the class I pathway, MHC class II protein *complexes* are loaded with *oligopeptides*, ranging from 8 to 13 residues in length, resulting from the action of above-mentioned proteases. Fully assembled MHC II-peptide *complexes* are *displayed* on the outer cell membrane, enabling the immune system to monitor their contents. A further similarity between the two pathways is the control exerted on the proteolytic activities. In the class I pathway, the *proteasomal catalytic* activities are shielded from the *intracellular* protein content thanks to the barrel-shaped architecture of the 20S core.

Protein substrates can reach the catalytic sites only through a series of *modification, recognition*, and *unfolding* events. Similarly, class II protein substrates are brought into contact with respective *proteolytic* activities by means of active sampling, by APCs, of their extracellular surrounding. *Proteolytic control* is exerted by the stepwise fusion of endocytic *compartments* with *vesicles* containing proteases.

Many of these proteases are synthesized in a proform and are activated upon removal of an *oligopeptide* stretch by either *autocatalytic* cleavage or by the action of another activity, and many of the *endosomal* and *lysosomal proteolytic* activities display a pH optimum. Of interest are the recent reports *describing* the existence of *cross-talk* between the class I and class II pathways.

Although *originally* thought of as separate routes dedicated to the *presentation* of either *endggenic* or *exogenic antigens,* it is now clear that oligopeptides derived from *extracellular* origin can end up being displayed by MHC class I assemblies The exact nature of the mechanism of this process, however, remains to be established: what events *trigger* the *occurrence* of this *"cross-presentation,"* and which *proteolytic activities* and cellular *mechanisms* are involved in the generation of *exogenous* class I antigenic peptides?

Cross-presentation is a relatively unexplored topic in the field of antigenic peptide presentation, and little is known about the role of proteolytic activities. The full extent to which proteolytic action is involved in cross-presentation remains to be established. Do additional aminopeptidases active in the class I pathway exist, both of *cytosolic* and ER origin, and what is the role of the individual cathepsins in the class II *processing* and *presentation*? Without much doubt, other yet-to-be-discovered *proteolytic* entities are involved.

These questions and the current state of the art in addressing these are amply reviewed elsewhere and are not the subject of this chapter. Here we discuss some recently emerged sets of *proteomics* techniques that provide ways to study and identify proteolytic activities with similar properties in one sweep, and based on their activity, rather than on their expression levels. Globally, these strategies are based on the design of *broad-spectrum*, covalent, and irreversible inhibitors of the targeted proteolytic activities in question.

These inhibitors, normally derived from oligopeptides and endowed with an electrophilic trap replacing the amide bond normally targeted by the proteases, can be tracked by means of an affinity tag. Through this tag, which can be either *incorporated* directly into the peptide-based probes (*biotin, fluorescent tags*) or attached at a later stage by means of a *bio-orthogonal* organic transformation, allows the study of targeted *proteolytic* activities by analytical *techniques* (gel electrophoresis, mass spectrometry).

In recent years, such techniques have become available through the endeavors of various research groups. Under Subheadings Procedures. the various *methodologies* are discussed in detail. Attention will be focused both on synthetic aspects (i.e., the *preparation* of the following probes: AdaK(Bio)Ahx_3L_3VS, Ada^{125}IYAhx3L3VS, AdaAhx((-N3)Ahx_2L_3VS, and DansylAhx_3L_3VS and application of these tools devised to study proteasome activity**.** The labeling results are obtained with EL4 or HeLa cells, but the methods are in *principle* applicable to the cell line of choice.

PROCEDURES

Requirements

1. Amino acid building blocks and peptide synthesis reagents (piperidine, diisopropy-lethylamine, hydroxybenzotriazole [HOBt], trifluoroacetic acid [TFA]) (Advanced ChemTech, Peptides International, Novabiochem, and American Bioanalytical).
2. Adamantane-acetic acid, triisopropylsilane [TIS], and iodoacetonitrile.
3. 4-Sulfamylbutyrylamimomethyl polystyrene resin (Novabiochem AG).
4. Iodo-gen (Pierce).
5. High-performance liquid *chromatography* (HPLC)-grade organic solvents dimethyl fonnamide [DMF], *dichloromethane* [DCM], *acetonitrile* [ACN], *N-methylpyrrolidinone* [NMP] (American Bioanalytical), and ethyl acetate [EtOAc] (Fisher) were used as received.
6. Sep-pak C_{18} columns (Waters).
7. Unless indicated otherwise, chemicals were obtained commercially of the highest available grade.

Multiplication and Destruction

1. EL4 (mouse thymoma), HeLa (human cervix epitheloid carcinoma), and U373 (human astrocytoma) cells lines were used.
2. RPMI 1640 medium (Gibco, Invitrogen Corp.), Dulbecco's modified Eagle's medium (DMEM),

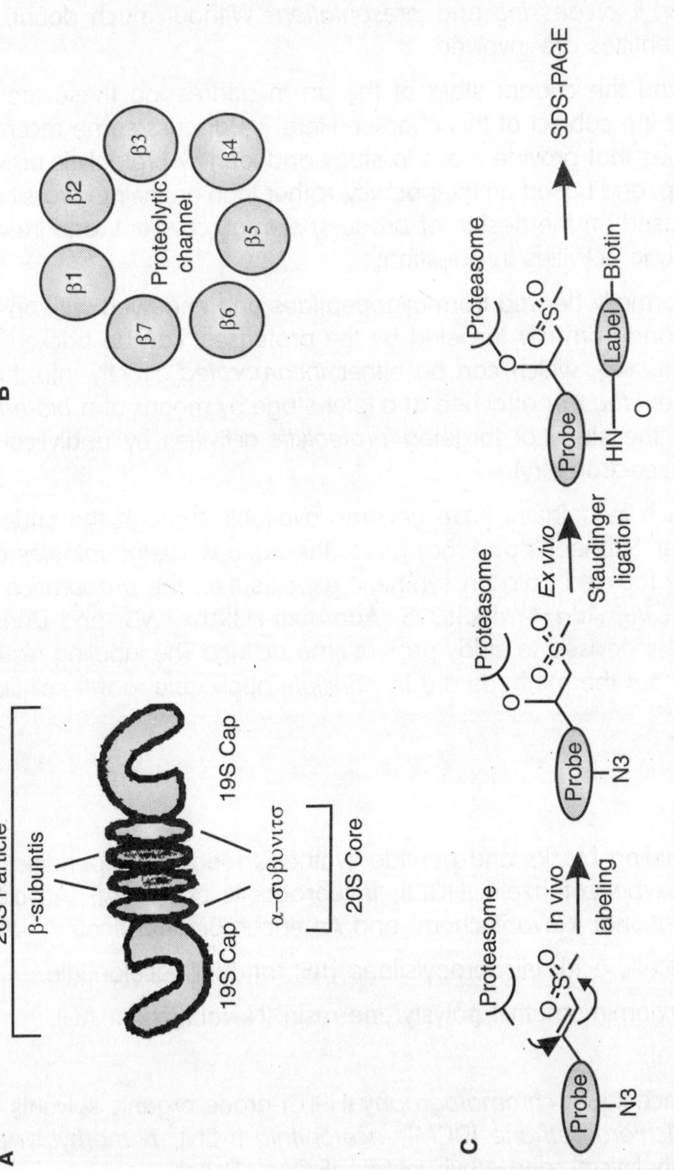

Figure 13.1: (A,B) Schematic representation of the proteasome. (C) Two-step labeling of the proteasome by means of the Staudinger ligation.

L-glutamine, 10% fetal calf serum (FCS), penicillin/streptomycin solution, phosphate-buffered saline (PBS), and trypsin-ethylenediamine tetraacetic acid (EDTA) solution.

3. Lysis buffer for protease active cell lysates: 250 mM sucrose, 50 mMTris-HCl pH 7.5, 5mM MgCl2, 1 mM dithiothreitol [DTT], 2 mM ATP in water.
4. Glass beads (140 μm).
5. NP-40 lysis buffer: 50 mMTris-HC1(pH 8.0), 150 mMNaCl, 1% NP-40 was used to lyse the inhibitor-treated living cells used in Subheadings esewhere in this chapter.

Electrophoretic Method for Two-Dimensional Studies

1. Rehydration solution: 8 M urea, 2% CHAPS, 0.5% immobilized pH gradient (IPG) buffer (Amersham Pharmacia Biotech) (pH 3-10), bromophenol blue, 18 mM DTT.
2. For two-dimensional, nonequilibrium pH gradient sodium dodecyl sulfate-poly-acrylamide get electrophoresis (2D NEPHGE SDS-PAGE): add 220 µL of rehydration solution to 30 µL of samples and mix.
3. Subject to first-dimension isoelectric focusing using a IPGphor isoelectric focusing system (Amersham Pharmacia Biotech) and second-dimension SDS-PAGE (12.5%). (The results obtained were comparable with conventional 2D NEPHGE SDS-PAGE methods.)

Western Blotting

1. SDS-PAGE gels of 12.5% polyacrylamide running gel and 4% stacking gel (BioRad).
2. Polyvinylidene difluoride (PVDF) membranes (Hybond P, Amersham).
3. Tris-buffered saline with Tween (TBS-T): prepare lox stock with l.37 MNaCl, 27 mM KCl, 250 mM Tris-HCl (pH 7.4), 1% Tween-20. Dilute 100 mL with 900 mL water for use.
4. Anti-dansyl-sulfonamidohexanoyl polyclonal antibody in TBS-T supplemented with 0.5% bovine serum albumin (BSA).
5. Horseradish peroxidase (HRP)-coupled goat anti-rabbit secondary antibody (Southern Biotech) in TBS-T supplemented with 0.5% BSA.
6. Streptavidin-HRP conjugate (Molecular Probes) in TBS-T (0.l% [v/v] Tween-20.
7. Enhanced chemiluminescence kit (ECL+, Amersham).

Biosynthesis of Probe A

1. Fmoc-Leu-Wang resin (0.5 mmol), Fmoc-Leu-OH, Fmoc-Ahx-OH, Fmoc-Ahx-OH, Fmoc-Ahx-Oh, Fmoc-Lys(biotin)-OH, and adamantaneacetic acid.
2. PyBOP (activating agent) and diisopropylethyl amine (base).
3. Piperidine (20%) in dimethylformamide (DMF), TFA/TIS/H20 95/2.5/2.5 v/v/v.
4. Leucine vinyl sulfone TFA salt, HBTU, and diisopropylethylamine (DiPEA) in DMF.
5. Ethyl acetate, diethyl ether, hexanes, 10% to 20% methanol in methylene chloride.

Biosynthesis of Probe B

1. Fmoc-Tyr(OtBu)-OH instead of Fmoc-Lys(biotin)-OH.
2. Iodo-gen (100 µg) was dissolved in phosphate buffer (50 mM, pH 7.4, 10 µL).
3. AdaYAhx3L3VS (20 µg) was dissolved in acetonitrile (30 µL).
4. Na^{125I} (10 L aqueous solution, 1 mCi), Sep-pak C_18 column, and acetonitrile.

Biosynthesis of Probe C

1. N-(α-Boc),N-(ε-Fmoc)-1-lysine (2.0 g, 4.2 mmol).
2. 50% v/v TFA/CH2C12, trifluoromethanesulfonyl azide, 1 *M* HCl, EtOAc.

3. Wang resin (l.0 g, 0.86 mmol), Fmoc-Leu-OH (1.2 g, 3.4 mmol) in CH_2C1_2 (25mL), diisopropylcarbodiimide (0.54 mL, 3.44 mmol), a catalytic amount of 4- (dimethylamino-)pyridine, CH2Cl2-MeOH, $CH2C1_2$, and Et2O.
4. Resin-bound AdaAhx(α-N3)Ahx_2L_2, TFA/H20 95/5 v/v, leucine vinyl sulfone TFA salt (1 eq), HBTU (1 eq), DiPEA (2.2 eq) in DMF, EtOAc.

Biosynthesis of the Staudinger Ligation Handle

1. Fmoc Rink amide resin (128 mg, 100 μcool), N-(a-Boc),N-(e-Mtt)-1-lysine (0.5 mmol, 312 mg) was coupled.
2. 1% TFA in $CH2Cl_2$ for 30 s, 10% DiPEA in DMF.
3. Biotin (0.5 mmol, 122 mg), PyBOP (1 eq), DiPEA (1.2 eq) in DMF.
4. Me3P (0.6 mL of a 1 M solution in toluene, 0.6 mmol, 6 eq) in dioxane/water (4/l v/v, 2 mL), dioxane, Fmoc-Ahx-OH (0.5 mmol, 180 mg).
5. Phosphine 2, Figure.elsewhere in this chapter(90 mg, 0.25 mmol), EDC (48 mg, 0.25 mmol), hydroxyben-zotriazole (HOBt) (41 mg, 0.3 mmol) in CH2Cl2 (2 mL)

Biosynthesis of Probe D

Wang resin, N-(3-dimethylaminopropyl)-N'-ethylcarbodiimide hydrochloride, (S,E)-5-methyl-l-(methylsulfonyl)hex-l-en-3-amine (leucine vinyl sulfone, LeuVS).

Proteasome Method Using AdaK(Bio)Ahx3L_3VS (A)

1. AdaK(Bio)Ahx_3L_3VS (A).
2. Cell lysate.
3. 3X SDS sample buffer: 0.67 mL 20% SDS, 1 mL 0.625 MTris-HCl(pH 6.8), 2.1 mL 87% glycerol, 0.4 mL β-mercaptoethanol, 0.l mL 10% phenol blue (N,N-dimethylindoaniline), 0.73 mL water.
4. Rehydration solution: 8 M urea, 2% CHAPS, 0.5% IPG buffer (Amersham Pharmacia Biotech) (pH 3.0-10.0), bromophenol blue, 18 mM DTT.

Proteasome Method Using AdaY(^{125}I)Ahx3L_3VS (B)

1. AdaY(^{125}I)Ahx_3L_3VS (B).
2. Cell lysate.
3. 3X SDS sample buffer: 0.67 mL 20% SDS, 1 mL 0.625 MTris-HCl (pH 6.8), 2.l mL 87% glycerol, 0.4 mL β-mercapto-ethanol, 0.l mL 10% bromophenol blue, 0.73 mL water.
4. Rehydration solution: 8 *M* urea, 2% CHAPS, 0.5% IPG buffer (Amersham Pharmacia Biotech) (pH 3.0-10.0), bromophenol blue, 18 mM DTT.

Proteasome Method Using Two-Step Labeling

1. Cells in culture.
2. AdaAhx(α-N3)Ahx_2L_3VS (C)

3. Glass beads.
4. Lysis buffer: 50 mMTris, 5 mMMgCl_2, 0.5 mMEDTA, 0.25 mMsucrose (pH 7.4).
5. Staudinger ligation handle.
6. 20% SDS solution.
7. 4X sample buffer: 1 mL 20% SDS, 1 mL 0.625 MTris-HCl (pH 6.8), 2.1 mL 87% glycerol, 0.4 mL β-mercaptoethanol, 0.1 mL 10% bromophenol blue, 0.4 mL water.

Proteasome Method Using DansylAhx$_3$L$_3$VS (D)

1. Cells in culture.
2. DansylAhx$_3$L$_3$VS (D).
3. Glass beads.
4. NP-40 lysis buffer: 50 mMTris-HCl(pH 8.0), 150 mMNaCl, 1% NP-40.

PROCEDURE

In this section, we describe several methods for the synthesis of selective proteasome inhibitors and their use in functional proteomics *methodologies*. In general, the protease inhibitors were dissolved in *dimethylsulfoxide* (DMSO) to 20 mM stock solutions that were further diluted in the buffer of interest. The active protease lysates should be kept on ice at all times to avoid protein *degradation.*

Lysates were frozen as 50-μL aliquots at –80°C and used only once, *preventing* repeated freeze/thawing cycles. In some cases, high concentration of chromatin, DNA, or RNA might influence the outcome of the experiments. Then, *postnuclear* lysate procedures should be explored either by using the NP-40 buffer or other mild methods like dounce *homogenization*, which selectively disrupts the cell *membrane*, leaving nuclei and other *organelles* intact.

General Technique

1. *Leucine* vinyl *sulfone* was prepared as reported.
2. *Solid-phase* peptide synthesis was carried out using a 180° Variable Rate Shaker (Peptides International).
3. Nuclear *magnetic* resonance (NMR) spectra were recorded on Varian (200 MHz, 500 MHz) and Bruker (200 MHz, 300 MHz, 400 MHz) spectrometers.
4. Mass spectra were *recorded* on an *electrospray* LCT *liquid chromatography* (LC)-*mass spectrometry* (MS) *instrument* (Waters).
5. Preparative reverse-phase HPLC *purifications* were carried out using a Waters Pre-pLC™ C 18 column (250 mm x 40 mm) with a solvent gradient *ranging* from 30 to 95% acetonitrile in water containing 0.1 % formic acid.

Multiplication and Destruction

1. The cell lines EL4 (*mouse thymoma*) was cultured in RPMI 1640 medium (Gibco, Invitrogen Corp.).

2. HeLa cells (human cervix *epitheloid carcinoma*) and U373 cells (*human astrocytoma*) were cultured in DMEM.
3. Both media were supplemented with L-glutamine, 10% FCS, penicillin, and streptomycin.
4. Cells were grown at 37°C and 5% CO_2 in a *humidified incubator*.
5. Cells were *harvested* and lysed by glass beads or *sonication* in lysis buffer or by *incubation* on ice for 30 min in NP-40 lysis buffer followed by 5 min *centrifugation* at 16,000g to remove membrane *fractions, nuclei* and *cell debris*.

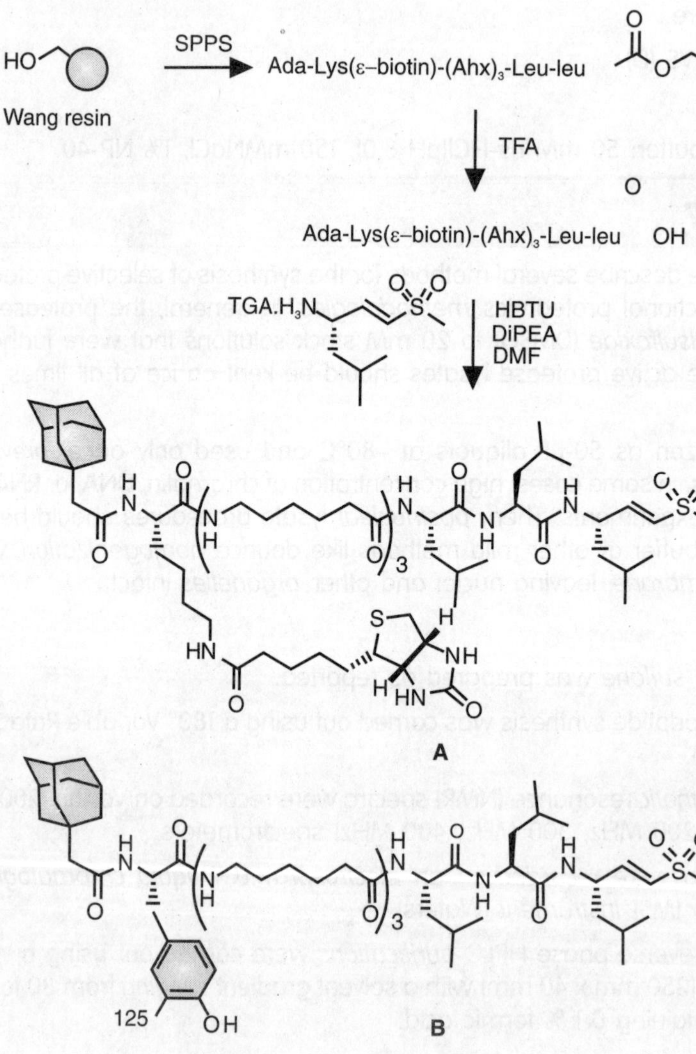

Figure 13.2: Synthesis of AdaK(Bio)Ahx$_3$L$_3$VS (A) and chemical structure of AdaY(^{125}I)AhX$_3$L$_3$VS (B).

Western Blotting

1. Electro-transfer protein from SDS-PAGE gel onto PVDF membranes (Hybond P, Amersham).
2. Dansyl: *anti-dansyl-hexanoylsulfonamido* immunoblotting was performed using a rabbit *antidansyl-sulfonamidohexanoyl polyclonal* antibody (1:7500, 1 h at room temperature) and HRP-coupled goat anti-rabbit secondary antibody followed by enhanced *chemiluminescence* detection.
3. Biotin: block the membrane with 0.5% (w/w) BSA and 0.1% (v/v) Tween-20 in PBS for 2 h. Incubate the blot at a 1:5000 dilution of s*treptavidin-HRP* conjugate in 0.5% (w/w), 0.1% (v/v) Tween-20 in PBS for 1 h. Wash the blot five times with PBS containing 0.1% (v/v) Tween-20 for 10-15 min for each washing step. Develop the blot by *chemoluminescense.*

Biosynthesis of Probe A

1. *Fmoc-Leu-Wang resin* (0.5 mmol) was elongated by standard Fmoc-based *solid-phase peptide* synthesis (SPPS) with Fmoc-Leu-OH, Fmoc-Ahx-OH, Fmoc-Ahx-OH, Fmoc-Ahx-Oh, Fmoc-Lys(biotin)-OH, and *adamantaneacetic acid,* each time 5 eq of the amino acid, 5 eq of PyBOP as the activating agent, and 6 cq of *diiso-propylethyl amine* as the base.
2. After each *condensation* step, the N-terminal Fmoc protective group was removed by treatment with 20% piperidine in DMF.
3. Treatment of the *immobilized* peptide AdaK(bio)Ahx_3L_2OH with TFA/TIS/H_2O 95/2.5/2.5 v/v and removal in vacuo of the solvents, was followed by solution phase block-coupling with *leucine vinyl sulfone* TFA salt (1q) under the agency of HBTU (1eq) and DiPEA (2.2eq) in DMF.
4. After evaporation of the solvent, the residue was dissolved in ethyl acetate.
5. Precipitation of the product was accomplished by *sonication* for 2 min. The *precipitate* was filtered and washed with *ethyl acetate, diethyl* ether, and *hexanes* to afford the title compound A in 40% yield with a purity of approx 90%.
6. The *peptide* can be purified to homogeneity by reversed-phase HPLC or by *silica gel chromatography* using a gradient from 10% to 20% *methanol* in *methylene chloride.* MS: found 1287.9 $(M+H)^+$. For ^{1}H-NMR (DMSO-d6, 500 MHz) data, see elsewhere in this chapter.

Biosynthesis of Probe B

1. AdaYAhx_3L_3VS: synthesized as compound A, but with Fmoc-Tyr(OtBu)-OH instead of Fmoc-Lys(biotin)-OH. MS: found 1097.1 (M+H)+. For ^{1}H-NMR (DMSO-d6, 500 MHz) data, *see* Note elsewhere in this chapter.
2. Radiolabeling: lodo-gen (100 μg) was dissolved in 50μL phosphate buffer.
3. AdaYAhx_3L_3VS (20 μg) in *acetonitrile* (30 μL) was added and the mixture was agitated for 2 min. $Na^{125}I$(10μL aqueous solution, 1 mCi) was added to the mixture and agitated for 10 min.
4. The *radioiodinated* proteasome inhibitor was separated using a Sep-pak C_{18} column. Free iodine was washed away with phosphate buffer (pH 7.4), and radioiodinated

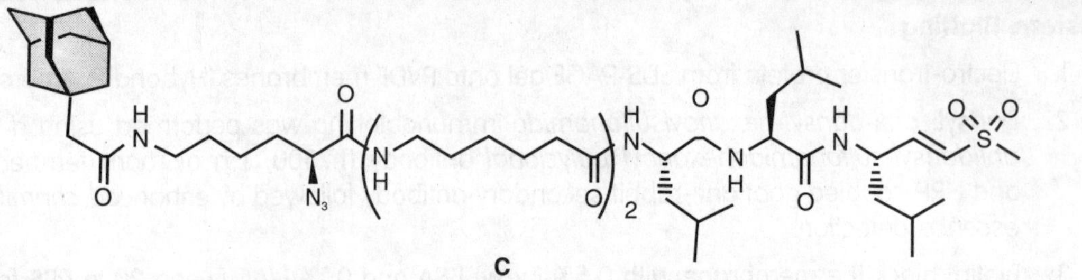

C

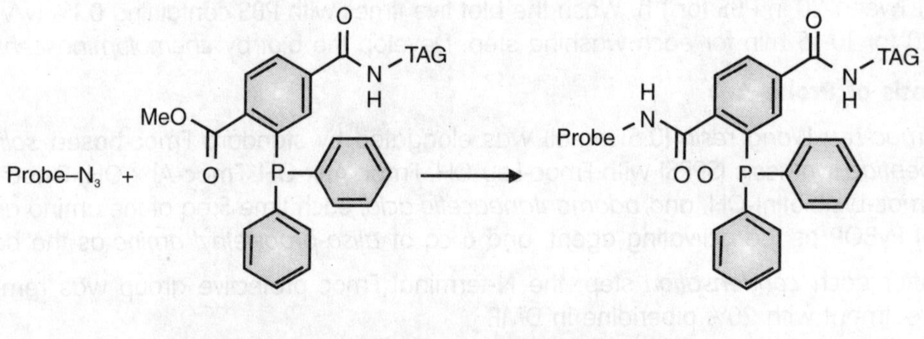

staudinger ligation

Figure 13.3: Chemical structure of AdaAhx(α-N_3)Ahx$_2$L$_3$VS (C) and reaction scheme for the Staudinger ligation.

AdaY(^{125}I)Ahx$_3$L$_3$VS B was eluted using *acetonitrile*. One-milliliter fractions were collected and the fractions containing most *radioactivity* (typically 0.3-0.7 × 10^6 cpm/μL) were used for proteasome *radiolabeling* experiments.

Biosynthesis of Probe C

1. Solid phase *peptide synthesis* was carried out essentially as described for Probe A.
2. Synthesis of Fmoc-Ahx(α-N_3)-OH: N-(α-Boc),N-(ε-Fmoc)-L-lysine (2.0 g, 4.2 mmol) was treated with 50% v/v TFA/CH2C12 for 40 min.
3. Solvents were removed *in vacuo,* yielding 1.9g (3.9 mmol, 92%) of crude TFA- salt of N-(ε-Fmoc)-L-lysine.
4. The crude product was treated with *trifluorocnethanesulfonyl azide* as described elsewhere in this chapter, followed by a modified work-up *procedure.*
5. After removal of the organic solvents, the *aqueous slurry* was *acidified* with 1 M HCl to pH 2.0 and extracted with EtOAc (4X). The combined organic phases were dried over $MgSO_4$, filtered, and concentrated *in vacuo*.
6. Purification of the crude product over silica gel (hexanes:EtOAc 2:l) yielded l.4 g (3.5 mmol, 89%) of a white foam. Electrospray ionization (ESI)-MS: m/z (%) = 417.2 (100) [M+Na+], 811.5 (20) [2M+Na$^+$]. For 1H NMR (200 MHz, $CDC1_3$, 25°C, TMS) and ^{13}C NMR (50.l MHz, $CDCl_3$, 25°C, TMS) data, *see* elsewhere in this chaptere.

7. Next, Wang resin (l.0 g, 0.86 mmol) was condensed with Fmoc-Leu-OH (1.2 g, 3.4 mmol) in CH_2Cl_2 (25 mL) under the agency of *diisopropylcarbodiimide* (0.54 mL, 3.44 mmol) and a *catalytic* amount of 4-(*dimethylamino*)*pyridine* for 2 h.
8. The resin was filtered off, washed (3X CH_2Cl_2-MeOH, CH_2Cl_2, and Et_2O), and air dried. Loading was determined by quantification (*ultraviolet*) of Fmoc cleavage, and proved to be 0.66 mmol/g.
9. The resin, 260 mg (0.17 mmol), was elongated using standard Fmoc-SPPS to give *resin-bound* AdaAhx(α-N3)Ahx_2L_2.
10. Treatment of the immobilized peptide with TFA/H_2O 95/5 v/v for 1 h, *filtration*, and removal of solvent *in vacuo* was followed by a solution phase block coupling with leucine vinyl sulfone TFA salt (1 eq) under the agency of HBTU (l eq) and DiPEA (2.2 eq) in DMF. After *evaporation* of the solvent, the residue was dissolved in EtOAc.

Precipitation of the product was *accomplished* by *sonication*. The precipitate was filtered and washed with EtOAc, Et20, and hexanes to yield 146 mg (0.15 mmol, 88%) of the title compound in 90% purity as judged by LC-MS.

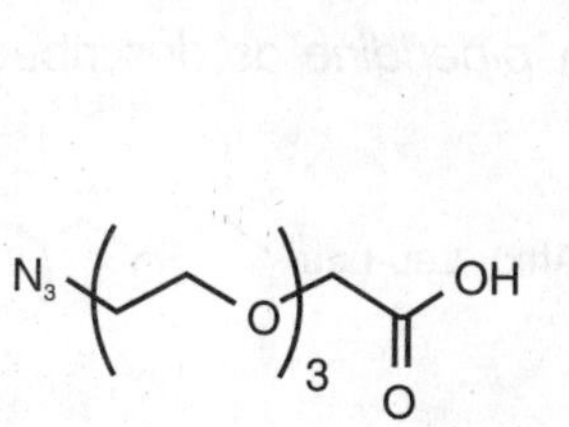

1

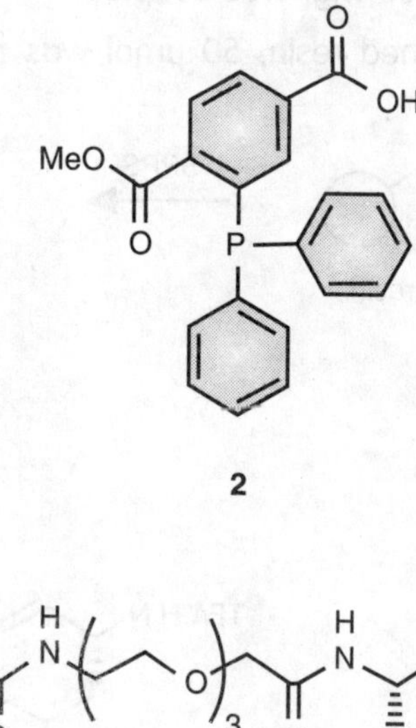

2

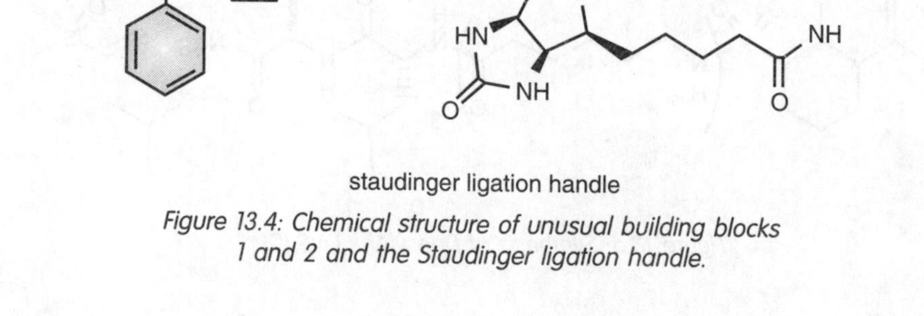

staudinger ligation handle

Figure 13.4: Chemical structure of unusual building blocks 1 and 2 and the Staudinger ligation handle.

11. Silica gel purification (0-10% MeOH/EtOAc) of 21 mg (22 µmol) of the crude product yielded 14 mg (14 µmol) of the title *compound* (58% overall yield) ready to use in *biological* experiments. LC-MS: m/z = 974.9 [M+H$^+$].

Biosynthesis of the Staudinger Ligation Handle

1. Fmoc Rink amide resin (128 mg, 100 µmol) was *deprotected* and N-(α-Boc),N-(ε-Mtt)-l-lysine (0.5 mmol, 312 mg) was coupled.
2. The Mtt *protecting* group was removed by treatment of the resin with 1% TFA in CH2Cl2 for 30 s. This *treatment* was repeated nine times, until no more yellow color was observed in the eluted solution. The resin was neutralized with 10% DiPEA in DMF.
3. Biotin (0.5 mmol, 122 mg) was coupled under the agency of PyBOP (1 eq) and DiPEA (1.2 eq) in DMF. After removal of the Fmoc group, the resulting free amine was condensed with 1(21) (0.5 mmol, 237 mg).
4. After washing of the *immobilized peptide* with *dioxane* (3X), the azide moiety was reduced by treatment with Me3P (0.6 mL of a 1 M solution in *toluene*, 0.6 moll, 6 eq) in *dioxane/water* (4/1 v/v, 2 mL) for 40 min (20), followed by dioxane washes (3X), and Fmoc-Ahx-OH (0.5 mmol, 180 mg) was coupled.
5. Of the obtained resin, 50 µmol was treated with *piperidine* as described *previously* to

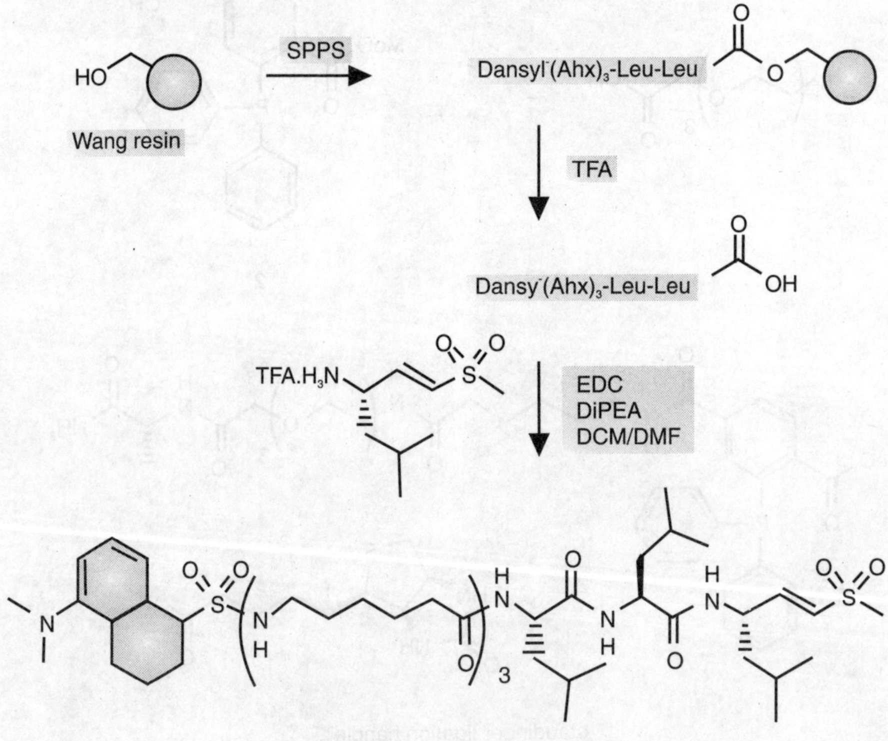

D

Figure 13.5: Synthesis of DansylAhx$_3$L$_3$VS (D).

remove the Fmoc *protecting* group, and the resin was then washed with CH_2Cl_2 (3X).

6. Phosphine 2 (90 mg, 0.25 mmol) was activated with EDC (48 mg, 0.25 mmol) and HOBt (41 mg, 0.3 mmol) in CH_2Cl_2 (2 mL) under an *argon atmosphere* for 5 min, and subsequently added to the resin.
7. The resin was agitated under argon atmosphere for 1 h, and the resin was washed (CH_2Cl_2, then DMF-MeOH *alternating* [3X], CH_2Cl_2-MeOH alternating [3X], and CH_2Cl_2), while keeping the resin under argon *atmosphere.*
8. The immobilized peptide was liberated from the resin by treatment with 50% v/v TFA/ CH_2Cl_2 for 1 h. Evaporation of the solvents *in vacuo* followed by HPLC purification of the crude product (linear gradient in B: 25-55% B in three column volumes) yielded 8.0 mg (8 µmol, 16%) of a white solid. LC-MS: m/z calculated = 1020 [M+H$^+$], m/z found = 1020 [M+H$^+$].
9. The product was kept as aliquots of a stock solution (1.6 *mM)* in degassed DMSO at -80°C.

Biosynthesis of Probe D

1. Dansyl-Ahx_3L_2-OH was synthesized on Wang resin using standard Fmoc-based solid-phase peptide synthesis *protocols* and *purified* by *silica gel chromatography* using a gradient of 10% → 20% of MeOH in CH_2Cl_2.
2. Dansyl-Ahx_3L_2-OH was coupled by N-β-dimethylaminopropyl)-N9-ethylcarbodi-imide hydrochloride (EDCI)-mediated condensation with (S,E)-5-methyl-l-(methylsulfonyl)hex-l-en-3-amine (leucine vinyl sulfone, LeuVS TFA salt) to afford D. MS(ESI): Calculated 990.5 $(M+H)^+$ 1012.5 $(M+Na)^+$. Observed 990.5 $(M+H)^+$ 1012.5 $(M+Na)^+$.
3. *Hydrogenation* and *deprotection* of Boc-LeuVS followed by condensation of the resulting (S)-5-methyl-l-(methylsulfonyl)hexan-3-amine Dansyl-Ahx_3L_2-OH afforded the negative control (mock). ESI-MS: Calculated 992.5 $(M+H)^+$ 1014.5 $(M+Na)^+$. Observed 992.5 $(M+H)^+$ 1014.5 $(M+Na)^+$.

Proteasome Method Using AdaK(Bio)Ahx_3L_3VS (A)

1. Incubate cell lysates (25 µg of total protein) with 3 µg AdaK(Bio)Ahx_3L_3VS (A) for 2 h at 37°C.
2. Add 3X SDS sample buffer (1X final concentration).
3. Analyze by one-dimensional SDS-PAGE (12.5%).
4. Western *blotting, probing* with *streptavidin-HRP*, and *chemoluminescence* detection.
5. For 2D NEPHGE SDS-PAGE, add 220 µL of rehydration solution to 30 µL of samples and mix.
6. Subject to first-dimension *isoelectric focusing* using a IPGphor isoelectric focusing system (Amersham Pharmacia Biotech) and second-dimension SDS-PAGE (12.5%). (The results obtained were comparable to conventional 2D NEPHGE SDS-PAGE methods.)

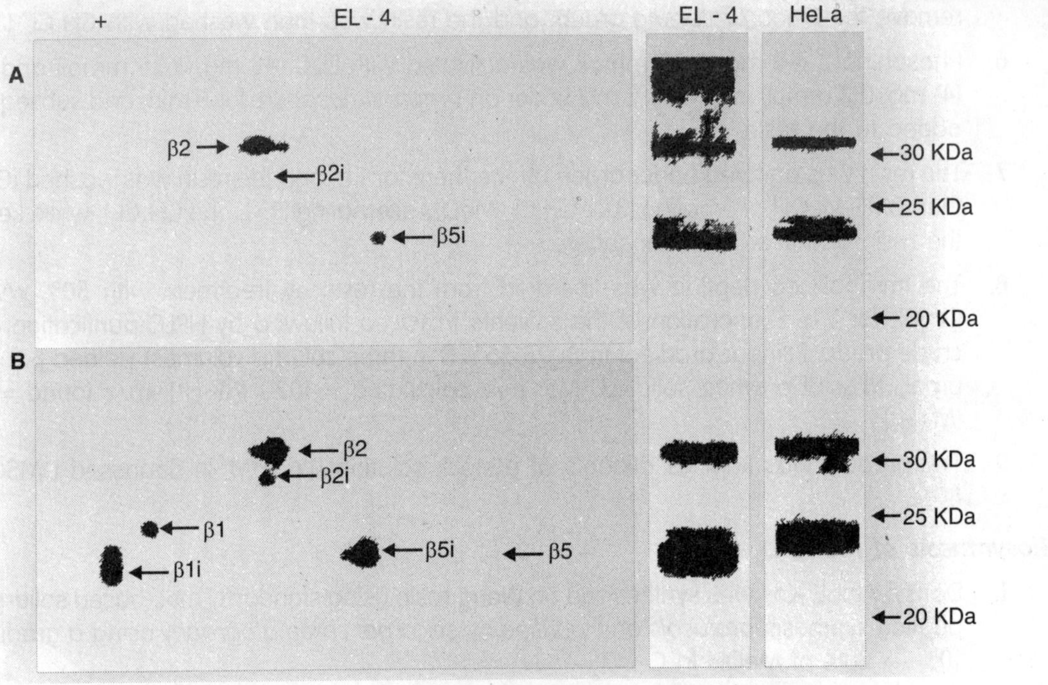

Figure 13.6: 1D- and 2D gel analysis of EL-4 and HeLa cell lysates labeled with either AdaK(Bio)Ahx$_3$L$_3$VS (A) or AdaY(^{125}I)Ahx$_3$L$_3$VS (B).

7. Western blotting, probing with streptavidin-HRP, and *chemoluminescence* detection.

Proteasome Method Using AdaV(^{125}I)Ahx$_3$L$_3$VS (B)

1. Incubate cell lysates (25 μg of total protein) with AdaY(^{125}I)Ahx$_3$L$_3$VS (B) (0.5 × 10^6 cpm) for 2 h at 37°C.
2. Add 3X SDS *sample buffer* (1X final *concentration*), heat to 95°C for 3 min.
3. Analyze by *one-dimensional* SDS-PAGE (12.5%).
4. Autoradiography.
5. For 2D NEPHGE SDS-PAGE, add 220 μL of rehydration solution to 30 μL of samples and mix.
6. Subject to first-dimension *isoelectric focusing* using a IPGphor isoelectric focusing system (Amersham Pharmacia Biotech) and second-dimension SDS-PAGE (12.5%). (The results obtained were *comparable* with conventional 2D NEPHGE SDS-PAGE methods.
7. Autoradiography.

Proteasome Method Using Two-Step Labeling With AdaAhx(α-N3)Ahx2L3VS (C)

1. Incubate 5 × 10^6 EL-4 cells cultured in RPMI *supplemented* with L-glutamine, FCS, penicillin, and streptomycin overnight with *concentrations* of AdaAhx(α-N$_3$)Ahx$_2$L$_3$VS C as indicated in Figure 13.7.

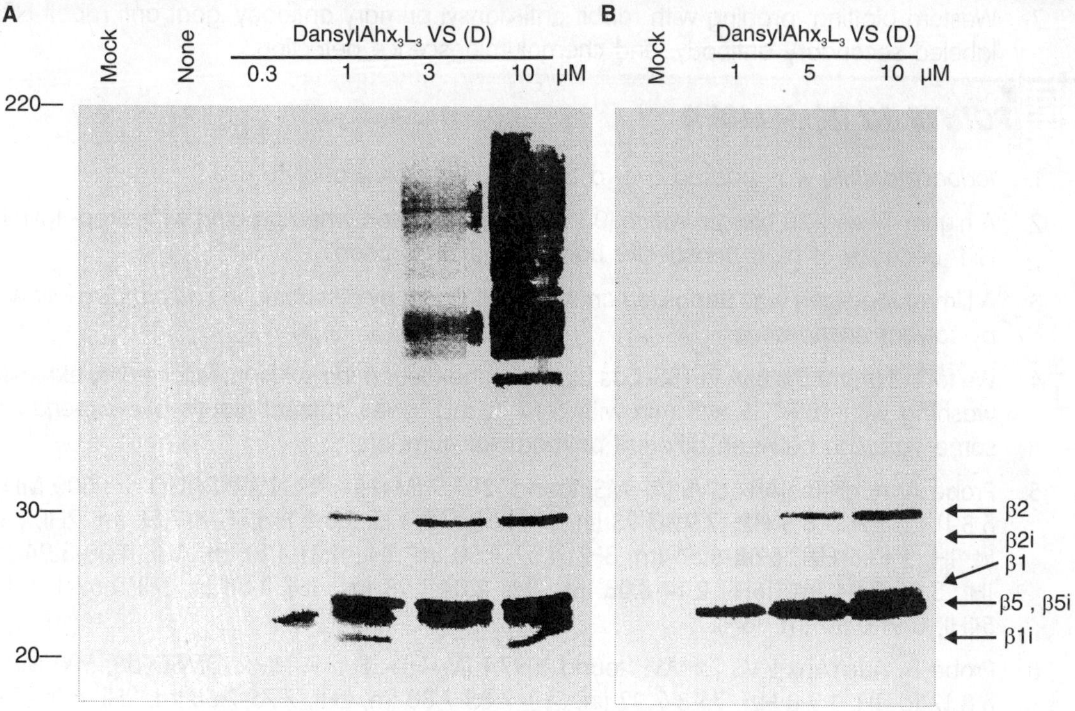

Figure 13.7: 1D gel analysis of A: EL-4 cell lysates and B: EL-4 cells labeled with DansylAhx$_3$L$_3$VS (D).

2. Centrifuge and harvest the cells.
3. After glass-bead lysis, resuspend 10 µg of protein in 33 µL of *lysis buffer* and *denatured* by the addition of 2 µL of 20% SDS followed by *brief boiling.*
4. Incubate the denatured sample with 10 µL (100 µW in DMSO/CH_3CN/H_2O 6/1/1 v/v/v) of *Staudinger ligation* handle for 1 h at 37°C. 5. Add 4X sample buffer.
6. Run SDS-PAGE (12.5%).
7. Western blotting, probing with *streptavidin-HRP,* and *chemoluminescence* detection.

Proteasome Method Using DansylAhx3L3 VS(D)

1. Incubate 10 × 10[6] EL4 cells in 10 mL RPMI 1640 with the indicated amounts of probe D or 10 tMmock, at 37°C for 2 h.
2. Harvest cells by *centrifugation.*
3. Lyse cells by glass bead lysis or by *incubating* them for 30 min in NP-40 lysis buffer and centrifuge 5 min to remove membrane *fractions, nuclei,* and *cell debris.*
4. Determine protein content by the *Bradford* method.
5. Denature equal amounts (typically 60 µg) of protein by boiling in *reducing* sample buffer.
6. Run SDS-PAGE (12.5%).

7. Western blotting, probing with rabbit anti-dansyl primary antibody, goat anti rabbit HRP-labeled secondary antibody, and chemoluminescence detection.

POINTS TO REMEMBER

1. *Iodoacetonitrile* was passed over a basic alumina plug prior to use.
2. A higher Tween-20 *concentration* (0.1 %) should be used when probing with strep- tavidin-HRP because of high *nonspecific* background *adsorption*.
3. A film of *iodo-gen* was deposited on the reaction vial by dissolving in *choloroform* followed by solvent *evaporation*.
4. We found that 0.5% BSA in TBS-T as blocking and *incubation* solution, followed by extensive washing with TBS-T (5 × 5 min with 5 to 10 mL), gives optimal results. We *experienced* some *variation* between different antibody lot numbers.
5. Probe A: AdaK(Bio)Ahx$_3$L$_3$VS (A): MS: found 1287.9 (M+H)+; ^{1}H-NMR (DMSO-d6, 500 MHz): δ 8.11 (d, 1 H, J 8.5 Hz), 7.99-7.93 (in, 2H), 7.84 (d, 1 H, J 8.0 Hz), 7.74-7.68 (m, 2H), 7.07 (d, 1H, J 10.5 Hz), 6.68-6.56 (m, 3H), 4.57-4.50 (m, 1H), 4.31-4.10 (m, 4H), 4.00-3.94 (m, 1H), 3.08-2.94 (m, 16H), 2.14-2.08 (m, 2H), 2.04-1.98 (m, 4H), 1.88 (s, 2H), 1.62-1.16 (m, 54H), 0.91-0.80 (m, 18H).
6. Probe B: AdaYAhx$_3$L$_3$VS (B): MS: found 1097.1 (M+H)+. For ^{1}H-NMR (DMSO-d6, 500 MHz): δ 8.12 (d, 1H, J 9.0 Hz), 7.98-7.92 (in, 3H), 7.85 7.80 (m, 2H), *7.75-7.68* (m, 2H), 6.99-7.97 (m, 2H), 6.67-6.58 (in, 4H), 4.58-4.50 (m, 1H), 4.41-4.35 (m, 1 H), 4.28-4.20 (m, 2H), 3.05-2.95 (m, 9H, 2.81 (dd, 1 H, J 5.0 and 13,5 Hz), 2.62 (dd, 1 H, J 9.8 and 13.5 Hz), 2.14-2.04 (m, 2H), 2.02-1.96 (m, 4H), 1/80 (s, 2H), 1.60-1.24 (m, 42H), 0.91-0.80 (m, 18H).
7. Fmoc-Ahx(a-N$_3$)-OH: ESI-MS: m/z (%) = 417.2 (100) [M+Na$^+$], 811.5 (20) [2M+Na$^+$]; ^{1}H NMR (200 MHz, CDC1$_3$, 25°C, TMS): δ = 7.75 (d, ^3J (H,H)=7.3 Hz, 2H; ArH), 7.57 *(d,* ^3J (H,H)=7.3 Hz, 2H; ArH), 7.40-7.25 (m, 4H; ArH), 4.41 (d, ^3J (H,H)=6.6 Hz, 2H; CHCH$_2$O), 4.19 (t, ^3J (H,H)=6.6 Hz, 1 H; CHCH$_2$O), 3.92 (t, ^3J (H,H)=6.7 Hz, 1H; αCH), 3.18 (m, 2H; εCH$_2$), 1.7-1.3 ppm (in, 6H; βCH$_2$, γCH$_2$, δCH$_2$); ^{13}C NMR (50.1 MHz, CDC13, 25°C, TMS): δ = 174.1, 156.8, 143.5, 140.9, 127.1, 126.7, 124.6, 119.6, 66.4, 62.1, 60.4, 46.7, 30.6, 28.8, 22.4 ppm.

Chapter 14

Rearrangements in Sex Chromosomes

X chromosome rearrangements usually convey *clinical manifestations* in the *hemizygous* males and are, thus, readily *ascertained*. They are found in all parts of the X chromosome and are associated with more than 20 disorders. Some of the rearrangements are the results of homologous *recombination* between low-copy repeats (LCRs) on the X chromosome or between large *homologous* regions on the X and Y chromosome, whereas others are caused by nonhomologous end-joining (NHEJ).

For most large *deletions* associated with contiguous gene syndromes, the deletion breakpoints remain *uncharacterized*. The deletions, as well as inversions and *duplications* on the X chromosome, occur mainly in male germ cells, indicating *intrachromatid* or sister *chromatid* exchange as the *underlying mechanism*.

Among all human chromosomes, the X chromosome seems to have more than its share of *genomic* disorder identified so far. There are two main reasons for the apparent prevalence of genomic disorders on the X chromosome. The *hemizygous* status of the X chromosome in males allows *phenotypic manifestation* of rare recessive genomic disorders that would have *escaped* detection if they were present on an autosome.

In addition, high homology shared between the X and the Y chromosome makes the X chromosome specifically prone to X/Y *translocation*, resulting in the deletion of the terminal sequences. To date, more than 20 genomic disorders have been identified on the X chromosome.

Deletions associated with the genomic disorders can be easily detected in the males by polymerase chain reaction or Southern blotting, whereas the carrier status of a female is usually determined by fluorescence *in situ hybridization* (FISH) analysis.

In this chapter, X-linked genomic disorders are grouped according to the *underlying mechanisms* such as nonallelic *homologous* recombination (NAHR) between direct LCRs, NAHR between inverted LCRs, and NHEJ. For most of the *Short Stature Homeobox (SHOX)* deletions and *contiguous* gene syndromes, the *breakpoints* have not been *characterized* and, thus, the mechanisms remain unclear.

LOSSES AND GAINS

Most cases of X-linked ichthyosis (XLI), color blindness, and incontinential pigmenti (IP) are caused by recurrent deletions resulting from NAHR between direct LCRs. In XLI the LCRs flank the steroid sulfatase *(STS)* gene, in color blindness the color *pigment* genes reside within the LCRs, and in IP the LCRs are one within and one outside the *nuclear* factor *(NF)-κB essential modulator (NEMO)* gene.

When NAHR occurs between two direct repeats on the same chromatid, it generates only deletion. If the recombination involves two X chromosome *homologs* or sister *chromatids,* it generates both duplication and deletion. *Reciprocal duplications* have been identified at the color *pigment* gene cluster and the *NEMO* locus.

Although duplications at the *STS* locus have also been reported, whether they involve the same LCRs as the recurrent deletion remains to be determined. Studies on *de novo NEMO* deletions in IP patients show that most cases occur in the paternal germ cells, indicating unequal crossover between sister *chromatids* or *intrachromatid* recombination as the *underlying* mechanism.

HETEROGENOUS CHANGES

XLI (MIM 308100) affects approx 1 in 5000 males worldwide. It is caused by a deficiency of the microsomal enzyme steroid sulfatase (STS) that *hydrolyzes* a variety of 3β-*hydroxysteroid* sulfates. The *STS* gene at Xp22.3 consists of 10 exons and spans approx 135 kb of genomic DNA. There is a *truncated* STS pseudogene at Yq11 that contains *sequences* homologous to five of the exons.

XLI is one of the first genomic disorders found to have recurrent deletions flanked by LCR elements. Up to 90% of XLI patients have the entire *STS* gene deleted . The deletions are heterogeneous in size and the *breakpoints* spread over several *megabases.* A *majority* of the deletions have common breakpoints within two direct LCRs, S232A and S232B, that are 1.6 Mb apart.

The common recurrent 1.6-Mb deletion is present in approx 87% of *STS* deletion patients in the Unites States and Japan, and 30% of the patient population in Mexico. One-third of the Mexican patients appear to have a smaller recurrent deletion with the distal breakpoint at S232A and the proximal breakpoint somewhere between S232B and another LCR G 1.3 (DXF22S 1).

The 1.6-Mb segment between S232A and S232B contains four genes, HDHDIA(GSI), *PNPLA4(GS2), VCX-8r/VCX-A,* and LOC392425 in addition to *STS.* Interestingly, patients with the 1.6-Mb deletion have the same *phenotype* as those with *mutations* within the *STS* gene, indicating that *removal* of the additional genes around *STS* has no apparent *phenotypic* consequences.

There are six copies of S232 repeats within the human genome, four at Xp22.3 flanking *STS* and two at Yq11. The X-linked S232 repeats are highly *polymorphic* in length. They span more than 2 Mb and are oriented in different directions. The 1.6-Mb deletion breakpoints fall within S232A and S232B, which are in the same direction and share more than 11 kb of 95% identity.

The structure of the S232 repeats is quite complex. Each S232 repeat contains 5 kb of unique sequence in addition to two elements, RU1 and RU2, that are composed of *variable* number of

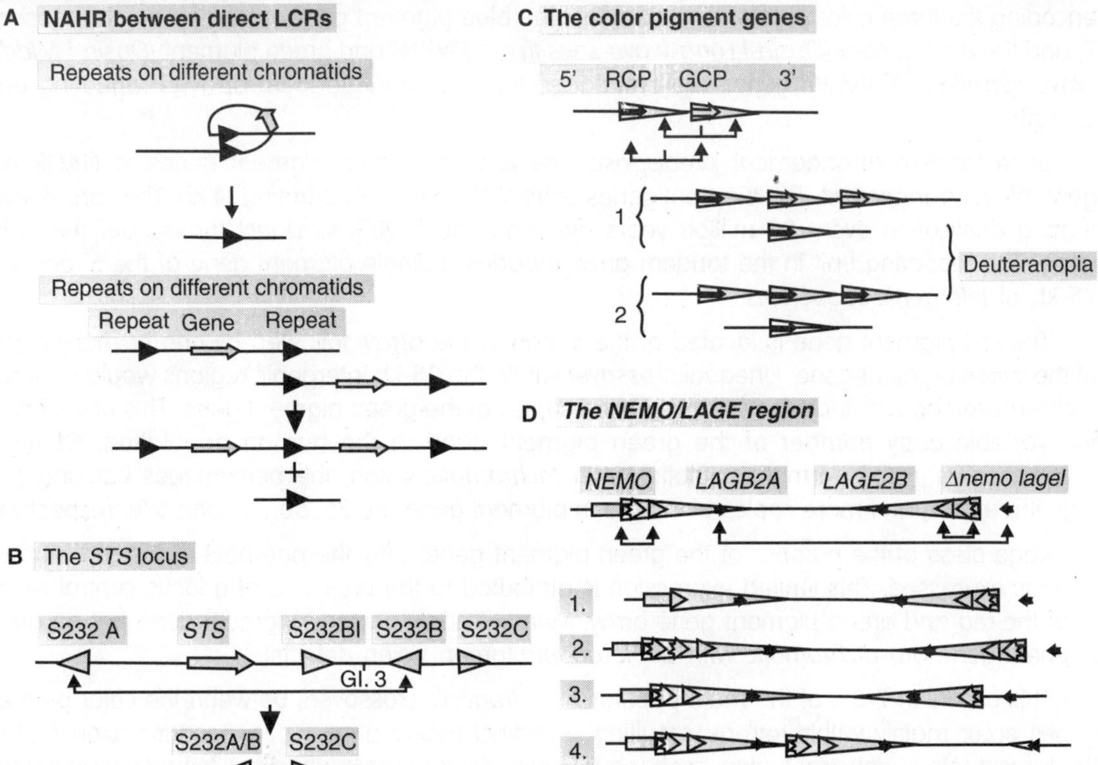

Figure 14.1: Nonallelic homologous recombination (NAHR) between direct low-copy repeats. Horizontal arrows depict genes in the 5' to 3' direction and arrowheads depict repeats. (A) Models of NAHR. (B) The steroid sulfatase (STS) gene and its flanking S232 repeats. (C) Tandem array of the color pigment genes. (D) Complex structure of the NEMO/LAGS *region.*

tandem repeats. The *repeating* unit of RU2 is of highly *asymmetric* sequence and in itself contains a variable number of a GGGA repeat.

The size of RU2 varies from 0.6 kb to more than 23 kb among different individuals, accounting entirely for the observed *polymorphism* at the S232 loci. RU 1 consists of a 30-bp repeat unit and is a part of a testis-specific gene, *Variable Charge X (VCX)* or *Variable Charge Y(VCY),* that is complete *embedded* in the S232 repeat. The VCX/Yfamily encodes basic proteins that appear to play a role in the regulation of *ribosome* assembly during *spermatogenesis*.

The RU2 VNTR sequences as well as the open chromatin structure associated with active *transcription* of the *VCX* genes may *facilitate* homologous *recombination* between S232A and S232B, though the exact locations of the deletion breakpoints within the S232 repeats have not been *determined.*

HUMAN COLOUR PIGMENTS

Color blindness affects approx 5% of the male population *worldwide,* and a vast majority of the cases are caused by defects in the color *pigment* genes on the X chromosome. Of the genes encoding the three color pigments in humans, the blue pigment gene is located on chromosome 7, and the red pigment *(Opsin 1 Long-wave-sensitive, OPN1LW)* and green pigment *(Opsin 1 Middle-wave-sensitive, OPN1MW)* genes are embedded in head-to-tail tandem-arrayed repeating units at Xq28.

Such tandem arrangement *predisposes* the red and green pigment genes to NAHR and genomic rearrangement. The pigment genes *consist* of six *exons spanning* 14 kb. They are derived from a duplication event 30 million years ago and share 98% similarity throughout the entire gene. The repeating unit in the tandem array includes a single pigment gene at the 5' end and 25 kb of *intergenic* sequences.

The red pigment gene is located at the 5' end of the array, followed by one or more copies of the green pigment gene. Unequal *crossover* within the 25-kb intergenic regions would generate X chromosomes with increased or decreased copies of the green pigment gene. This accounts for the variable copy number of the green pigment gene in the human population. Of the X *chromosomes* in the normal population with *trichromatic* vision, the percentages carrying one, two, three, four, and more copies of the green pigment gene are 25, 50, 20, and 5 %, respectively.

Regardless of the number of the green pigment gene, only the one next to the red pigment gene is expressed. This limited expression is attributed to the presence of a locus control region 5' of the red and green pigment gene array. Men with only one red pigment gene but no green pigment gene are *dichromatic* with *deuteranopia* (green vision defects).

This occurs in 2-3% of the male population. *Intragenic* crossovers between the color pigment genes occur mainly within introns, resulting in distinct hybrid gene types. Because exon 5 plays the largest role in spectral tuning, men would have *deuteranopia* when the *hybrid* genes contain exon 5 of the red pigment gene, and protanopia (red vision defects) when the hybrid genes contain exon 5 of the green pigment gene.

X-LINKED DOMINANT DISORDERS

Familial IP (MIM 308310) is a rare X-linked dominant disorder affecting approx 1 in 50,000 newborns. It is characterized by *abnormalities* of the skin, hair, nails, teeth, eyes, and central nervous system, with a distinctive pattern of *hyperpigmentation* and dermal scarring. Affected males die *in utero,* whereas affected females have *extremely* skewed X-inactivation owing to the death of cells carrying the mutation on the active X chromosome.

The disorder is caused by defects in the *NEW* gene that encodes a regulator of the NF-KB signal pathway. The gene spans 23 kb and consists of 10 exons. A recurrent 10-kb deletion that removes exons 4-10 accounts for 70-80% of the *mutations* identified in IP patients. This deletion is *flanked* at the ends by two identical 878-bp MER67B repeats (termed *int3h* repeats), one located within intron 3 of the gene and the other in the 3' flanking region.

In most of the *de novo* cases, the deletion occurred in the paternal germline, indicating *intra-chromosomal* recombination between the repeats. In addition to the recurrent deletion previously described, there are other arrangements around the NEMO gene that are promoted by the complex structure in the region.

The 3' portion of NEMO is in fact included in a 35.5-kb segmental duplication that occurred

10-15 million years ago after divergence of the *great ape lineage*. The *duplicated* regions are present in opposite orientation and include exons 3-10 of NEMO (designated D*NEMO) and the LAGE2 gene. The inverted repeats share more than 99% sequence identity that is believed to be *maintained* through frequent gene *conversion*.

Immediately distal to the repeats is the LAGE1 gene that is related to LAGE2. Molecular *analyses* of normal controls and IP patients had identified rare *rearrangements* in the region, including int3h duplication in NEMO, int3h deletion in D*NEMO, and LAGE1-LAGE2A inversion. Both the int3h duplication and LAGEI-LAGE2A inversion are nonpathogenic because they were found in normal individuals.

It is unclear whether int3h deletion in D*NEMO is also nonpathogenic because the rare mutation has only been identified in IP patients with a mutated NEMO gene. However, D*NEMO appears to be a nonfunctional pseudogene by several criteria.

INDIRECT REPEATS

NAHR between indirect repeats on the same chromatid results in inversion of the intervening sequence. When one or both of the repeats are within a gene, the inversion would disrupt the gene structure and result in a *phenotype*, such as seen in the X-linked hemophilia A and Hunter *syndrome*. When both repeats are situated outside a gene, the inversion would generate a *polymorphism*, such as the one associated with the emerin (EMD) gene.

The inversions usually create altered *restriction* fragments that are easily detected by Southern blotting. It is interesting that all three genes associated with recurrent inversions are located at Xq28, close to the end of the X chromosome long arm. Studies on de novo mutations of hemophilia A showed that the inversions occur almost exclusively in male germ cells.

It is suggested that pairing of the X chromosomes in female germ cells inhibits mispairing of the inverted repeats. The proximity of the locus to the free end of the chromosome could also facilitate chromosome looping required for the alignment of the *inverted* repeats.

Disorders of Vision and Coagulation

Hemophilia A (HEMA, MIM 306700) is a common X-liked coagulation disorder affecting 1 in 5000 males. It is caused by defects in the factor VIII (F8) gene, located approx I Mb from the Xq telomere. The F8 gene consists of 26 exons and spans more than 141 kb. Within its intron 22 reside two genes, factor VIII-associated gene A (F8A) and factor VIII-associated gene B (F8B), that are *transcribed* in the opposite and the same direction, respectively, as F8.

F8A is a small 1.8-kb intronless gene that is ubiquitously expressed. There are two additional copies of F8A situated approx 300 and 400 kb, respectively, 5' and telomeric to F8. These genes are transcribed in the same direction as F8. The homology between F8 intron 22 (int22h-1) and the 5' regions (int22h-2 and int22h-3) extends beyond the F8A genes, and the inverted repeats share 9.5 kb of 99.9% sequence similarity.

About 45% of severe hemophilia A cases (25% of all cases of hemophilia A) have an inversion of the F8 gene caused by NAHR between int22h-I and int22h-2 or int22h-3. The inversions are detected by Southern blotting that gives different fragments for inversions involving int22h-2 or

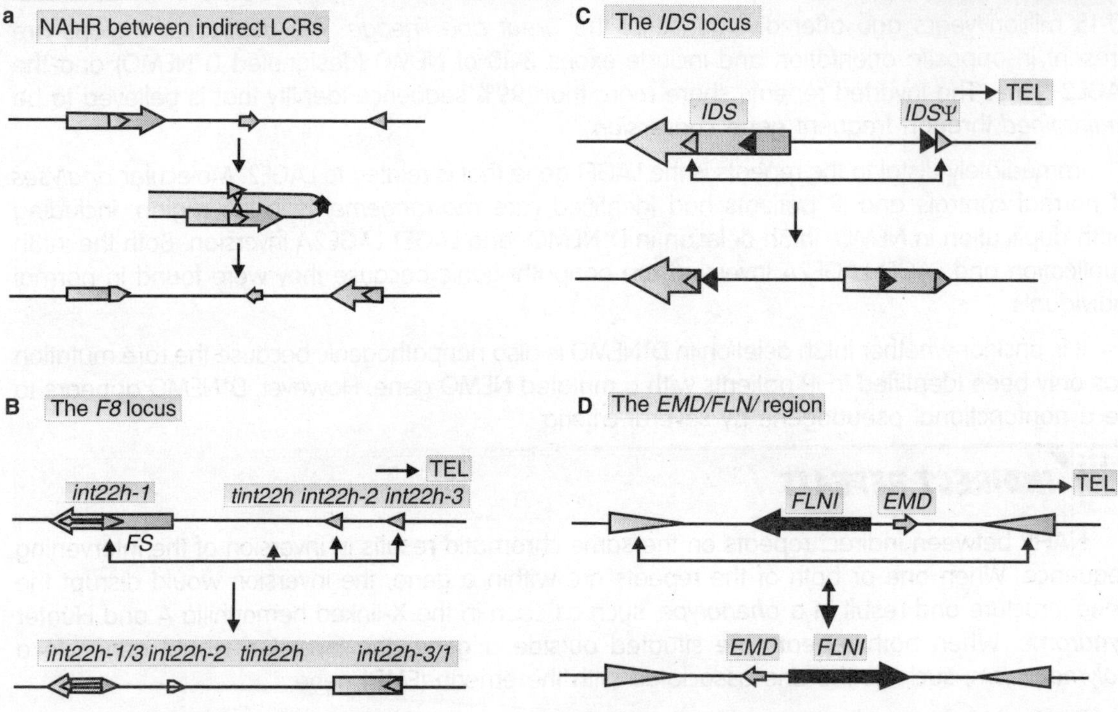

Figure 14.2: Nonallelic homologous recombination (NAHR) between inverted low-copy repeats. (A) A model of NAHR. (B) The F8 *gene and its flanking repeats. (C) The* IDS *gene and its pseudogene IDSψ. (D) The* EMD/FLN1 *region and its flanking repeats.*

int22h-3. No DNA is lost or *gained* during the inversion process. *Interestingly, inversions* involving the more distal int22h-3 are four times more common than those involving the more proximal *int22h-2*. It *is* possible that the longer distance between *int22h-1* and *int22h-3* favors the looping *of* the DNA strand that brings the LCRs together.

In addition to inversions involving *int22h-2* and *int22h-3,* a *hemophilia* A patient was found to have a third kind of inversion disrupting *F8.* This inversion involves an extra-truncated *copy of int22h (tint22h)* located between *int22h-2* and *F8*. The *tint22h* repeat contains only 1.9 kb *of* the 9.5 kb *int22h* sequence. It is detected only in the patient's *family* and does not appear to be a common *polymorphism* in the UK population studied.

X-Linked Lysosomal Storage Disorder

Hunter syndrome (or mucopolysaccharidosis type II, MPS II, MIM 309900) is a rare X-linked lysosomal storage disorder that affects about 1 in 132,000 male live births. It is caused by iduronate-2-sulfatase (IDS) deficiency, which results in the deposits of *muco*-polysaccharides in the *lysosomes* and the *excretion* of large amounts of chondroitin sulfate B and *heparitin sulfate* in the urine. The *IDS* gene consists of 9 exons spanning 23 kb.

About 20 kb 5' and *telomeric* to the *IDS* gene is an *IDS pseudogene (IDSy*)* arranged in the opposite direction. *IDSyr* is *homologous* to two discontinuous segments of *IDS; a* 1.3-kb segment with 98.4% homology to the 3' portion of exon 2 through the first half of intron 3, and a 1.6 kb segment with 96% homology to intron 7. About 13% of Hunter syndrome patients have a recurrent inversion caused by NAHR between *IDS* and *IDSψ*.

Of the *IDS/IDSψ* inversions characterized to date, all have the breakpoints mapped within the intron 7 homologous regions. Therefore, similar to the recurrent inversion at F8, recombination at *IDS* also prefers the homologous regions more distant from each other. Because of the rarity of the disorder, it has not been studied whether the *IDS inversion* also occurs exclusively in male germ cells.

X-Linked Degenerative Myopathy

Emery-Dreifuss muscular dystrophy (EMD, MIM310300) is an X-linked *degenerative myopathy*, characterized by progressive weakening and atrophy of muscle and cardiomyopathy. It is caused by defects in the emerin *(EMD)* gene, which encodes a nuclear lamina-associated protein involved in the *anchorage* of nuclear *membrane* to the *cytoskeleton*.

The *EMD* gene consists of six exons and spans 2 kb. *EMD,* together with its neighboring filamin *(FLN1)* gene, are flanked by two 11.3-kb inverted repeats that share more than 99% sequence similarity. A polymorphism consisting of inversion of the 48-kb *FLN1/EMD* region, likely caused by NAHR between the two long inverted repeats, was discovered during characterization of the molecular defects in EMD patients.

The inversion is present in 18% of the X chromosomes in the Caucasian population and does not seem to increase the risk of DNA rearrangement in the region. However, it could dramatically reduce recombination in females *heterozygous* for the inversion and is thought to account for the *discrepancy* between the physical map and the *genetic* map of the region.

DUCHENNE AND PELIZAEUS DISEASES

Duchenne/Becker muscular dystrophy and Pelizaeus-Merzbacher disease are two additional X-linked disorders with *deletions/duplications* as the major molecular defects. However, *deletions/duplications* in these disorders lack repetitive *sequences* at the breakpoints, suggesting that they result from *nonhomologous* end joining. This chapter *focuses* on Duchenne/Becker muscular dystrophy and the dystrophin gene at Xp22.

Pelizaeus-Merzbacher disease is covered elsewhere in this book. Duchenne muscular dystrophy (DMD; MIM 310200) is the most common form of muscular dystrophy, affecting approx l in 3500 newborn males. About one-third of the cases represent *de novo* new mutations. The disorder is characterized by a progressive muscle weakening, and patients rarely survive beyond the age of 20 years.

Becker muscular *dystrophy* is a milder allelic form of DMD with a later age of onset and slower progression of the disease. The gene underlying both disorders, *DMD,* is the largest gene in the *human* genome. It consists of 79 exons spanning more than 2.2 Mb. Its protein product, dystrophin, is associated with integral *membrane glycoproteins* on the inner surface of muscle fibers and links the inner cytoskeleton to the *extracellular matrix.*

Deletion of one or more exons of the *DMD* gene accounts for 60% of mutations in DMD/ BMD patients in the European and North American populations and much less in other populations.

The *deletions* are *heterogeneous* in size and they cluster at two hotspots. About 30% of the deletions are located in the 5' region centered around exon 7, and the remaining 70% are mapped to the middle of the gene centered around exons 40-50'.

However, the deletion breakpoints scatter throughout the "hotspots" and no extensive homology is present at the ends of the majority of the deletions. The *DMD* deletions, thus, appear to occur mainly by NHEJ. The unusual high incidence of deletion at *DMD* is thought to be promoted by the high recombination rate within the gene. *DAD* has a *recombination* rate about four times the rate that would be expected for a gene of its size, and the major meiotic *recombination hotspots* coincide with the *deletion hotspots*.

Partial duplication of the *DMD* gene accounts for approx 6% of the mutations. Similar to the deletions, no extensive *homology* is present at the ends of the duplicated segments in most cases. In all cases studied, the *duplications* were found to occur in the paternal germline, indicating unequal *sister-chromatid* exchange as the *underlying* mechanism.

OSTEOGENIC GENE

The *SHOX* gene, also known as the *pseudoautosomal homeobox-containing osteogenic gene (PHOG), is* another gene with a high deletion frequency, yet the underlying mechanisms remain unclear. The gene is located within the *pseudoautosomal* region of the X chromosome short arm (PAR 1), approx 500 kb proximal from the telomere. It consists of seven exons *spanning* approx 40 kb, and encodes a member of the paired-like *homeodomain* protein family that acts as a *transcriptional activator* in osteogenic cells and plays an important role in limb *development.*

SHOX mutations have been found in three disorders: *idiopathic* short stature (SS: MTM 604271), Leri-Weill dyschondrosteosis (LWD; MIM127300), and *Langer mesomelic dysplasia* (MIM 249700). In addition, SHOX haploinsufficiency is responsible for the skeletal anomaly in Turner *syndrome patients.* Approximately 2-7% of *individuals* with *idiopathic* SS and 60-100% of LWD patients have *SHOX mutations.*

The rare and more severe Langer mesomelic dysplasia has homozygous or compound heterozygous mutations of the *SHOX* gene. A striking feature of *SHOX* mutations is that there is considerable *phenotypic heterogeneity* among the patients, and no *correlation* exists between the type and position of a mutation within the gene and the resulting phenotype.

Complete gene deletion or the same point mutations could result in SS or LWD, and there is a significant inter- and *intrafamilial* variation in the severity of the phenotype, indicating the involvement of modifiers.

Deletions involving the entire *SHOX* gene account for a vast majority (>70%) of *SHOX* mutations *regardless* of the clinical *manifestation.* Most of the deletions were detected by FISH, however, and the extents of the deletions have not been determined.

For those that have been characterized, they were both terminal and interstitial deletions, and their sizes ranged from 150 kb to more than 9 Mb. Deletions that extend beyond the PAR1 boundary are associated with contiguous gene syndromes, to be described in the next section. The high frequency of *SHOX* deletion has been attributed to the prevalence of repetitive sequences within the PAR1 region, although none of the deletion breakpoints have been isolated and sequenced.

Table 14.1: Contiguous Gene Syndromes on the X Chromosome

Location	Genes deleted	Clinical features
Xp22.3	*SHOX, ARSE, VCX-A, STS, KAL1, GPR143, MID1, HCCS*	Short stature/Leri-Weill dyschondrosteosis, condrodysplasia punctata, mental retardation, ichthyosis, hypogonadism/anosmia, ocular albinism, microphthalmia/linear skin defects
Xp21	*IL1RAPL1, NR0B1, GK, DMD, XK, CYBB, OTC*	Mental retardation, adrenal hypoplasia, glycerol kinase deficiency, muscular dystrophy, McLeod phenotype (acanthocytosis, compen-sated hemolytic anemia), chronic granulom-atous disease, ornithine transcarbamylase deficiency
Xp11.3	*NDP, MAOA, MAOB*	Norrie disease (congenital blindness, hearing loss, mental retardation), disruptive behavior, abnormal sexual maturation, seizures, hypotonia
Xq12	*AR, OPHN1*	Androgen insensitivity, mental retardation
Xg21.1	*REP1, POU3F4, RPS6KA6*	Deafness, mental retardation, choroideremia
Xg21.3	*BTK, DDP*	Agammaglobulinemia, sensorineural deafness
Xq22.3	*COL4A6, COL4A5*	Alport syndrome (hematuria, hearing loss), leio-myomatosis
Xq22.3	*COL4A5, FACL4, AMMECRI, KCNE1L*	AMME: Alport syndrome, mental retardation, midface hypoplasia, elliptocytosis
Xq28	*ABCD1, BCAP31 (DXS1357E)*	Adrenoleukodystrophy, onset before 3 years of age
Xq28	*MTMI, F18*	Myotubular myopathy, intersexual genitalia

X-LINKED CONTIGUOUS GENE SYNDROMES

A large fraction of contiguous gene syndromes (CGSs) identified to date are located on the X chromosome. The CGSs map to all regions of the X chromosome. Their clinical manifestations depend largely on the gene contents of the deletions, though some genes can be deleted with little consequences.

With some exceptions, most CGS patients are males, and female carriers for the same deletions are phenotypically normal. Some of the CGSs are quite rare and only a few cases have been reported. The more common ones at Xp22.3 and Xp21 are described in some detail next.

X-Linked Contiguous Gene Syndromes

More than 40 patients with contiguous gene syndromes at Xp22.3 have been reported. These patients have clinical manifestations associated with two or more disorders mapped to Xp22.3, including SS, LWD, X-linked chondrodysplasia punctata (CDPX, MIM 302950), mental retardation (MR), XLI, X-linked Kallmann syndrome (KALI, MIM 308700), ocular albinism type 1 (OA 1, MIM 300500), and microphthalmia with linear skin defects (MLS, MIM 309801). CDPX, KAL 1, OA 1, and MLS are caused by defects in the *aryls ul fatase E (ARSE), KALI, G protein-coupled receptor 143 (GPR143),* and *holocytochrome c-type svnthetase (HCCS)* gene, respectively.

A mental retardation locus *(MRX)* was mapped to a 15-kb critical region between *ARSE* and STS based on phenotype-genotype correlation. However, the only gene identified within the critical region is *VCX-A,* a member of a gene family with testis-specific expression. It is possible that an unidentified nonprotein-coding or microRNA gene within the region is responsible for the mental retardation phenotype.

With the exceptions of *SHOX, STS,* and *HCCS* mutations, complete gene deletion accounts f'or only small fractions of the mutations identified in patients with isolated disorders within the region. Patients with CGSs at Xp22.3 have large deletions that span several megabases, and their phenotypes correlate with the genes that are eliminated by the deletions. Most of the patients are males.

However, MLS is found almost *exclusively* in females because of embryonic lethality for the *hemizygous* males. All of the affected MLS individuals are monosomy from Xpter to Xp22 owing to *unbalanced translocations* or terminal deletions of the X chromosome that remove more than 10 Mb of Xp terminal sequences. In addition to MLS, these females have SS because of haploinsufficiency of the *SHOX* gene.

A large fraction of the deletions associated with Xp22.3 CGSs is terminal, resulting from X/Y or X/autosomal translocation. X/Y translocations occur rarely in the human population, and the majority have breakpoints at Xp22 and Ygl1. These regions share 89-94% sequence similarity over an extended area. *ARSD, ARSE, VCX-A, STS,* and *KALI* at Xp22 all have homologous sequences at Yq 11

Mapping and characterization of the breakpoints indicate that some of these X/Y translocations result from recombination between homologous sequences at these regions. No breakpoints of the interstitial deletions associated with Xp22.3 CGSs have been isolated and sequenced, and it is not known whether these deletions are formed through NAHR or NHEJ. Two families of LCR are known to be present at Xp22; the S232 family that are involved in the recurrent deletion of *STS,* and the G1.3 family.

There are four clusters of G1.3 related sequences at Xp22, one between *STS* and *KAL1,* two between *KALI* and *GPR143,* and one proximal to *Midline I* (MIDI). Similar to the S232 repeats, the G1.3 repeats also contain a gene family, *fami/y with sequence similarity 9 (FAM9),* that is expressed exclusively in the testis. There may be additional LCRs in the region waiting to be discovered.

Xp21 Contiguous Deletion Syndromes

Close to 100 patients with Xp21 contiguous deletion syndromes have been reported. About two-thirds of the patients have DMD, glycerol kinase deficiency (GKD; MIM 307030), and adrenal hypoplasia congenital (AHC; MIM 300200), 25% have AHC and GKD, and the rest have GKD and DMD, or the more proximally located disorders such as the McLeod phenotype (MIM 314850),

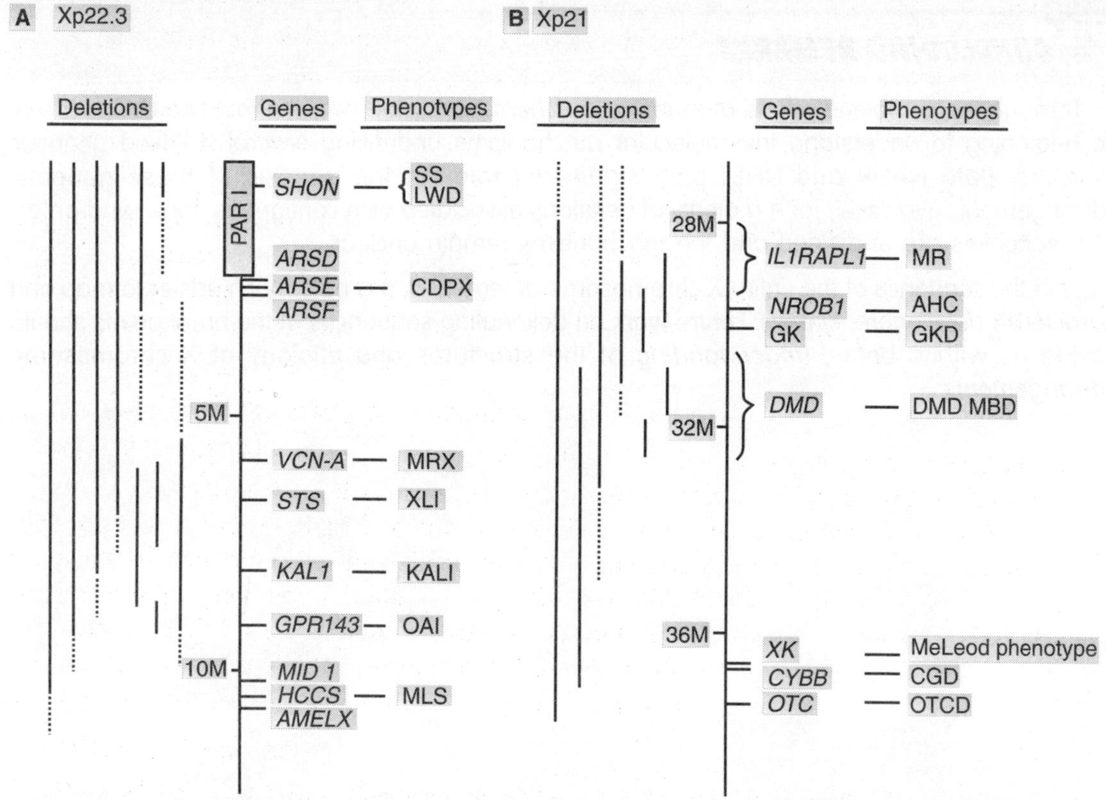

Figure 14.3: Contiguous gene syndromes at two regions of the X chromosome. The maps show the distances (in megabases) of the genes from Xp telomere, taken from the NCBI map viewer. PAR1, pseudoautosomal region 1 ; SS, short stature; LWD, Leri-Weill dyschondrosteosis; CDPX, X-linked chondrodysplasia punctata; MR, mental retardation; XLI, X-linked ichthyosis, KALI, X-linked Kallmann syndrome; OA 1, Ocular albinism type 1; MLS, microphthalmia with linear skin defects; AHC, adrenal hypoplasia congenital; GKD. glycerol kinase deficiency; DMD, Duchenne muscular dystrophy; BMD, Becker muscular dystrophy; CGD, chronic granulomatous disease; OTCD, ornithine transcarbamylase deficiency.

chronic granulomatous disease (CGD; MIM 306400), and ornithine transcarbamylase (OTC) deficiency (MIM 311250).

AHC is caused by defects in the *nuclear receptor subfamily 0 group B, member I (NROB1)* gene, also known as *DAX1,* which when present in double dosage causes XY sex reversal. CGD is caused by defects in the *cvtochrome bβ subunit (CYBB)* gene.

The sizes of the deletions range from less than 1 Mb to more than 14 Mb, and the deletion breakpoints spread over a wide range.

A boy with AHC and mental retardation was reported to have a 2-Mb deletion extending from the *NR0B1* gene to the distal *interleukin-1 receptor accessory protein-like (IL1RAPL)* gene, which is 1.4 Mb in size, implicating *IL1RAPL* in nonspecific mental retardation. None of the deletion breakpoints have been isolated and the mech anism(s) underlying the *deletions* remains unclear.

CONCLUDING REMARKS

Through the combined efforts of many *researchers* worldwide over the last two decades, we are beginning to understand the molecular mechanisms underlying several X-linked genomic disorders. Both NAHR and NHEJ play *significant* roles in the genesis of these genomic rearrangements. However, for a majority of deletions associated with contiguous gene syndromes, the breakpoints are undefined and the mechanisms remain unclear.

With the sequence of the entire X chromosome determined, it is now much easier to map and characterize *deletionbreakpoints.* Future work on delineating sequences at the breakpoints should provide us with a better *understanding* of the structures and etiology of X chromosome *rearrangements.*

Chapter 15

Rearrangements in Y-Chromosomes

Approximately 0.03% of men carry a Y-chromosomal defect that leads to *azoospermia,* the absence of sperm cells from semen. Deletion mapping of the Y chromosomes of *azoospermic* or *oligozoospermic* men suggested that loss of three *nonoverlapping* regions, *AZFa, AZFb,* and *AZFc,* could be responsible. When the finished *Y-chromosomal* reference sequence became available, the recurrent deletion of each of these intervals could be explained largely by non-allelic *homologous* recombination between direct repeats.

However, in contrast to the conclusion from deletion mapping, *AZFb* deletions were found to overlap with *AZFc* deletions. In addition, a background level of *nonhomologous* recombination was found to generate a minority of deletions of these intervals. USP9Yappears to be the critical gene underlying the *AZFa* phenotype, but the critical genes lost in the *AZFb* and *AZFc* deletions have not yet been identified.

Inspection of the sequence allowed additional duplications, inversions, and partial deletions of the *AZF* intervals to be anticipated, and many of the predicted structures have subsequently been identified in the population. The *phenotypic* consequences of these additional rearrangements of the *AZFc* region are unclear.

High levels of gene conversion homogenize duplicated sequences in both direct and inverted *orientations* on the Y, which could *potentiate* subsequent rearrangements. The Y chromosome provides an *excellent* model for *understanding genomic* disorders; however, more finished *sequences* and new *methodologies* are needed.

"*Azoospermia*" is the absence of sperm cells (*spermatozoa*) from the semen. It is found in as many as 1-2% of men and is a heterogeneous condition with diverse causes. It has been classified in a number of ways, for example, according to whether there is a failure of sperm cell production, or a failure of transport from the testes *tubules* to the *semen fluid*. Among the underlying causes are genetic factors, and this chapter focuses on one subset of these: *microdeletions* of *Y-chromosomal* DNA loci.

After *Klinefelter syndrome* (approx 0.1 % of births), such deletions are the second most frequent cause of *spermatogenic* failure and may be present in approx 0.03% of men. The link between *azoospermia* and Y defects has been *recognized* for almost three *decades*.

We outline some aspects of the *historical* development of the field in the next section, but because the finished DNA sequence of most of the *euchromatic* portion of the Y chromosome is now available, in the remainder of the chapter we emphasize two themes: the way that known *deletions* can be explained by the DNA sequence, and, in a *complementary fashion*, the way that the *sequence* can suggest additional possible deletions or other *rearrangements* that can then be *experimentally* verified and *investigated* further. We end with a discussion of findings that are not so readily explained and outline some possible future directions for the field.

CONDITION WITH AZOOSPERMIA

The complete *absence* of sperm cells from semen may seem like a simple *phenotype*, *unambiguously* ascertained by an *examination* of semen, but sperm count can be affected by many factors (such as time since the last *ejaculation*, substance abuse, frequency of hot baths, and even dental caries) and can, thus, vary *considerably* in a single *individual*.

A discussion of this variation lies outside the scope of the present chapter; here we note that sperm count can be regarded as a continuous trait with *azoospermia* grading into *oligozoospermia* (few sperm) and the *normozoospermic* state, and will not distinguish between the first two, so that the *phenotype* under *consideration* here would more accurately be described as "*azoo-* or *oligozoospermia*."

Although *azoospermia*, in the strict sense, leads to *infertility* under natural conditions (this can sometimes be overcome by assisted reproduction), the relationship between *oligozoospermia* and fertility (producing *offspring*) is complex but can only be discussed briefly in this chapter.

Y-Linked Azoospermia

The idea that *azoospermia* was linked to the Y chromosome was initially established by the cytogenetic *observations* of *Tiepolo* and *Zuffardi* in 1976, who detected visible deletions of the distal euchromatic region of Yq in 5/1170 infertile males and an additional male of unknown fertility; four of the six were *de novo* events.

The authors, therefore, presciently suggested that factors (in the plural, later designated *AZFs* for AZoospermia Factors) controlling *spermatogenesis* were contained within the deleted region. The availability of DNA *hybridization* probes and subsequently *sequence-tagged* sites (STSs) detecting Y-specific loci allowed the construction of deletion maps from rearranged chromosomes.

Deletion mapping provides a clear ordering of loci into deletion intervals when the underlying sequences are unique, but the results can be *misleading* when the sequences are present in *multiple* copies.

Some regions of the Y chromosome are entirely *multicopy* and *difficulties* in deletion map *interpretation* underlie some of the problems discussed below. *Individuals* with *cytogenetically* normal Ys and idiopathic infertility were characterized using this approach, and *microdeletions* were identified in a subset of them; however, these did not all overlap.

An extensive compilation of data from 370 men with azoospermia or oligozoospermia led to

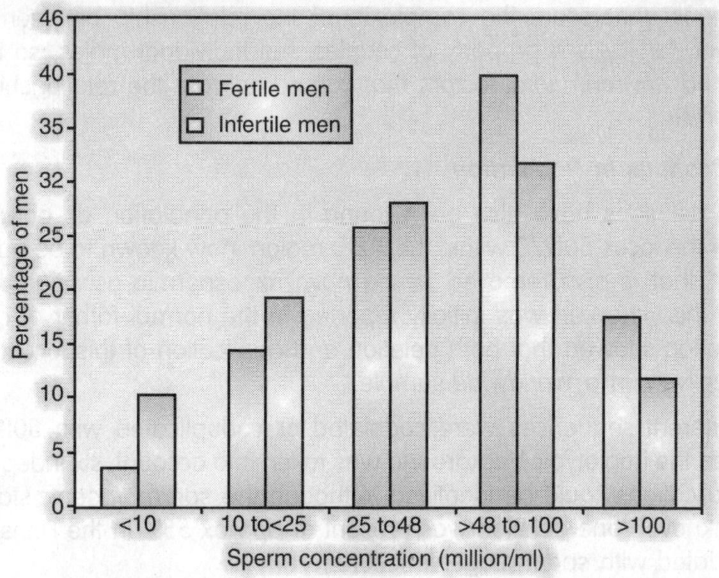

Figure 15.1: Distribution of sperm counts in fertile and infertile men. Note the broad ranges and overlap between the two samples.

a widely held view that three nonoverlapping deletions, designated *AZFa, AZFb,* and *AZFc* in the order centromere to Yq telomere, could be defined. It was expected that further work would refine the locations of the individual genes within these intervals that were responsible for the *spermatogenic* failure, using a *combination* of finer deletion mapping and the subsequent detection of point mutations.

Although this expectation has been largely fulfilled for *AZFa,* the *AZFb* and *AZFc* intervals have proved more *refractory* to analysis and the identity of the key genes within them remains unclear.

Sporadic Deletion

As attempts to identify *AZF* genes progressed, it became clear that several lines of evidence could not be incorporated into a simple picture of an intact "*normal*" Y chromosome with sporadic deletion mutations leading to *azoospermia.*

Visible Variability

The sperm count associated with any Y-chromosomal genotype is variable. The variation within a single individual has been previously mentioned, so it should not be surprising that considerable variation is also seen in the *phenotypes* of individuals with deletions that are *indistinguishable* at the molecular level.

Nevertheless, it is striking that men with *AZFc* deletions, where independent events lead to a common structure (*see* The *AZFc* Deletion) and are usually associated with *spermatogenic* failure, can occasionally conceive children naturally. In one family, an AZFc-deleted man *fathered* four sons, the last at the age of 38, although at the time of analysis when he was aged 63, he was azoospermic; the three sons tested were all oligozoospermic.

These observations underscore the complexity of the relationship between *microdeletions* and fertility. Moreover, fertility is a property of couples, not individual males, so there are many additional genetic and *environmental* factors that can complicate the relationship between low sperm count and *fertility*.

Deletion and Duplications in Population

Y-chromosomal deletions have also been found in the population as polymorphisms. For example, deletion of the locus 50f2/C within the *AZFc* region (now known to be part of the single-copy sequence "u3" that is also removed *by de novo* azoospermia-associated deletions; see Figure elsewhere in this chapter) was initially reported in the normal father of a 46,XY female, and further investigation showed that both deletion and duplication of this locus were common: 6.4 and 1.4%, respectively, in a *worldwide* sample.

Furthermore, different sequences were codeleted or coduplicated with 50f2/C in different individuals, and when the haplotypic background was taken into account, six independent deletion events and four duplications could be identified. Although the *spermatogenic* status of the men investigated was unknown. one deletion was present at approx 55% in the Finns, and thus was unlikely to be associated with spermatogenic failure.

Surveys of infertile patients and controls (either fertile or normozoospermic individuals) provided further examples of deletions that were not associated with azoospermia, either because they were found only in controls, or were *transmitted* by the fathers of *infertile* patients. In one large-scale evaluation using 48 STSs and 920 fertile individuals, 11 men (1.2%) were found to show *deletions* of single STSs that removed one of four different *loci*.

One of these was Amelogenin on Yp, but the other three (sY207, sY269, and sY272) are repeated sequences with copies lying within *AZFh* or *AZFc,* as defined at that time. These four deletions were assumed to represent *polymorphisms,* but in the same study, deletions of other single STSs (e.g., sY75, sY126) were found only in infertile patients. Are these related to the patient's infertility or not?

Mapping

Interpretation of the meaning and order of the intervals defined by deletion mapping proved to be difficult, irrespective of the phenotype associated with any particular deletion. These complexities are illustrated by the proposed *designation* of a fourth deletion interval, *AZFd*. This was identified by Kent-First et al. in a study of 514 infertile males.

Six men who retained DAZ lacked only one STS, sY152, and eight lacked only the adjacent STS, sY153; one of the latter *deletions* was *a de novo* event. This was taken as evidence for a distinct deletion interval located between *AZFb* and *AZFc*.

Yet on the finished sequence, all copies of sY152 and sY153 are located withinAZFc as defined by the common b2/b4 deletion. Current European guidelines for molecular diagnosis of Y-chromosomal *microdeletions* consider that "a putative fourth *AZFd* region *postulated* by Kent-First et al.... does not exist". The deletion maps and yeast artificial chromosome contigs in many repeated parts of Yq that were available at the time have not been *scrutinized* in such detail, but may be equally *unreliable.*

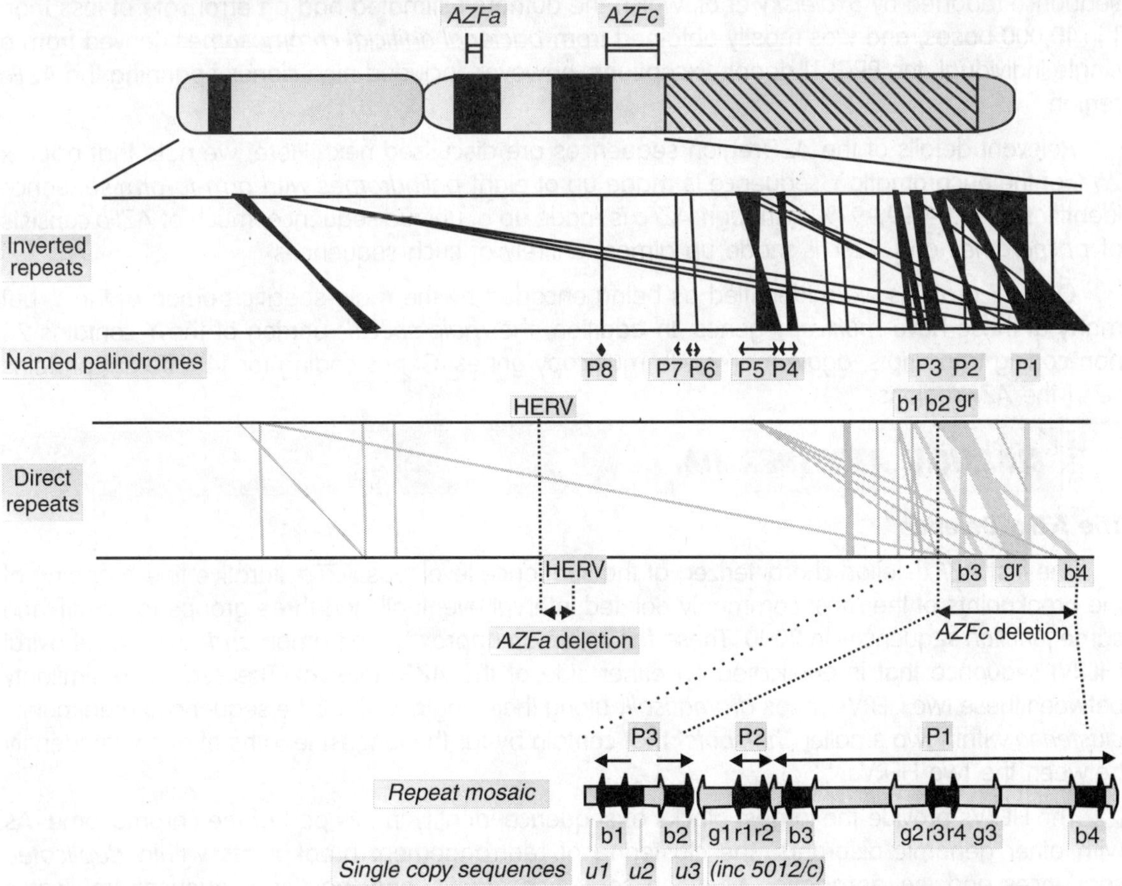

Figure 15.2: Classical AZF *deletions mapped onto the finished Y chromosome sequence. (Top) Approximate locations of the* AZFa *and* AZFc *deletion intervals on the Y chromosome. Middle) Visualization of duplicated sequences with greater than 900 by of identity in inverted (upper middle) and direct (lower middle) orientation on the Y chromosome-specific euchromatic reference sequence. This image wasgenerated using the GenAlyzer software and places diagonal lines between aligned chromosomes to show the location of inverted and direct repeats. Underneath the inverted repeats are shown the eight palindromes identified within the Y sequence. The visualization of direct repeats is annotated with the positions of the repeats promoting the* AZFa *and* AZFc *deletions, HERV-HERV and b2/ b4 elements,respectively. (Bottom) Expanded view of* the AZFc *region and adjacent sequences, showing the mosaic of smaller repeats referred to in the text and including the nonrepeated segment u3 that contains 50f2/C.*

Significance

Lessons to be drawn from the preceding section are that: reliable sequence information is essential for interpreting the structures of deletions and other rearrangements involving repeated Y regions; (2) the sequence should be based on the *Y chromosome* of a single man in order to avoid confusion between polymorphisms and variants in different copies of repeated sequences within the same *chromosome*; but (3) sequence may be needed from more than one individual in order to understand the extent of normal *polymorphic variation*.

The first two of these requirements have largely been met by the near-complete *euchromatic* sequence reported by *Skaletsky* et al, which the authors estimated had an error rate of less than 1 in 10,000 bases, and was mostly obtained from *bacterial artificial chromosomes* derived from a single individual, the RPCI-11 donor. Exceptions, however, included nine clones spanning the *AZFa* region.

Relevant details of the *AZF* region sequences are discussed next. Here, we note that approx 25% of the euchromatic Y sequence is made up of eight *palindromes* with *arm-to-arm* sequence identities of 99.94-99.997%. Although *AZFa* is made up of unique sequence, much of *AZFb* consists of *palindromes* and *AZFc* is made up almost entirely of such sequences.

Only 27 proteins were identified as being encoded by the male-specific portion of the Y, but many of these have *multicopy* genes. In addition, the *male-specific* portion of the Y contains 28 non-coding transcripts, again many with multicopy genes. Genes coding for 14 different proteins lie in the *AZF* regions.

CLASSICAL AZOOSPERMIA

The AZFa Deletion

The first *AZF* deletion characterized at the sequence level was *AZFa.* Iterative fine-mapping of the breakpoints of the most commonly deleted interval eventually led three groups to identify the same *junction* sequences in 2000. These fall within an approx 10-kb human *endogenous* retroviral (HERV) sequence that is duplicated on either side of the *AZFa* interval. The sequence similarity between these two HERVs varies *dramatically* along their length, with all the sequenced breakpoints *clustering* within two smaller "*hotspots*" that contain by far the longest lengths of absolute identity between the two HERVs.

The HERVs provide the largest blocks of sequence identity in this part of the chromosome. As with other genomic disorders, the clustering of rearrangement breakpoints within *duplicated* sequences and the *association* between sequence *identity* and *junction* sequences implicates nonallelic homologous recombination (NAHR) between these two HERVs as the *mechanism* underlying the deletion event.

The *AZFa* deletion removes two protein-coding genes, *USP9Y* and *DBY,* and one non coding *transcript,* and results in a *complete* absence of germ cells known as the sertoli cell only phenotype. A single infertile male with a *de novo* null *mutation* (a 4-bp deletion) that produces a truncated USP9Y protein has the less severe, spermatid arrest, phenotype, which implicates one or more *additional* sequences, most likely *DBY* , in the more severe *sertoli* cell only phenotype.

The AZFc Deletion

Deletions of the *AZFc* interval are the most common Y-chromosomal lesion seen in infertile males. The highly duplicated nature of *the AZFc* interval-only 7% of which is single copyhindered efforts to refine the position of the breakpoints of the common *AZFc* deletion using STSs. Only when the entire region had been sequenced using clones derived from a single male could the position of these *breakpoints* be *localized.*

Although most of the duplicated sequence in distal Yq resides in *nearidentical palindromes,* there are a number of long *highly* similar sequences in direct *orientation,* including some within

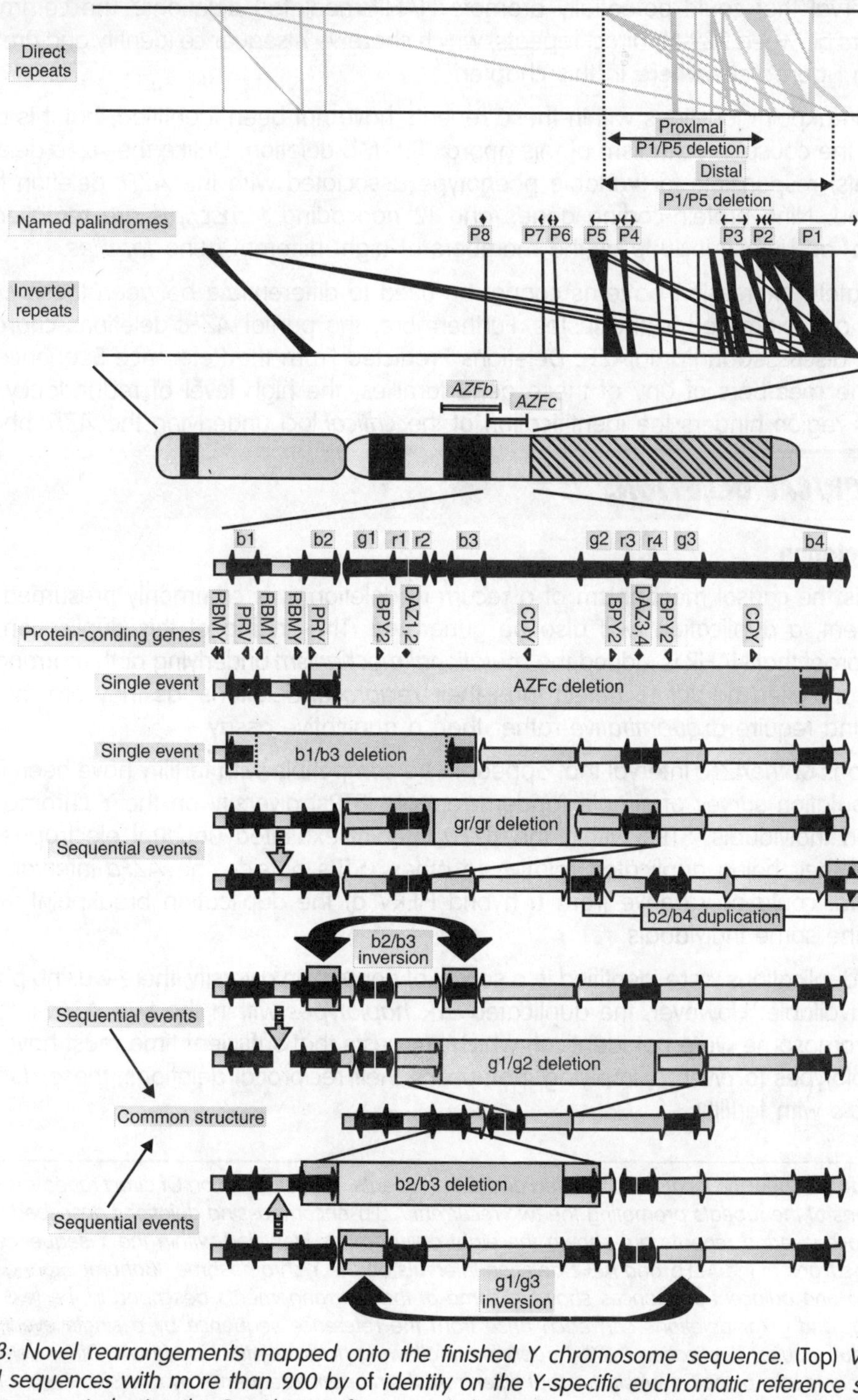

Figure 15.3: Novel rearrangements mapped onto the finished Y chromosome sequence. (Top) *Visualization of duplicated sequences with more than 900 by of identity on the Y-specific euchromatic reference sequence. This image was generated using the GenAlyzer software and places diagonal lines between aligned chromosomes to* ***(Caption Contd.)***

the *AZFc* interval that could potentially promote NAHR-mediated deletions. The common *AZFc* deletion occurs between 229-kb direct repeats, which share 99% sequence identity and are denoted b2 and b4 in Figure's elsewhere in this chapter.

Precise breakpoint junctions within these repeats have not been identified, but it is assumed that NAHR is the causal mechanism of this approx 3.5-Mb deletion. Unlike the *AZFa* deletion, the specific gene(s) responsible for variable phenotype associated with the *AZFc* deletion have not been identified. Nine protein-coding genes and 12 noncoding *transcripts* are removed by the *AZFc* deletion, and these include all the members of eight different gene *families*.

Unfortunately, expression patterns cannot be used to differentiate between these *candidate* genes as all are expressed in the testes. Furthermore, the partial *AZFc* deletions characterized thus far (and discussed in PartialAZFc Deletions Predicted From the Reference Sequence) do not remove all the members of any of these gene families; the high level of redundancy of gene copies in this region hinders the identification of the *critical* loci underlying the *AZFc* phenotype.

HISTORICAL DELETIONS

Recurrent Deletion

If NAHR is the causal mechanism of a recurrent deletion, it is commonly presumed that the *reciprocal* event, a duplication, will also be generated. The finding of this duplication is often regarded as proof that NAHR is indeed the *mutational mechanism* underlying both rearrangements. *Duplications* are often harder to detect than their *reciprocal* deletions. as they are more rarely *pathogenic* and require a *quantitative* rather than a qualitative assay.

Duplications of *theAZFa* interval that appear to be compatible with fertility have been identified during a population survey of simple tandem repeats (STRs) diversity on the Y chromosome. In two unrelated individuals, STRs within the *AZFa* interval exhibited unusual electropherograms suggestive of their being duplicated relative to other YSTRs outside the *AZFa* interval. Junction sequences that could only derive from a hybrid HERV at the duplication breakpoint were also identified in the same individuals.

As these duplications were identified in a survey of population diversity, there was no phenotypic information available. However, the duplicated STR *haplotypes* within the two *AZFa* intervals on the same chromosome were not identical, which suggests that sufficient time must have elapsed for these haplotypes to *diverge*, implying that, unlike their reciprocal deletions, these *duplications* are *compatible* with fertility.

Caption 15.3 Contd.: *show the location of inverted and direct repeats. The visualization Of direct repeats is annotated with the positions of the repeats promoting the two recurrent* AZFb*-encompassing deletions described in the text, P1 and P5. Above inverted repeats are shown the eight palindromes identified within the Y sequence. (Middle) Approximate locations of* the AZFb *and* AZFc *deletion intervals on the Y chromosome. (Bottom) Expanded view of the* AZFc *region and adjacent sequences showing some of the rearrangements described in the text. The* AZFc *(b2/b4), b1 /b3, and gr/gr deletions can each arise from the reference sequence by a single event, and one example* of *a\subsequent duplication (b2/b4 duplication following gr/gr deletion)* is *shown. Other deletions (g 1/ g3, b2/b3) appear to take place on substrates that differ from the reference sequence by inversion events. Alternative inversion-deletion routes leading to the same structure are shown.*

It is worth noting that on a constitutively haploid chromosome such as the Y, duplications must be caused by NAHR between sister chromatids. In addition to the deletion and duplication events within the *AZFa* interval, NAHR has been involved in the *homogenization* of the HERV sequences that promote the rearrangements.

In two independent lineages of the Y chromosome phylogeny, the maximal length of *absolute* identity that is shared between the HERVs has been expanded fivefold as a result of homogenization of the approx 4.5 kb of sequence between the two NAHR hotspots. This homogenization process most likely occurred because of an *AZFa* deletion on a chromosome that had previously undergone an *AZFa* duplication.

This dramatic polymorphism for the absolute length of sequence identity between the HERV s, and the apparent relationship between the length of sequence identity and hotspots of NAHR activity, together suggest that some males may have higher rates of *AZFa deletion/ duplication* than others.

Partial Deletions

Although early studies and several subsequent analyses had provided evidence for partial deletions and duplications within the *AZFc* region, their structures and phenotypic consequences were unclear. The availability of the sequence and an understanding of the mechanism of the *AZFc* (b2/b4) deletion led to the prediction that two partial deletions might arise by homologous recombination between other direct repeats: b1/b3, or the various green and red repeats, now collectively abbreviated as gr/gr.

A search strategy based on the presence or absence of STSs spanning amplicon boundaries and thus providing some specificity in this highly repeated region was used to characterize 689 men. STS patterns indicative of both of the predicted deletions were found: one b1/b3 deletion and 22 gr/gr deletions, and all those that could be tested (1/1 and 20/22, respectively) were confirmed to be deletions by *fluorescent in situ* hybridization.

In addition, two of the gr/gr deletions showed duplications interpreted as b2/b4 events. The spermatogenic phenotype of the man with the b l /b3 deletion was not reported, but gr/gr deletions were found in 9/246 men with total sperm counts less than 2×10^7 (or $<10^7$/mL) compared with 0/148 with total sperm counts more than 4×10^7, a difference that achieves statistical significance ($p < 0.014$, Fisher exact test).

This finding led the authors to suggest that the gr/gr deletion, which removes some but not all members of three protein-coding and five non-coding gene families, exists within the population at a lower than expected frequency as a result of mutation/ selection balance. However, statisticai significance would be lost if the 149th man with normal *spermatogenesis* were to carry the deletion (9/246 vs 1/149,$p > 0.05$), a conservative consideration often applied in the statistical interpretation of *forensic* evidence.

This factor and the prevalence of the deletion in the general population mean that the association of' the gr/gr deletion with spermatogenic failure must be considered uncertain until more data are available, particularly in the light of results from other deletions discussed next. Irrespective of the phenotype associated with the gr/gr deletions, their genomic properties are of considerable interest.

The Y chromosome outside the pseudoautosomal regions evolves along stable lineages that are not disrupted by recombination and can be identified and organized into a phylogenetic tree using neutral polymorphisms. gr/gr deletions were found in 14 branches of a tree that resolved 43 lineages and had apparently arisen independently in each of these branches, although the locations of the recombination events within the gr repeats are unknown and could differ between lineages: for example, some could remove *DAZI/DAZ2,* wereas others could remove *DAZ3/DAZ4.*

The deleted chromosomes made up a small proportion of most lineages, but in one, D2b, all 12 chromosomes examined carried the deletion, and this lineage is known to be present in approx 30% of men in Japan. Such a frequency can be achieved by a neutral marker by genetic drift, but is unlikely to be attained by a lineage that is selected against even to the extent of 1%, the selective disadvantage estimated by the authors.

Possible explanations are that: (1) some or all gr/grdeletions have no significant effect on spermatogenesis; (2) they reduce sperm numbers but have no significant effect on fertility, or (3) a compensatory mutation has arisen. Such a mutation would be most effective if it was located on the Y chromosome and was, thus, permanently linked to the deletion, and the b2/b4 duplication reported in 2/30 gr/gr deletions provides a candidate for such a mutation since it restores gene copy numbers to their original levels or higher.

Unfortunately, it is not yet known whether or not D2b gr/gr deletions carry a b2/b4 duplication. Ironically, although the gr/gr deletion may not be in mutation/selection balance, the *AZFc* deletion characterized by the same authors provides a clear example of such a balance.

The incomplete penetrance of the *AZFc* deletion leads to its infrequent but recurrent transmission from father to son *(see* The *AZFa* Deletion) and results in the *AZFc* deletion reaching a population frequency higher than its *de novo* mutation rate, but being kept in check by the deletion's substantial fitness costs.

Deletion in a Population

The detection in the population of deletions predicted from the sequence provides a powerful illustration of the usefulness of sequence data. Further surveys, however, have discovered additional partial *AZFc* deletions that were not predicted. A major class lacks two apparently discontinuous regions, one including u3 (and thus 50f2/C) and the other the *DAZ3* and *DAZ4* genes, which lie in the distal *DAZ* cluster.

Neither deletion could arise by *homologous* recombination from the reference sequence, but a *combination* of two events, an inversion followed by a deletion, involving both the b2/b3 and gl/g3 repeats, could remove all the missing sequences by either of two pathways. This deletion product has been identified in four *lineages,* but most deleted chromosomes belong to one of them, N, where it appears to be fixed.

The N lineage is even more widespread than D2b, predominating in northern Asia and making up 12% of Y chromosomes in one worldwide survey. The proposed mechanism of deletion gains credence from the detection of both of the possible intermediate inverted structures in nondeleted chromosomes, with b2/b3 inversions showing at least two independent origins in a 44-branch tree and gr/gr inversions 5; additional duplications were found in 3 of the 14 deleted chromosomes.

Genes removed by the deletion (known either as g1/g3 or b2/b3) include some but not all members of three protein-coding and five non-coding gene families, but with the loss of more family members than the gr/gr deletion. The success of the N lineage is easily understood under the hypothesis that such *partial deletions* are neutral eventsit is entirely owing to chance genetic *drift*-but requires more convoluted *explanations* involving compensating mutations other than the known duplications under the *hypothesis* that such partial deletions are associated with spermatogenic failure.

Mapping Deletions

The relatively simple relationship observed for *AZFa* and *AZFc* between a deletion *interval* mapped using STSs and a single common deletion structure does not hold true *forAZFb.* There are at least two recurrent deletions that remove the *AZFh* interval defined by STS mapping, as well as additional *nonrecurrent* deletions.

All of these deletions also remove part of the *AZFc* region, suggesting that the *AZFb* interval is not a discrete block as was *originally* supposed. NAHR is again the most likely mechanism underlying the recurrent deletions, although the lack of homology at the breakpoints of a nonrecurrent deletion indicates a low background level of nonhomologous *recombination.*

As with *AZFc,* the direct repeats responsible for both of the recurrent deletions lie within different palindromes in distal Yq. The 6.2-Mb recurrent deletion that removes only a small portion of the *AZFc locus* in addition to the *AZFb* region occurs between approx 100-kb direct repeats that reside in the proximal copies of palindromes P I and P5.

The 7.7-M b recurrent deletion that removes a much greater proportion of the *AZFc* region occurs between the corresponding repeats in the distal copies of palindromes P1 and P5. There is evidence that within the approx 100-kb repeats lies a 933-bp hotspot in which the deletion breakpoints are preferentially located, although more AZFb-deleted chromosomes need to be characterized at the sequence level for unambiguous identification of hotspot activity.

Twenty-one protein-coding genes and 11 non-coding transcripts are removed by the proximal P1/P5 deletion, which represents all the family members of 15 different genes. In the larger distal P1/P5 deletion, an additional four protein-coding genes and six non-coding transcripts are deleted, with the result that 19 gene families are lost in their entirety.

When overlapping deletions elsewhere in the genome result in similar *phenotypes,* it is usual to assume that the critical genes *underlying* the phenotype map to the region of overlap. This *viewpoint* has not been adopted for the overlapping *AZFb/AZFc* deletions causing male *infertility* on Yq, probably for historical reasons.

The overlap of the *deletions* contains only members of gene families and not entire families. Is loss of these members primarily responsible for the phenotype, with *modifications* by *flanking* genes, or do similar *phenotypes* arise from the loss of genes in the *nonoverlapping* regions?

AZF_B Rearrangements

Perhaps unsurprisingly in view of the *AZFc* complexity, there are hints of a greater variety of *AZFb* rearrangements. One case of a duplication involving both 50f2/C and a second locus approx 1.4 Mb proximal, 50f2/E, was detected in a population screen of 595 normal individuals, and may represent an example of one of the predicted P1/P5 duplications. It seems likely that systematic searches would reveal a wealth of further inversions, duplications, and deletions.

AZOOSPERMIC GENES

Detection of Azoospermic Genes

The recurrent *AZF* deletions previously discussed account for most of those understood at a molecular level, but there are exceptions for all three regions. Partial deletions of *AZFa* have been described; out of a sample of 11 deletions removing *AZFb, 2* could not be explained by homologous recombination; and I out of 48 *AZFc* deletions was smaller than the common size shared by the other 47.

Although these findings require further investigation and interpretation in the light of the large variety of possible deletion substrates, they do suggest that there is a low level of more diverse and sometimes nonhomologous deletion. Although of less clinical significance because of their rarity, they may be highly informative in pinpointing the functionally important genes in the large regions identified by the recurrent mutations.

An alternative way of identifying the critical genes is offered by some of the variant structures. Despite searches, point mutations have not so far been found in *AZFb or AZFc,* and one possible explanation for this failure is that inactivation of a single copy of any of the genes present, which are all *multicopy* in the reference sequence, would be insufficient to lead to *spermatogenic* failure.

In some *lineages,* however, such as N, the genes *BPY2* and *CDY1* are present in single copies, and *DAZ* is reduced to two copies. Men belonging to lineage N might therefore experience *spermatogenic* failure as a result of *inactivation* of the last copy of one of these genes by a point mutation, so a screening program focusing on this lineage might prove *fruitful.*

Significance

Independent replication of the association study between gr/gr deletions and spermatogenic failure is needed, and further work should identify the deletion breakpoints and thus the individual members of gene families that are lost. In addition, more thorough investigation of the effect of the common partial deletions within *AZFc* on spermatogenesis is desirable, taking into account both the phylogenetic background and any associated duplications.

Lineages Db2 and N would be of particular interest. Population genetic approaches might also provide information: have these *lineages* experienced negative selection and expanded less rapidly than expected for a neutral lineage'?

Methods that compare the age of a lineage with its frequency and determine whether the observed pattern is compatible with neutrality could be used to address this question. If some partial deletions are associated with fitness costs, these might indicate the underlying selective pressures that have operated to favor multicopy genes on the Y chromosome.

Polymorphism

It should be apparent from the previous discussion that the mosaic pattern of direct and inverted repeats capable of promoting rearrangements generates substantial structural polymorphism on the long arm of the Y chromosome. The principle that it is possible to predict potential variants is well established and we have not cataloged them further here.

However, it is worth emphasizing that not every structure observed in the population can be explained by a single mutational step from the reference sequence, and that in many cases there are alternative recombinatorial routes to the same structure.

Because most rearrangements are expected to occur more frequently than one in 3 billion meioses, we can reasonably expect that most, if not all, of these structures could be found in the current global population of more than 3 billion males. However, we would like to know how prevalent these different events and structures are, the locations of the recombination events within the repeated regions, and what their effects are on both spermatogenesis and further Y-chromosomal rearrangements.

Some of the rearrangements may predispose to further pathogenic rearrangements, but others are likely to protect against them: for example, the b1/b3 deletion removes the b2 repeat and presumably prevents the common *AZFc* deletion. In view of the likely complexity of rearrangement events, it will be important to understand them in their phylogenetic context.

It is worth considering the consequences of the choice of reference sequence: if this had been derived from lineage N, our view of the *AZFc* region would be simpler and we would have no sequence data from sections like u3. In the same way, the current reference sequence may lack regions present on other chromosomes. Additional sequences and/or structural maps of other Y-chromosomal lineages are therefore desirable.

Selection of high-priority lineages could be directed by the phylogeny, perhaps choosing the most divergent-known as haplogroup A *-or* by preliminary mapping experiments to identify the most distinct structures. Unfortunately, such duplicated structures cannot be sequenced efficiently using current high-throughput methods that rely on the assembly of short reads.

Hemizygous Chromosomes

In many respects the constitutive haploidy of the Y chromosome renders it a useful model for understanding chromosomal dynamics. There is no need to generate somatic cell hybrids to separate a rearranged chromosome away from its unaltered homolog. This saves time and effort in the study of individual rearrangements.

The unusual evolutionary history of the Y chromosome also facilitates studies of the origins of rearrangements. The lack of allelic recombination over the entire male-specific portion of the Y chromosome makes it possible to relate all extant chromosomes with a single phylogeny. Structural polymorphisms can be shown to have multiple independent origins if they lie on different

lineages of this well-characterized phylogeny. In principle, the origin of a specific polymorphic rearrangement can be placed in both time and space, albeit with a substantial degree of imprecision. Having emphasized the positive qualities of the Y chromosome, it is necessary to add the caveat that it is presently unclear how good a model the Y chromosome will be for other regions of the genome enriched for segmental duplications.

The generality of the Y-chromosome is called into question by its paucity of genes, its high repeat content and especially its absence of a meiotic pairing partner over most of its length. As a consequence of all of these factors, rearrangement is the dominant mode of pathogenic mutation on this chromosome, an observation that is unlikely to hold true for other chromosomes.

Nevertheless, the lessons learned from studying distal Yq may well be informative for the interpretation of other complex mosaics of inverted and direct repeats, where rearrangements of different types and sizes are associated with different phenotypic outcomes, and the concept of a single wild-type structure breaks down.

Y Chromosome of Primate

Much can be learned about the nature of mutational processes from the comparative analyses of homologous sequences between species. Thus, the prospect of a finished Y-chromosomal sequence from the common chimpanzee represents an opportunity to infer the evolutionary history of the structure and sequence of the human Y chromosome, and to explore how mutational processes differ between duplicated and single-copy sequences.

Already, limited sequencing of palindromes and their flanking sequences have revealed extensive gene conversion between the palindromes in distal Yq. Similarly, comparative sequences from chimpanzees and gorillas have revealed localized concerted evolution in and around the rearrangement hotspots within the HER Vs that promote *the AZFa* rearrangement.

Gene on Y-Chromosome

As previously described, studies of both sequence variation within species and sequence divergence between species have revealed that much of the duplicated sequence on the Y chromosome undergoes sequence homogenization. In principle, two different mechanisms could account for this observation.

The first is gene conversion, which represents one resolution of a homologous recombination intermediate in which no rearrangement results, but sequence information is transferred in a nonreciprocal fashion from one repeat to another. The second process is unequal crossover in which successive duplication and deletion events generate hybrid repeats that are more similar to one another than the ancestral repeats.

Both of these processes have been observed at Y-chromosomal loci, although only gene conversion is capable of homogenizing repeats in inverted orientation. It has been suggested that the palindromic nature of distal Yq is functionally important because it allows gene conversion to maintain the integrity of critical genes.

However, this suggestion ignores the fact that gene conversion does not require a palindromic structure; it also occurs between direct repeats. In addition, gene conversion is blind to the ancestral state of a mutation, so would not be biased toward maintaining integrity. It is more likely that the Yq palindromes owe their predominance to either a mutational bias in the duplication mechanism toward inverted duplications, or to a selective bias toward repeats in inverted orientation because these are more stable structures.

It has been observed that gene conversion between duplicated sequences is capable of altering patterns of sequence divergence between orthologous sequences. In the case of the Yq palindromes, gene conversion seems to reduce sequence divergence beneath that observed at single-copy Y-chromosomal sequences. The reverse is true at the *AZFa*-HERVs, where gene conversion elevates sequence divergence between humans, chimpanzees, and gorillas.

More theoretical modeling and data from elsewhere in the genome are required to explain these opposing outcomes. Sequence similarity between duplicated sequences potentially influences the rate of NAHR between them. Gene conversion generates variation in the degree of sequence similarity between duplicated sequences.

Consequently, it will be interesting to see whether past gene conversion events are capable of *potentiating* future *rearrangement* events. Such studies will require rates of NAHR-promoted *rearrangement* to be accurately ascertained in *individuals* with varying levels of *duplicate* sequence *similarity*.

Advancement

As befits the emergence of a new *discipline*, there are many questions that remain *unanswered*. Why do some pairs of repeats promote NAHR much more readily than others? What underlies the existence of rearrangement hotspots? What *phenotypic* consequences result from the altered copy number of which genes? Some of these questions can be answered using existing methods, for example the functional characterization of candidate genes in model organisms.

However, many of the questions relating to duplicated sequences in the human genome require the development of a whole new set of techniques. Among these desired methods are those that will allow us to rapidly characterize megabasesized chromosomal structures, to efficiently capture long paralog-specific sequences, to type inversions without resorting to fluorescent *in situ* hybridization and to quantify rates of NAHR in different individuals.

The haplotyping of single molecules of DNA derived from sperm and other tissues are likely to be central to the development of many of these methods.

CONCLUDING REMARKS

The abundance and chromosome-wide distribution of segmental duplications on the Y allows an enormous variety of further duplications, deletions, and rearrangements to take place by NAHR. The nonessential nature of the chromosome (it is absent from females) and its low gene content mean that many of these rearrangements are compatible with life and may persist in the population as polymorphisms.

A subset of rearrangements removes genes required for spermatogenesis and therefore leads to the recognizable genomic disorders of azoospermia. The finished sequence of the *chromosome* has allowed the molecular basis of several AZF deletions to be *elucidated* and led to the discovery of additional structural variation.

It high-lights the need for structural information from diverse Y chromosomes and further studies of both the mechanisms that generate rearrangements and the *phenotypes* associated with each structure.

Chapter 16

Parasitic Fragmented Chromosome

Chromosome *fragmentation* (CF) constitutes one means of manipulating eukaryotic genomes and provides a powerful tool for examining the *structure* and function of *chromosomes*. In particular, it is an essential step toward constructing artificial chromosomes through a better understanding of chromosome functional elements such as *centromeres*, origins of *replication*, and *telomeres*.

During the past 15 yr, CF has been widely used in yeast and mammals to *elucidate* the functional elements required for normal chromosome maintenance. It has also proved a useful tool for refining *physical mapping* of large genomes (in particular human) through yeast artificial chromosome (YAC) fragmentation and the *creation* of deletion derivatives of large DNA *fragments* cloned in YACs.

The different approaches used in CF fall into three broad categories: (a) telomere-mediated chromosome fragmentation, in which *telomeres* are integrated at different positions along the length of the chromosome, resulting in breakage and production of a truncated chromosome; (b) large (several hundred kilobases) internal targeted chromosomal deletions; and (c) induction of gross recombination events leading to *de novo* minichromosome formation. The construction of sets of deletion derivatives obviously is very useful for such studies.

These approaches are based on the use of transfection as a means of introducing CF vectors that target a specific site within the genome. Therefore, the construction of appropriate CF vectors as well as a successful *transformation* system are essential steps toward *mastering* this *technology*.

After the fragmentation experiment, three main steps are necessary to obtain interpretable results: (a) *determining* the nature of the chromosomal *rearrangements* and *karyotype* changes induced; (b) precisely analyzing the recombination events that occurred, and (c) examining the behavior of the truncated or artificial chromosome in cells cultivated in vitro.

CF has only been used recently in parasites. Early transfection experiments in *Trvpanosorna brucei* using small plasmidic constructs produced more or less unstable minichromosomes that integrated into unidentified genomic sequences, but this was not true CF. Basically, only those parasites in which homologous recombination after transfection is highly efficient and for which genome knowledge is well advanced, e.g., *Leishmania, Trvpanosorna,* and *Plasmodium*, have benefited from CF developments Different model chromosomes have been used, addressing the identification of potential centromeric sequences, putative replication origins, and telomere elongation.

This chapter describes basic methods for CF experiments and analysis in *Leishmania.* Current limitations of this *methodology* are the lack of knowledge of the nature of centromeres and *autonomously replicating* sequences (ARS) in this and other *protozoa,* the poor understanding of recombination mechanisms, and the fact that the deletion of unknown genes essential to parasite *survival* may interfere with recombination events and chromosomal rearrangements when selecting the *transfectants.*

The latter factor, combined with the now well-documented genome plasticity in this *protozoon,* may be responsible for the *numerous unexpected* and *aberrant* events obtained when attempting CF.

REQUIREMENTS

PCR Method

1. lox polymerise chain reaction (PCR) buffer.
2. 25 m*M* $MgCl_2$.
3. 5 mM dNTPs.
4. Goldstar *Taq* DNA *polymerase* (*Eurogentec*).
5. Standard primers: T7 promoter and SP6 (MWG Biotech).
6. 0.5-mL *microfuge tubes.*
7. *Thermocycler.*
8. Mineral oil for PCR.

Role of Restriction Enzymes

We recommend using *highquality* restriction enzymes (e.g., New England Biolabs).

1. Restriction enzymes *Kpn*I, *Xba*I, *Hpa*I, and *Mfe*I and corresponding 10X buffers. 2. Homing endonuclease *Eco57*I and buffer (10X buffer + 0.5 mM S-adenosyl-homo-cysteine) (MBI Fermentas or Eurogentec).
3. T4 DNA ligase (400 U/mL) and 10X buffer (New England Biolabs).

Electrophoresis and Electroclution

1. High-quality agarose (Seakem GTG) for preparative gels (DNA purification).
2. Agarose (Seakern LE) for analytical gels.
3. 0.5X TBE buffer (stock solution 20X: 1.8 M Tris-borate, 0.04 M EDTA).

4. Ethidium bromide, 0.1 mg/mL (stock solution 10 mg/mL).
5. Standard double-chamber electroelution block (e.g., CBS Scientific).
6. Dialysis membrane.

DNA Extraction and Precipitation

1. Phenol.
2. Chloroform/iso-amyl alcohol (24:1 v:v).
3. Ethanol, 100% and 70%.
4. 2.5 *M* ammonium acetate (stock solution *7.5 M)*.
5. Vacuum apparatus.

Minipreparation of DNA

1. Solution I: 50 mM glucose, 25 mM Tris-HCl, pH 8.0, 10 m*M* EDTA, pH 8.0. 50-µg/mL RNase A. Store at +4°C.
2. Solution II: 0.2 *N* NaOH freshly diluted from a 10 N stock, 1 % sodium dodecyl sulfate (SDS); prepare immediately before use.
3. Solution 111: 5 *M* potassium acetate (60 mL), glacial acetic acid (11.5 mL), distilled H_2O (28.5 mL). Store at +4°C.

Maxipreparation of DNA

1. Affinity columns: e.g., Qiagen Plasmid Maxi Kit (Qiagen).

Culture of Bacterial Cells

1. Luria Bertani (LB) medium: 10-g/L peptone 140, 5-g/L yeast extract, 10-g/L NaCl.
2. LB-agar: LB + 15-g/L agar.
3. Ampicillin (200-mg/mL stock solution).
4. JM109-competent *Escherichia coli* cells (Promega).

DNA Sequencing

1. Licor DNA sequencer Long Readir 4200 (Amersham) apparatus.
2. Sequence gels (66 cm): Long Ranger gel solution (BioWhittaker Biomol. Appl.) (4.6 mL), Long Ranger TBE 10X (5 mL), H_2O(32 mL), dimethyl sulfoxide (DMSO) (50 µL), Urea (21 g), TEMED (50 µL), 10% ammonium persulfate (freshly prepared) (350 µL).
3. Labeled primers: T7 IRD700 and Sp6 IRD800 (MWG Biotech).
4. Sequencing kit: Sequitherm EXCEL™ II DNA Sequencing Kit-LC (Amersham).
5. Sequence analysis software: Sequencher (Amersham).

Laboratory Production of Parasites

1. "Complete" RPMI 1640 medium (all reagents are from Invitrogen Life Tech., and the concentrations given here are final concentrations): RPMI 1640, 20% heat-inactivated fetal

calf serum (FCS), 1 % v/v nonessential MEM amino acids (100X), 1% v/v MEM amino acids (50X), 1. mM sodium pyruvate (IOOX), 2-mg/mL dextrose, 25 mM HEPES, pH 7.4, in bidistilled H_2O, 2 MM L-glutamine, 100-U/mL penicillin, 100-μg/mL streptomycin.

2. Selection antibiotics (Euromedex): stock solutions are 50 mg/mL for hygromycin B, puromycin dihydrochloride, and geneticin disulfate, and 20 mg/mL for phleomycin. Aliquot and store at -20°C.
3. Sterile glass bottles.
4. Pipetting aid and sterile pipets.
5. 25-cm^2 and 75-cm^2 sterile culture flasks.
6. 96-well and 24-well microplates.
7. Sterile-flow cabinet.
8. 26°C incubator with a 5% CO_2 concentration and a humidified atmosphere.

Transfection

1. Electroporation buffer: 21 mM HEPES, pH 7.5, 137 mM NaCl, 5 mM KCl, 0.7 mM Na_2HPO_4, 6 mM glucose. Filter-sterilize and keep at +4°C.
2. Electroporator Easyject Plus (Equibio).
3. Electroporation cuvets (0.4-cm gap).
4. Plasmid DNA resuspended in 50 μL sterile H_2O at a concentration of 80-200 μg/mL.
5. Sterile 50-mL Falcon conical centrifuge tube.
6. Hemocytometer (Thoma cell).
7. Phosphate-buffered saline (PBS), pH 7.2.
8. l.5-mL microfuge tubes.
9. Micropipet, sterile tips.
10. Ice.
11. SeaPlaque agarose GTG.
12. 10-cm-diameter Petri dishes.
13. 15-mL Falcon polypropylene round-bottomed centrifuge tubes.

Methods of Transfectants

1. 15-mL conical Falcon centrifuge tubes.
2. Low-melting-temperature agarose (SeaPlaque GTG).
3. PBS.
4. Plug molds (Bio-Rad).
5. Lysis buffer: 0.5 MEDTA, pH 8.0, 1% N-lauroyl-sarcosine. Add 1 mg/mL of proteinase K just before use.

6. 50°C water bath.
7. Complete pulse-field gel electrophoresis (PFGE) apparatus.
8. 1X TE buffer: 10 mM Tris-HCl, pH 8.0. 1 mM EDTA.
9. 0.5X TBE buffer: 45 mM Tris-HCl, pH 8.3, 45 mM boric acid, 1 mM EDTA.
10. Agarose (Seakem GTG).
11. HCl, NaCl, NaOH.
12. Nylon-membranes: Hybond N+ (Amersham).
13. Chromatography paper: Whatman 3MM.
14. Random hexanucleotides pd(N)6.
15. Deoxynucleotides triphosphate dATP, dGTP, dTTP.
16. Radiolabeled a*^{32}P-dCTP.
17. DNA polymerase 1 Klenow fragment.
18. 37°C water bath.
19. 20X SSPE: 3 mM NaCl, 200 mM NaH_2PO_4, 20 mM EDTA, pH 7.5.
20. SDS.
21. 65°C shaking water bath.
22. Hyperfilm (Amersham).
23. Autoradiography cassettes and amplifying screens.
24. Restriction enzymes (NEB) and corresponding buffers.
25. PCR reagents.

PROCEDURE

Chromosome Fragmentation Vector

Here, we describe the utility of a family of versatile vectors termed pVV (GenBank Accession no. AF315645) that has been successfully used for internal CF.

They should also be useful for telomere-mediated CF. Indeed, the telomere mediated CF vector can simply be viewed as a deletion vector in which one of the two sequences *homologous* to the target genome consists of *telomeric repeats*. A telomere-mediated CF vector based on telomeric and *subtelomeric* sequences of *Trypanosoma brucei* has been shown to be functional in *Leishmania.*

Vector pVV was created from the pGEM3Zf *bacterial plasmid* (Promega). Three essential features were introduced that allow (a) interchange of selection marker cassettes, for *sequential targeting*; (b) *modification* of the nature and orientation of *homologous* sequences; and (c) *linearization* by *restriction* within the sequences homologous to the targeted genome.

Four different selectable marker genes can be used, conferring resistance to hygromycin *B*

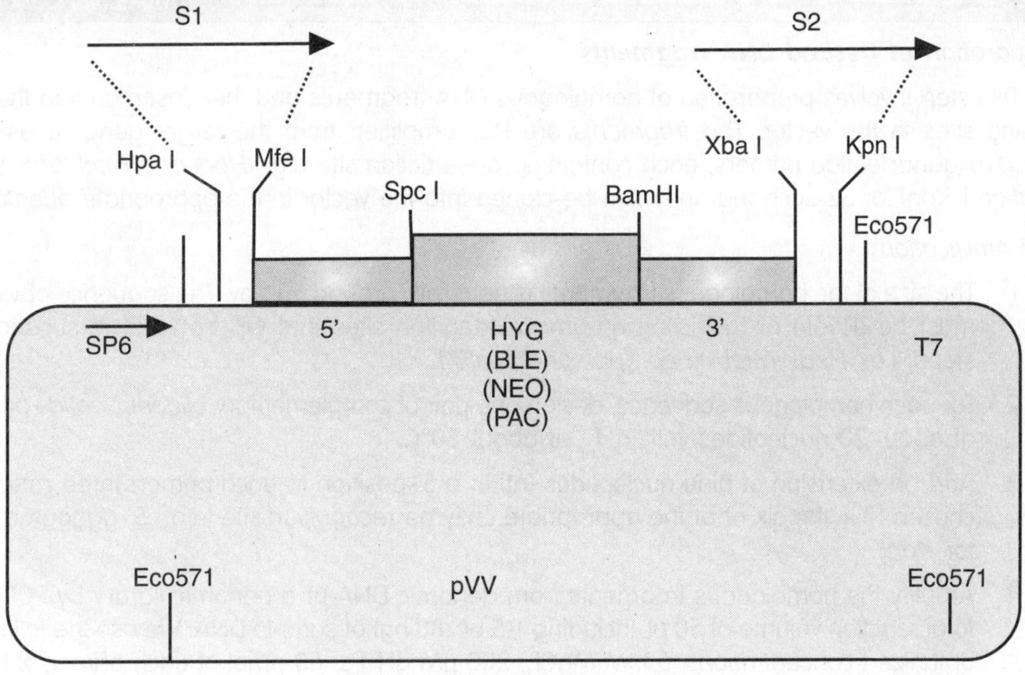

Figure 16.1: Schematic representation of the "versatile" transfection vector pVV. *See text for description. S 1 and S2 = sequences homologous to the targeted chromosomal DNA.* HYG, BLE, NEO, *and* PAC *= resistance marker gene cassette (see text for details). 5' and* 3' *= parts of the 5' and* 3' *Leishmania major* DHFR-TS *gene flanking regions. SP6 = SP6 RNA polymerase promoter, T7 = T7 RNA polymerase promoter.*

(HYG gene), the aminoglycoside G418 *(NEO),* puromycin *(PAC),* or phleomyein *(BLE).* These *resistance* gene cassettes can be readily *interchanged* by *SpelIBarnHI* double digestion, as shown in Figure 16.1. 5'- and 3'-DNA sequences (953 and 1738 bp, respectively) have been inserted on both sides of the selectable marker to drive the *expression* of the *drug-resistance* gene.

They derive from the flanking regions of the *Leishmania major* dihydrofolate *reductase-thymidilate synthase (DHFR-TS)* gene (GenBank Accession no. U59231). Each *cassette* (5'-region-drug-resistance marker gene- 3'-region) is flanked by two different cloning sites, one containing the restriction sites *HpaI* and *MfeI,* and the other *XbaI* and *KpnI.*

These enzymes are rare-cutters in the GC-rich *Leishmania* genome and offer a wide potential for choice of DNA sequences homologous to the targeted locus. When using *telomere-mediated* CF, one of the homologous sequences consists of telomeric repeats, whereas the other one defines the targeted locus on the chromosome to be deleted.

For example, the vector used by Tamar and Papadopoulou included approx 1.5 kb of subtelomeric repeats from *T. brucei* adjacent to the telomeric repeats. Finally, a restriction site for the homing endonuclease Eco571 (CTGAAGN$_{16}$/GACTTCN$_{14}$) has been introduced adjacent to each *cloning* site.

Thus, the vector can be linearized by cutting within the homologous sequences, providing free ends for invasive recombination into the target genome.

Preparation of Desired DNA Fragments

This step involves preparation of homologous DNA fragments and their insertion into the two cloning sites in the vector. The *fragments* are PCR-amplified from the target genome using a pair of oligonucleotide primers, each containing a restriction site-e.g., *HpaI* and *MfeI* for S 1 and *XbaI* and *KpnI* for S2-such that they can be cloned into the vector in the appropriate *orientation.*

PCR Amlification

1. The size of the homologous fragments is generally around 500 by. The sequence obviously must be devoid of restriction enzyme recognition sites that will be used in subsequent steps, i.e., *HpaI, MfeI, XbaI, KpnI,* and *Eco57I.*
2. For each homologous sequence, design one pair of complementary *oligonucleotide* primers of about 20 nucleotides with a T_m at about 60°C.
3. Add an extension of nine nucleotides (nt) in a 5'-position to each primer: three *randomly* chosen nt + the six nt of the appropriate enzyme *recognition* site (e.g., 5'-gggggtacc...-3' for *KpnI*).
4. Amplify the homologous fragments from genomic DNA or a genomic library by PCR, in a total reaction volume of 50 µL including 1-5 pL (80 ng) of sample DNA. We use the following optimized *concentrations*: 5 tnM $MgCl_2$, 200 µM dNTPs, 50 pmol of each primer, *2 U Taq* DNA polymerase.

   We use the following cycling conditions: 95°C for 2 min 30 s, 62°C for I min, 30 cycles of 72°C for 1 min then 94°C for 30 s, 62°C for 30 s, and 72°C for 10 min. Perform two to three replicates of each reaction to have *sufficient* material for the following steps.

Purification

1. Pool the contents of the two or three replicate PCR reactions mixtures and load all into a large slot (made by sealing comb teeth if necessary) of a 1 % agarose minigel (10 × 10 cm) (Seakem GTG agarose in 0.5X TBE buffer + 0.1-mg/L *ethidium bromide*). *Electrophorese* for 1 h at 10 V/cm in 0.5X TBE buffer.
2. Visualize the band of interest (at the expected size) under UV light. Using a sharp *scalpel blade,* cut out a slice of *agarose containing* the band of interest.
3. Place a double-chamber *electroelution* block in an electrophoresis tank. Cover the bottom of each chamber with a *dialysis membrane,* previously boiled for 10 min in distilled HO. Transfer the slice of agarose to the *electroelution* block. *Electroelute* at 6 V/cm (120 V) for 1 h in 0.5X TBE buffer.
4. Recover the DNA in 200 µL of buffer in a 1.5-mL *microfuge* tube; rinse the block with 200 µL of 0.5X TBE.
5. Extract the DNA using a standard *phenol-chloroform* extraction method (1 vol phenol + 1 vol chloroform). Add *ammonium* acetate to 2.5 M and 2 vol ethanol to precipitate the DNA. Wash the pellet with 70% *ethanol* once. Dry the pellet, then dissolve the DNA in 20 µL of distilled H_2O.
6. Load 1 µL on a 1% *agarose minigel* (0.5X TBE + 0.1-mg/mL ethidium bromide) along with

a DNA size standard of known *concentration* (e.g. 200 ng of PstI-restricted ?,DNA). This allows you to estimate of the DNA concentration for each amplified fragment by comparison of fluorescence intensity.

Hydrolytic Removal of the PCR Products

The PCR-amplified homologous sequences must be subjected to a double digestion (with *HpaI-MfeI* for SI and *KpnI-XbaI* for S2) to be inserted into the vector.

1. Digest all of the PCR-amplified DNA fragments (19 ML) in 200 µL of 1X digestion buffer with 1-2 U of enzyme/pg of DNA. Perform a control *digestion* of 500 ng of plasmid DNA containing the corresponding *restriction* sites (e.g., pBluescript) with one-fifteenth of the 200 µL of restriction digestion solution. *Incubate* both tubes at 37°C for 1 h 30 min.
2. Load 1-5 µL of the contents of the control tube onto a 1% *agarose* minigel (0.5X TBE + 0.1-mg/mL *ethidium bromide*) adjacent to a DNA size standard to verify that the *digestion* is complete.
3. Extract the digested fragment using *phenol-chloroform* as in step 5 of the head purification in this chapter.
4. Dissolve the pellet in 10 µL of distilled H_2O. Load 1 µL on a 1 % agarose *minigel* next to a DNA size standard as in step 6 of the head purification in this chapter.

Insertion of Fragment S2 Into pVV

High Hydrolysis of PVV

It is essential to carry out both *digestions* stepwise and to perform separate control reactions. Partial restriction of the vector DNA must be avoided.

1. Digest 6 µg of pVV DNA with *KpnI* in 200 µL of 1X enzyme buffer + 1–2 U/µg *KpnI* of DNA, for 2 h at 37°C.
2. Load 5 µL on a 1 % agarose minigel to assess whether the digestion is complete.
3. Extract the digested product with phenol-chloroform.
4. Dissolve the DNA pellet in 20 µL of distilled H2O. Load 1 µL on a 1 i agarose minigel to assess DNA concentration.
5. Proceed with the second digestion using *XbaI* as in step 1. Perform a control digestion.
6. Load 5 µL of the control digestion on a 1 % agarose minigel and verify that the digestion is complete.
7. Extract the digested vector with phenol-chloroform.
8. Dissolve the DNA pellet in 10 µL of distilled H_2O Load 1 µL on a 1 % agarose minigel with a DNA size standard (e.g., *X/PstI)* as in Described elsewhere in this chapter.

Ligation of S2 Into PVV

1. Set up a ligation reaction in a microfuge tube containing 1X T4 DNA ligase buffer, 50-75 ng of the double-digested pVV vector, 100-150 ng of the double-digested *homologous*

sequence S2, and 1 µL (400 cohesive end units) of T4 DNA ligase (New England Biolabs) made up to 10 µL with distilled H_2O.

2. Set up an *additional* control reaction containing the same elements except S2.
3. Incubate both tubes overnight at 15-16°C, then store at +4°C.

Transformation

We generally transform competent *E. coli* by the "heat-shock" method.

1. Place both ligation reaction tubes on ice.
2. Remove an aliquot of preprepared competent cells (home-made or JM I 09 Promega) from the -80°C freezer and place immediately on ice for 5-10 min.
3. Using a sterile pipet tip, quickly transfer 50 µL of cells into both ligation tubes on ice. Gently *homogenize* tube contents by stirring with pipet tip (but without *pipetting*).
4. Leave on ice for 30 min. then transfer the tubes to a *circulating* water bath pre-heated at 42°C for exactly 1 min.
5. Immediately transfer the tubes to ice and allow the cells to chill for 2 min.
6. Remove the tubes from ice. Using a *sterile* pipet tip, add 450 µL of LB medium without *antibiotic* to each tube. Gently mix by inverting tubes four or five times, then *incubate* at 37°C for 45-60 min. Invert the tubes every 15 min to mix.
7. From each tube, transfer 50, 100, and 200 µL of the transformed cells onto three 90-100-mm plates containing LB-agar + 70-µg/mL ampicillin. *Gently* spread over the surface of the agar plate. Leave plates to dry for 10-15 min, then invert and *incubate* overnight (16 h) at 37°C.

Marking The Recombibant Clones

To screen for recombinant clones, we perform both a PCR and the isolation of the clone by transfer of a single *colony* to *culture* medium (LB + *antibiotic*).

1. Estimate the number of recombinant clones by comparing the number of *colonies* present on the six plates derived from the control and test ligations.
2. Using a sterile pipet tip, gently remove a well-separated colony from an agar plate. Transfer some of the colony from the tip onto the bottom of an empty PCR tube. Immediately transfer the same pipet tip into a well of a 96-well *microplate* containing 200 µL of LB medium + 70-µg/mL ampicillin. (We usually screen 10 clones in this manner.)
3. Process one or two *clones* from the control plate in the same way, to be used as positive *controls* for the PCR.
4. Incubate the microplate overnight at 37°C.
5. Amplify by PCR to verify insertion of the desired fragment. For this, prepare a PCR mix (1X buffer, 5 mM $MgCl_,$, 200 pM dNTPs, 50 pmol of each primer, 1 U Goldstar Taq DNA polymerase, Eurogentec). Transfer 50 µL of the mix into each of the PCR tubes containing

the colony smears Cover the mix with mineral oil. Cycle as follows: 95°C for 2 min 30 s, 62°C for 1 min; 30 cycles at 72°C for 1 min 30 s, 94°C for 30 s, and 62°C for 30 s; finally, 72°C for 10 min.

6. Load 8 μL of each PCR product on a 1 % agarose (SeakemGTG) gel (0.5X TBE + 0.1-mg/mL *ethidium bromide*) adjacent to a DNA size standard (e.g., λ/*PstI*).

   Electrophorese at 7 V/cm for 2 h and *visualize* under UV. The size of the amplicons from nonrecombinant clones is 200 bp, whereas that from *recombinant* clones is 200 by + the size of the *inserted* sequence.

DNA Minipreparation of DNA

The integrity of the cloned *homologous* fragment in the pVV vector must be verified by sequencing. For this, the DNA from recombinant clones is purified by a *minipreparation* method. The protocol given below is adapted from a standard *protocol* but yields the high-quality DNA *necessary* for sequencing reactions.

1. Select two to three recombinant clones. Inoculate 3 mL of LB medium + *ampicillin* (70 μg/mL) with 20 μL of culture from each selected clone in loosely capped 15-mL *polypropylene* tubes. Incubate overnight at 37°C with *vigorous* shaking.
2. Centrifuge at 1800g for 20 min at room temperature, then discard the *supernatant* and leave the tubes inverted on a paper towel. Place the tubes at -80°C for 5 min.
3. Remove the tubes and thaw for 2-3 min at room temperature.
4. Add 200 μL of solution I to each tube. Vortex briefly and gently until the pellet is completely redissolved.
5. Transfer the 200 μL, to a 1.5-mL microfuge tube. Add 300 μL of lysis solution II. Mix the contents by inverting the tubes five to six times, and leave 5 min at room temperature.
6. Add 250 μL of ice-cold *neutralization* solution III. Invert the tubes five to six times, and leave 20 min on ice.
7. Centrifuge at 12,000g for 30 min at +4°C, then transfer 600 μL of the supernatant to another *microfuge* tube and centrifuge again at 12,000g at +4°C for 20 min.
8. Transfer 500 μL of the *supernatant* to another *microfuge* tube and precipitate with 2.5 times the volume of *ethanol* at -20°C for 20 min.
9. Centrifuge at 12,000g at +4°C for 20 min. Discard the ethanol, wash once with 70% ethanol and *resuspend* the *pellets* in 20 μL of *distilled* H_2O. Store at -20°C.

The DNA prepared in this manner should be sequenced. This step is important, to ensure that the selected clones contain the desired DNA fragment before *proceeding* to the second stage of cloning.

Use the primer T7 IRD800 (MWG, Biotech), located downstream of the *KpnI* site on pVV, to determine the *nucleotide* sequence of the cloned S2.

Cloned S2-Containing Vector DNA

1. Select one of the recombinant clones and perform six separate *minipreps* of this clone.

2. Pool the six minipreps (6 × 20 pL) in a *microfuge* tube and add 280 μL of distilled H_2O.
3. Purify the DNA with phenol-chloroform and ethanol precipitation as in Described elsewhere in this chapter
4. Resuspend the DNA pellet in 20 μL of distilled H_20. Visualize 1 μL on a gel described already in this chapter.

Insertion of the Sequence S 1 Into pVV

Here, the purified DNA fragment S 1 is introduced into the *HpaI* and *MfeI* sites of pVV-S2. Follow the same protocol as used to clone fragment S2, except for a few technical *variations* at steps 4 and 6:

Step 4. Marking of The Recombibabt Clones

Use S 1-specific PCR primers: standard M 13-reverse (5-aacagctatgaccatg-3'), located on pVV upstream of the *HpaI* site, and another (5'-gtgctactacttagct-3') *homologous* to the *5'-DHFRTS* region.

Step 5. Sequencing

Use the labeled primer is Sp6 IRD800 (MWG, Biotech), located upstream of the *HpaI* site.

Maxipreparation of the CF Vector DNA

Having selected the appropriate clone to be used as the complete vector pVV-S2-S 1, you should now perform a maxi preparation of this clone using a commercial kit (we recommend affinity columns such as Maxiprep, QiaGen).

1. Inoculate 3 mL of LB medium containing 70-μg/mL *ampicillin* with 10 μL of the clone culture that was retained in a 96-well microplate. Incubate *overnight* (>8 h) at 37°C with *vigorous* shaking.
2. Inoculate 400 rnL of LB medium + 70-μg/m, ampicillin with 2 ml, of the culture. *Incubate* overnight at 37°C with vigorous shaking.
3. Extract the vector DNA using a commercial kit according to the manufacturer's instructions.
4. Redissolve the DNA in 150 μL of distilled H_2O. Estimate the DNA *concentration* on an agarose gel described already in this chapter.

Extraction of the Transfection Vector pVV-S2-S 1

Digestion With Eco57I

1. In a 2-mL microfuge tube, digest 60 pg of vector DNA with *Eco57I* (0.4 U/pg of DNA) at 37°C overnight in a final volume of 500 μL.
2. Verify that the digestion is complete by loading 1 μL of the digest on a 1% agarose minigel.
3. Precipitate the DNA in 2.5 M ammonium acetate and 2 vol of ethanol at -20°C for 20 min. Spin at 12,000g at +4°C for 30 min. Wash the pellet once with ethanol 70% and dry.
4. Resuspend the DNA precipitate in 400 μL of distilled H_2O.

Extraction of The Vector DNA

1. Cast a 0.7% agarose (Seakem GTG) gel (0.5X TBE + 0.1-mg/mL ethidium bromide) containing a single 15-cm-long well. Load the whole digest (400 µL) into the well and migrate at 3 V/cm overnight.
2. Visualize the band of interest (at 4.5-5.5 kb) under UV. Using a sharp scalpel blade, cut out the 15-cm-long slice of agarose containing the band of interest. Place it on another gel plate and directly cast a 0.7% agarose gel (as above) around the slice. Electrophorese at 3 V/cm overnight, followed by 5 V/cm for 4-5 *h*.
3. Again, using a sharp scalpel blade, cut out the slice of agarose containing the band of interest (at 4.55-5 kb) and chop up into eight pieces.
4. Electroelute the DNA from the eight agarose pieces in four electroelution blocks at 5 V/cm for 4 h.
5. Recover the DNA in 4X 200 µL in a 2-mL Eppendorf tube; rinse the four blocks with the same 200 µL of 0.5X TBE, and pool the 1000 µL.
6. Purify the DNA with phenol-chloroform and precipitate with *ethanol* as under Described elsewhere in this chapter.
7. Resuspend the precipitate in 50 µL of sterile distilled H_2O. Check the amount and quality of the DNA by loading 1 µL on a 1 % agarose gel. The DNA yield is generally between 6 and 10 pg. This amount of *linearized* vector DNA is necessary for *transfection*. The DNA is now ready for use.

Transfection of Leishmania Cells

The following protocol has been adapted for transfection of *L. major* and *L. donovani*.We use RPMI 1640 medium enriched with 20% FCS and other *complements* to favor *efficient* growth.

1. Cultivate the cells in complete medium at 26°C with 5% CO_2 in a humidified atmosphere. When the selection drug is required (after *transfection* of the *parasites*), add it to the medium just before *inoculation* of the cells.
2. Examine the cells under a *microscope* on a regular basis. Passage the cells *triweekly* with a one-tenth dilution into 25-cm^2 sterile plastic tissue culture flasks (6 mL/flask), just before they reach late log phase (approx 1.5-2 × 10^7 cells/mL).

Observing Sensitivity

It is necessary to determine the *minimal concentration* of selection drug at which no survivors are observed in the wild-type population. This test should be performed as a two-step process in liquid culture medium and in *agarose* plates, because *transfected* cells will be *cultivated* on both and the effective concentrations are not the same.

1. Test the selection antibiotics in liquid medium at increasing concentrations. generally a *twofold* dilution series from 1 to 100 µg/mL for hygromycin B, 0.5 to 50 µg/mL for *phleomycin*, 10 to 150 µg/mL for *puromycin*, and 0.5 to 100 µg/mL for *geneticin*. Include a drug-free control. For each *concentration*, inoculate in duplicate 3 mL of medium at 1-2 × 10^5 *Leishmania* cells per mL in loosely capped 15-mL *polypropylene* tubes.

2. Count the cells in a *haemocytometer* daily for 1 wk. The lowest *concentration* at which no *survivors* are observed is chosen for the subsequent selection of *transfectants* in liquid culture.
3. Estimate the approximate EC_{50}. Then test four drug concentrations (from the EC_{50} to several times the EC_{100}) in agarose plates. For this, pour RPMI 1640-agarose plates at the selected drug concentrations, already described in this chapter. Choose the minimal drug concentration yielding no colonies for subsequent selection on agarose plates

Transfection

For highest transfection efficiency, it is essential to use particularly "*healthy*" parasites, cultivated in optimal conditions.

Electroporation

All steps below must be performed in *sterile* conditions and as quickly as possible.

1. Two days before transfection (d "-2"), inoculate an adequate volume of medium with promastigotes at 10^6/mL. For example, if three *electroporations* (including a control) are *foreseen, inoculate* 50 mL.
2. On d 0, estimate the number of live cells by counting with a hemocytometer.
3. Place the *electroporation* cuvets (0.4-cm gap) containing 50 µL of vector DNA (6-10 µg of purified *linearized* vector pVV-S2-S1 DNA) on ice.
4. Spin down *Leishmania* cells in a sterile 50-ml.- Falcon tube at 900q for 10 min at room temperature.
5. Remove the supernatant and resuspend pellet in 10 mL of *electroporation* buffer by gently *pipetting* up and down.
6. Centrifuge as above and carefully remove the *supernatant*, without *disturbing* pellet, and *resuspend* in 10 mL EP buffer.
7. Repeat centrifugation as above, then gently remove the supernatant and resuspend the cells at a concentration of 10^8/mL in EP buffer.
8. Incubate the cells on ice for 10 min.
9. Set up the electroporator (Easyject plus"', Equibio). We routinely use the following conditions: one pulse with 2 kV/cm, 2310 S2, and 25 pFarrads. With these settings, we get a 57.7-ms pulse time.
10. In each electroporation cuvet, mix 800 µL of the *Leishmania* cell suspension (8×10^7 cells) with the 50 pL of ice-cold vector DNA (= test sample [TS]).
11. Prepare a "mock cuvet" containing the same amount of cells but no vector DNA (= control without DNA [CWD]). This will serve as a control for the *mortality* of *nontransfected* cells in the presence of the selection drug.
12. Carefully dry the outside walls of the cuvet before *placing* it in the chamber. Zap (subject to electric field) *immediately*, then place the cuvets on ice for 10 min.

13. Transfer the cells from each cuvet into a 50-mL tube containing 10 mL of drug-free complete medium.
14. Also process a control *electroporation* with 800 pL of nonelectroporated cells (NEC). 15. Incubate all tubes for 24 h in the 26°C CO_2 incubator until you commence drug selection on d 1.

Obtaining Selective Plates

Two methods may be used to isolate the transfectant cells: either in liquid culture or on RPMI medium-agarose plates. The latter obviously has the advantage of *isolating* clones in one step, whereas the former requires subsequent cloning in liquid medium. We describe the second method below. All steps must be performed in sterile conditions.

1. Melt and sterilize 6% (10X) LMT agarose (SeaPlaque GTG) in double-distilled H_2O by *autoclaving* (120°C for 20 min). Place the flask in 65°C water bath until use.
2. Prepare complete medium and keep at 37°C.
3. In a sterile-flow cabinet, pour medium into the 10X agarose (9:1 v/v) and mix gently while maintaining the flask at 42°C until completely *homogenized.* (You need about 25 mL/ plate).
4. Immediately pipet 25 mL onto one or two plates for use as drug-free controls.
5. Add the selective drug at appropriate concentration to the agarose-medium mix. Rapidly *homogenize* and pipet 25 mL of the mix onto each plate. Carefully avoid air bubbles. Do two to three plates for each *electroporation* cuvet (TS and CWD).
6. Leave the plates uncovered in the hood for 1 h 30 min to solidify.

Plating Techniques

1. Vigorously shake the cultured *Leishmania* cells to break down cell aggregates. Count the cells in the control tubes NEC and CWD to assess cell viability after *electroporation.*
2. For each sample (TS and CWD), transfer 50 µL of cells to 1 mL of drug-free medium in a 24-well microplate to test viability of electroporated cells in liquid medium (= viability control [VC]).
3. For each sample, transfer I mL of cells to a 25-cm² culture flask containing 8 mL of complete medium with the selection drug at the appropriate concentration. We routinely use 30-pg/mL *hygromycin* B for *L. major* and 50-µg/mL hygromycin B for *L. donovani,* 8-µg/mL geneticin for *L. major* and 50-µg/ mL geneticin for *L. donovani,* 10-µg/mL *phleomycin* and 100-µg/mL puromycin. Incubate in the 26°C CO_2 incubator with the cap loose. This selective liquid culture (SLC) is maintained as a *safeguard* in case plating efficiency is too low.
4. Centrifuge the remaining 9 mL at 900g for 10 min at +4°C, then gently remove the *supernatant* with a pipet, carefully avoiding the cell pellet.
5. Resuspend the pellet in 600 µL of sterile medium by flicking the tube.
6. Transfer 10 µL of this suspension to 5 mL, of medium. mix, and transfer 100 µl, to 1 mL of medium. Plate 200 µL of this dilution on a nonselective plate to evaluate plating efficiency.

7. Divide the remaining 600 μL of TS and CWD samples between three plates each by depositing 200 μL of cells onto the center of the plate and *spreading* them by gently tilting the plate. Leave uncovered for a few minutes, then place the closed plates in the 26°C CO-, *incubator*. It is not necessary to wrap the plates in plastic film or *Parafilm*''

Separation of Transefectants

1. On the drug-free plates, colonies begin to emerge after 4-7 d. Count these to assess plating efficiency.
2. On the CWD plate, many *aggregates* containing live cells are visible during the first 4 d but then disappear progressively. On d 10, no *survivors* should be observed.
3. On the selective plates, colonies start to emerge after 7-14 d. They should be transferred to liquid culture as soon as possible. Identify reasonably sized and well-separated colonies. Using a sterile pipet tip, transfer each selected colony to 200 μL of drug-containing complete medium in a 96-well *microplate*. Rinse the tip by pipetting up and down. Change tip for each colony. Maintain *sterility* at all times.
4. Wrap the plates with Parafilm and store at room temperature: growth will slow down, and the plates may prove useful at a later stage.
5. After 1-2 d, transfer the contents of each well to separate wells of a 24-well *microplate* containing 1 mL of drug-containing complete medium.
6. After 1-2 d more, transfer each well to 2 mL of selective medium in a 25-cm^2 flask with the cap loose. Scale up the cultures by adding 3 mL of *selective* medium 1-2 d later and again 4 mL after 1-2 d more.
7. When cells reach log phase, remove 9 mL of culture for extraction of DNA. Add fresh selective medium to the *flasks*, and maintain the *cultures* until the DNA has been analyzed.

Cultural Technique

These flasks are used only to assess the success of different stages of the process. Check cell viability in the VC drug-free cultures for a few days after *electroporation* and discard when they reach *stationary* phase. In the SLC cultures, the cell density should decrease considerably after 3-7 d. In the control CWD, no *survivors* should be observed after 10 d, showing that the drug *concentration* was sufficient to kill all *nontransfectant* cells.

In the TS preparations, a low proportion of cells survive and *multiply*. This culture may be discarded when *numerous* colonies have been isolated from *selective* plates.

Basis of Transfectants

Transfectants with the desired CF event are identified by analyzing the molecular *karyotype* by PFGE. This *necessitates* preparation of *chromosomal* DNA. To screen a high number of clones *simultaneously*, we harvest the cells from a small volume of cell *suspension*. This is usually performed 3 wk after the electroporation.

Preparation of Chromosomal DNA

1. Prepare 2% LMT agarose (SeaPlaque GTG) in PBS and keep in water bath at 12°C.

2. Remove 9 mL of cell suspension containing 1.2-1.6 × 10^8 parasites in late log phase. This will be enough to prepare two DNA plugs.
3. Spin down the cells at 900g at +4°C for 10 min and *resuspend* the cell pellet in PBS buffer. Repeat this *process* twice to wash the pellet.
4. Prior to the final resuspension, count the cells using a *hemocytometer*. Resuspend the cells in PBS at a *concentration* of 1.6 × 10^9/mL (in about 0.1 mL).
5. Add an equal volume of LMT *agarose* kept at 42°C. Mix quickly and pour immediately into the plug molds (100 µL per plug). Leave to solidify 15 min at +4°C.
6. Remove the *agarose* plugs from the molds using a fine *spatula*, and drop the plugs into labeled microfuge tubes containing 1 mL of *lysis buffer* with *proteinase* K (1 mg/mL).
7. Incubate overnight at 50°C, then store at +4°C until required.

Gel Electrophoresis

1. Wash the agarose plugs 4 times in 1X TE buffer for 15 min with gentle shaking.
2. Prepare a 1.5% agarose (Seakem GTG) gel in 0.5X TBE, then place the plugs in the wells and the gel in the tank in freshly prepared 0.5X TBE.
3. Electrophorese using appropriate parameters for the chromosome size class to be separated.
4. Stain the gel in 0.5X TBE containing 0.5-µg/mL *ethidium bromide* for 1 h. Photograph the gel under UV light, with a ruler placed on the gel to aid *karyotype* analysis.

Method of Southern Blotting

Depurinate the gel in 0.4 N HCl for 15 min twice and then transfer onto a *nylon membrane* in denaturing conditions following a standard alkaline *Southern blotting meth.*

Hybridization

The verification of the deletion of the targeted region and of its replacement by the drug-resistance gene is done by hybridization to a Southern blot of DNA probes *homologous* to the resistance gene and to chromosomal segments that are within and flanking the targeted *chromosomal locus*. This step allows you to select the clones of interest and discard the rest.

1. Label 30 40 ng of purified DNA probe by random priming labeling and *hybridize* to the DNA on the nylon filter.
2. Wash the filters at high stringency (at 65°C in O.1X SSPE + 0.1% SDS).
3. Nylon blots may be *stripped* and *rehybridized* 5-10 times.

Multiplication of the Transfectants

This method has been described in several publications.

1. Perform cloning of parasites at the beginning of *log-phase* growth, for example, 1-2 d after a one-tenth dilution of the cells.
2. Transfer the cell sample into a sterile 15-mL *conical tube* and allow the cell aggregates to sediment for 1 h.

3. Transfer I mL from the higher phase (which contains the most motile cells) to another tube containing medium without FCS. Homogenize *vigorously* by pipetting up and down to ensure that no more cell *aggregates* remain.
4. Count the cells precisely with a *hemocytorneter* (mean of two countings at two *dilutions*).
5. Prepare tubes with one-tenth serial dilutions of the cells in medium without FCS. The last dilution should be 2.5 cells/mL in 40 mL of complete medium containing the selection drug. Mix by pipetting.
6. Pipet 0.2 mL of the final dilution into the wells of two 96-well microplates. This should result in a mean distribution of 0.5 cells per well. Continually swirl the tube while distributing the cells. *Incubate* in the 26°C CO_2 *incubator.*
7. From d 5, examine the plates using an inverted *microscope.* The clones will grow by forming *rosettes.* Therefore, discount any well *containing* more than one rosette. Using the 0.5 cell/well *concentration,* the cloning efficiency usually is between 50 and 90%, i.e., about 25-45 clones will grow on each of the two microplates.
8. After 2-3 wk, when dense growth is *observed,* transfer each sample thought to contain a single cloned parasite population to 1 mL of *drug-containing* complete medium in a 24-multiwell plate.
9. After 1-2 d, transfer the contents of each well separately to 2 mL of complete medium with selective drug in a 25-cm' flask. After 1-2 d, add a further 3-4 mL of medium. Because the limiting dilution method cannot guarantee the purity of each subclone (particularly with *Leishmania),* it is recommended to perform a second subcloning using the same protocol.

Applications of the Recombination Events

The recombination events that have occurred in each selected clone must he *characterized* and the rearranged chromosomal region should be analyzed further. This is performed by *enzymatic restriction* of total chromosomal DNA, as well as *PCR* and sequencing *of* the junction regions. The correct or, conversely, incorrect insertion of the vector into the chromosomal locus, as well as *aberrant* events, can be detected in this way.

Increase the culture volumes of the selected transfected *Leishmania* clones up to 100 mL. From these cultures, prepare chromosomal DNA. This will yield about 20 *agarose* plugs, which can be kept at +4°C for > 1 yr.

Enzymatic Hydrolysis of Chromosomal DNA

We digest the chromosomal DNA directly in the agarose plugs prepared for PFGE.

1. Place half an agarose plug (representing 3-4 × 10^7 cells. i.e.. 2—3 pg of total 1)N A in a 50-mL polypropylene tube.
2. Wash in 20-40 mL of sterile cold IX TE for 30 h at +4°C, changing the buffer six times during this period.
3. Transfer to a 15-mL round-bottomed *polypropylene* tube. *Equilibrate* in 1 mL of IX digestion buffer for 2 h.

4. Remove the buffer and replace with 200 µL of new buffer or as small a volume as possible (just enough to cover the plug). Add 80 U of enzyme per tube (it is essential to use high-quality *restriction* enzymes) and incubate overnight at the *recommended* temperature.
5. Separate the restriction *fragments* by *electrophoresis*. For fragments <10 kb, load the plugs onto a 0.7-0.9 % agarose gel (Seakem GTG). For fragments >10 kb, it will be necessary to perform PFGE (test PFGE conditions prior to the *experiment* to determine the best conditions for the size of fragments to be resolved). Include appropriate DNA size standards such as *lambda* DNA restricted with *PstI, HindIII,* and/or *EcoRI*.
6. After electrophoresis, proceed to *Southern blotting* and *hybridization* described already in this chapter.

Integration by PCR and Sequencing

This step involves PCR *amplification* of both junctions between the targeted chromosome and the inserted vector, and determination of the *nucleotide* sequence of the amplified *fragments*.

1. For each junction, design a pair of oligonucleotide primers, one homologous to the inserted vector sequence (in the 3'- or *5'-DHFRTS* region, depending on the *junction* tested) and the other to the chromosome sequence immediately upstream or downstream (depending on the junction tested) of the *insertion* locus.
2. Melt an agarose plug containing the DNA of the clone of interest in a water bath at 65°C for 10 min.
3. Extract the DNA with *phenol-chloroform* and *precipitate*.
4. Resuspend the *DNA pellet* in *50* µL of distilled H_2O. Estimate the DNA concentration on a I % agarose gel.
5. PCR-amplify each of the junctions in two separate tubes using 80 ng of template DNA and the parameters described in Described elsewhere in this chapter but with annealing temperatures appropriate to the T_m of the *primers*.
6. Load the PCR products on a 1% agarose gel. After *electrophoresis,* excise and purify both PCR fragments.
7. Clone the PCR products into a TA-cloning type vector (e.g., pGEMT-Easy, *Promega*) according to the manufacturer's instructions.
8. Sequence the products already described in this chapter.

Chromosomal Stability

To determine whether the chromosome fragmentation has affected *mitotic stability* of the truncated chromosome, the parasite clone is cultivated in the *absence* of drug pressure for 100-150 generations. Samples of cells are collected at intervals and the persistence of the *fragmented chromosome* is assessed both by PFGE and drug *susceptibility* testing as follows:

1. At various time *intervals,* collect three samples of each culture in *mid-log phase* and *process* in three ways:
   a. Transfer the first sample into both drug-containing and drug-free medium at a

concentration of 2×10^5 cells/mL. Count *promastigotes* daily with a hemocytometer from days 2 through 7. The percentage of *drug-resistant* cells in a given line is estimated by comparing the cell concentrations in both cultures in the mid-log phase (days 2-3).

b. Clone the second sample by limiting dilution as in Described elsewhere in thils chapter. Select 20-30 subclones. For each subclone, (1) assess the *drug-resistant phenotype* by double culture as in Described elsewhere in this chapter and (2) test for the presence of the truncated chromosome by PFGE.

c. Assess the presence of the *truncated* chromosome in the population by PFGE analysis of the third sample.

POINTS TO REMEMBER

1. The size of the *homologous* sequences is usually between 200 and 1000 bp. It may exceed this size but should not be <200 bp. Below this threshold, the efficiency of *integration* into the *Leishmania* genome is low or null.
2. It is necessary to add 2-3 nt *upstream* of the restriction sites, so they are not located right at the ends of the PCR products. Most restriction *enzymes* will only recognize their cutting site if it is *flanked* by a minimal number of nt.
3. The cycling conditions, and particularly the *annealing temperatures,* may be optimized according to the *theoretical* T_m of the primers and the thermocycler, but note that the annealing temperature of the *primers* should not be modified becuase of addition of *restriction* sites.
4. Minimize the time of exposure to *avoid radiation* damage to the DNA. Also, remember that UV light is *dangerous,* particularly to the eyes. Wear a full safety *protective* mask and *gloves* for this procedure.
5. It is possible to use a commercial DNA extraction kit to purify the PCR products. However, it is not practical because several columns must be used to process the contents of the two to three PCR tubes, and the elution volume must be kept to a minimum.
6. We use New England Biolabs® enzymes with buffer 2 for *KpnI-XbaI* and buffer 4 for *HpaI-MfeI.*
7. It is essential to ensure that purified *amplification* product is completely restricted for *high-efficiency* insertion into the vector. As restriction sites have been added to the S2 and S1 homologous fragments, double digestion will yield a fragment 12 by smaller than the *undigested* PCR product.

   This size difference is too small to be visible on a simple agarose gel. By adding one-fifteenth of the DNA to be digested to the plasmid DNA control digestion, it is possible to monitor the progress of the sample DNA digestion by following that of the control DNA.

   To do this, use a *plasmid* in which the double digestion will yield a fragment detectable on an agarose gel (>100 bp).

Otherwise, you should include one control tube for each of both enzymes. Note that this procedure tests the purity of the reaction medium and not merely the enzyme efficiency. An alternative, more lengthy procedure is to clone the PCR product into a TA-cloning type vector (e.g., pGEMT-Easy, Promega), carry out a double digestion on the *recombinant plasmid*, and assess the quality of the *digestion* by the presence of both the insert and plasmid bands on an *agarose* gel.

8. If the digestion is not complete, incubate for a further 1 h 30 min. Be careful to avoid overdigestion, which may alter the ends of the homologous sequences and *considerably* lower cloning efficiency. Ideally, follow the manufacturer's *instructions*.
9. The *XbaI* and *KpnI* sites are too close on the vector to be able to verify the second digestion on a gel, hence the need for a control as described elsewhere in this chapter. It is essential that the digestion be complete to avoid recircularization of the vector during the subsequent ligation step. Recircularization leads to considerable background (false positive clones) during selection of transfectants, making the latter much more cumbersome.
10. It is recommended to include a control consisting of the vector alone to check for the recircularization of the vector, and thus the quality of the double *digestion*, as pVV is lacking the a-complementation system used for identifying recombinant clones after transformation (blue/white colonies). By comparing the number of colonies obtained after transformation with the test and control *ligations*, respectively, one can estimate the percentage of *recombinant* clones.
11. For optimal results, it is critical that the cells be kept at < +4°C during the whole procedure until the "heat shock."
12. It is preferable to carry out these steps in sterile conditions. preferably *in a flow* cabinet.
13. The yield may be considered *satisfactory* if the ratio of clones obtained from the control *ligation* compared to that obtained from the test ligation is lower than 0.25. This restricts the number of clones to be screened to isolate four to five recombinant clones.
14. PCR amplification aims to detect insertion of the fragment into the cloning site of the vector. The primers are *located* on pVV on each side of the *KpnI* and *XbaI* sites: downstream of the *KpnI* site (5'-gtaaaacgacggccag-3') and in the *3'-DHFRTS* region (5'-cctcggcgtctcgtcatca-3'). Note that different primers are used to verify insertion of Sl and S2 sequences.
15. We use a Licor© DNA sequencer Long Readir 4200 (Amersham). Sequencing reactions are performed using the Sequitherm EXCEL" 11 DNA Sequencing Kit-LC (Epicentre Tech.) and read using the Sequencher program (Amersham). Instructions for performing sequence reactions may be found in any of the standard kits.
16. Most often, such miniprep DNA is not completely devoid of impurities. It is essential to purify to achieve complete restriction digestion-a crucial step for the construction of the vector.
17. We recommend digesting 1 μL of the purified DNA with a restriction enzyme that *linearizes* the *vector*, and loading several dilutions of this digest adjacent to varying concentrations of a DNA size standard (e.g., *klPstl).* The DNA yield obtained from 400 mL of culture

generally approximates 400 pg (i.e., l. mg/mL).

18. Four *Eco571* restriction sites are present in pVV-S 1-S2. Thus, *Eco571* digestion will yield three fragments of 800, 900, and 1200 by and one fragment of 4500-5500 bp, depending on the size of fragments S 1 and S2. The larger fragment represents the linearized vector DNA that will be introduced into the *Leishmania* cells.

19. This double purification in agarose allows a considerable reduction in the presence of circular forms of the vector, which constitute a severe limitation in transfection success.

20. The in vitro cultivation of *Leishmania* promastigotes is generally straightforward in various culture media. However, variations in growth efficiency are observed among different species and strains and different media. When a strain is to be cultivated in a new culture medium, it should be adapted progressively by frequent passaging at higher cell densities.

    Transfected cells, perhaps because they are exposed to drug selection, are more difficult to grow than wild-type cells. If the medium is not adequate (e.g., not rich enough), the plating and cloning efficiencies after transfection may be considerably reduced.

21. Culture media should be prepared in sterile conditions and stored at +4°C. Avoid storing medium more than 3-4 d. Sterility should be tested by incubating an aliquot at 37°C for a few days, followed by microscopic examination to detect microorganisms, in particular fungi, which are slower to grow and more difficult to detect.

22. Different batches of FCS may influence parasite growth as well as transfection and cloning efficiencies. It may be necessary to test several batches before a suitable batch can be purchased.

23. If no CO_2 incubator is available, use a hermetically sealed container with a CO_2 source such as Alka-Seltzer or a candle jar.

24. The culture may be observed directly on an inverted microscope without manipulation. This allows a good estimate of the cell density, morphology, and motility, and therefore vitality.

25. Regular passaging is necessary to avoid overcrowding and decline of cultures. On the other hand, be cautious not to overdilute the cells, especially transfectant clones. For species that grow faster than *L. major* and *L. donovani,* passage more frequently or with a greater dilution.

26. This test is critical. Indeed, a too high or too low drug concentration may lead to loss of transfectants or to false positives, respectively. Moreover, the test must be performed for each strain, because the lethal dose of antibiotics varies among strains.

    Aim to test the drug sensitivity of each strain on a regular basis,. as progressive inactivation of drugs may be induced by aging or light and heat sensitivity of stock solutions. We recommend to aliquot and store stocks at -20°C.

27. After thawing, culture cells for 2-3 wk before transfection. Also, do not perform transfection with cells that have been cultured for a long time.

28. These concentrations may have to he modified depending on the species and the culture medium used. We recommend establishing a growth curve for the model in use.

29. The cell concentration is critical for best transfection efficiency. Cells should be taken in mid- to late log phase, and concentration should not exceed 8 × 10^6/mL (for *L. major)* or 10^7/mL (for *L. donovani).*
30. These conditions are indicative. It is necessary to test electroporation conditions for each type of parasite and apparatus. Our conditions yield 30-50% of survivors after electroporation, shown to be adequate for efficient transfection.
31. It is important to leave the transfected cells without drug selection for 24 h, to allow expression of the introduced resistance gene.
32. Careful preparation of the plates is critical. We recommend preparing both culture medium and agarose plates on the day of use (day 1).
33. If drug selection can be obtained only in liquid medium, divide contents of each cuvet into four samples immediately after electroporation, and transfer to 1 mL of drug-free medium in a 24-well microplate for 24 h.

    On the next day (d 1), transfer the contents of each well to 8 mL of complete medium containing the selection antibiotics in 25-cm'- culture flask. Allow to replicate, then isolate clones by the limiting dilution method.
34. It is important to let the plates dry sufficiently before plating, to avoid movement of the motile *Leishmania* cells between colonies.
35. Be very careful at this stage, as the agarose is semisolid, hence particularly soft. We do not recommend using a solid spreader for this step.
36. Using our conditions, the transfection efficiency varies from 0.2 to 1× 10^{-4} transfectants/ electroporated cells for *L. major* and *L. donovani* and an efficient transfection should yield about 100-150 colonies. If the colonies are not picked soon enough, they will rapidly merge into a continuum, making selection of individual colonies impossible.
37. It is important to use high-quality agarose for long migrations such as PFGE and for enzymatic DNA digestion and purification.
38. To analyze a large number of transfectant clones simultaneously. it is useful to have access to several PFGE apparatuses.
39. PFGE migration conditions vary considerably among apparatuses. We use a homemade apparatus and migrate at 7.5 V/cm for 1-3 d, depending on the chromosome size class. The pulse times for different size classes can be found in Wincker et al. (24) for our apparatus and on the *Leishmania* Genome Network website. www. ebi.ac.uk/parasites/ leish.html, for the Bio-Rad CHEF apparatus.
40. Use standard protocols.
41. Despite all precautions taken in the purification steps, recircularization as well as *multimerization* of the transfection vector may occur during or after *electroporation*. This will give rise to false positives, i.e., drug-resistant clones that have not integrated the resistance gene. These can only be detected by hybridization analysis of Southern-blotted PFGE gels.
42. *Promastigotes* have a tendency to grow as "rosettes," increasing with cell density, and it

is important to minimize this in the sample to be cloned. Avoid using late growth phases in which density is higher.

43. Cloning efficiency varies depending on the strain and the drug used. In some experiments, the 0.5 cell/well dilution may not yield any clones. For these "fastidious" strains, we recommend inoculating one additional plate with 1 (and sometimes 2) cells/well.
44. This step aims to remove all traces of Sarkosyl-EDTA and proteinase K. To avoid nucleases in the reaction buffers, use double-distilled H_2O and sterile tubes. Wash at +4°C.

Chapter 17

Micromolecular Genome

The involvement of *politicians* on both sides of the *Atlantic* in the *announcement* of the *completion* of the first draft of the human genome heralded a lot of *speculation* about the use and abuse of the information that had been produced, or could be produced, when the genome was completed. As always with major developments of this kind, more problems are thrown up by the data than can be immediately solved.

After all, the sequence is merely a string of letters representing one-dimensional information of the genes. How this information is to be translated into more useful knowledge was a more *complex* task and has led to an *explosion* of interest in bioinformatics. Many financially advanced nations have *devoted* considerable fractions of their science research budgets to new initiatives in bioinformatics; no one wants to be left behind, because the *perceived* commercial value of genomics is believed to be massive.

Bioinformatics is an exciting young discipline aimed mainly at sorting and *categorizing* data. The big challenges are twofold: the development of novel methods for automated *linkage* of *disparate* data silos, and the shift from *one-dimensional* genome sequence data into *three-dimensional* (3D) protein structure data.

Automated linkage methods seek to find *relationships* such as possible protein-protein interactions, that can then be resolved by physical experiments. Once those transitions have been overcome, the genomic data will have been *manipulated* into knowledge that can be used as input for rational and *automated* methods for drug design.

Primarily, we are hoping to develop new methods that will be exquisitely selective for specified proteins. Drug design on the genome scale is a huge challenge. However, if we can foresee a way through the challenge, even though the *computing* power needed is massive, then that way will surely be found.

Data History

Although the size of the human genome is not as large as *initially envisaged* (some 30-40,000 genes rather than the 140,000 genes predicted before the sequence of the first draft) there are still about 3 billion nucleotide bases in the sequence. The *conversion* of this gene information into protein sequences is more *problematic*, because there are many splice *variants* that lead to different proteins.

The size of the human *proteome* may therefore be an order of magnitude more complex than the genome. Furthermore, the old dogma of one gene/ one *enzyme* has been shown to be untrue, some proteins being *multifunctional* depending on where in the body they are expressed.

For example, *phosphoglucose isomerase* is a multifunctional protein; in the cell it plays a part as an enzyme in both *glycolysis* and gluconeogenesis; however, outside the cell it acts as a *cytokine* with autocrine motility factor activity. Similarly, the m-RNA *splicing* factor Prp19p also functions as a *ubiquitin* ligase.

A subset of the proteome contains the targets for drugs and can be categorized into three principal groups:

- Receptors, where the drug modulates the activity of the receptor without the necessity of another small ligand being involved,
- Enzymes, where the drug interferes with the activity of the enzyme in turnover of *substrate* to product, and
- Protein-protein *interactions*, where the drug interferes with the formation of a protein *complex* necessary for the *propagation* of a biological effect.

If the size of the proteome is *P*, there will be no more than P^n possible n-mer interactions. However, as far as drug design is concerned, the size of the interacting surfaces that could be interfered with by a small molecule is small and unlikely to include three proteins; this limits the potential number of interactions to P^2.

If the human proteome contains 10^5 proteins, the number of *homodimer interactions* is potentially 10^{10}. *Chemical* space is thought to be massive: 10^{180} different structures with molecular weights <500; the number of drug-like molecules is thought to be much smaller at 10^{60}, whereas the number of chemotypes (novel core *chemical scaffolds*) may be 10^{40}. These numbers can be put into a context of the mass of the visible universe, 10^{54}g.

Placing these numbers within a computing *perspective*, a dedicated cluster of 10,000 teraflop machines would provide only 10^{23} flops per year. Clearly, if computing methods are to be used either to *generate* or to search these large chemical spaces, they would have to be efficient *combinatorial* optimizers to complete the job in a reasonable time.

The task for drug discovery is to reduce these numbers to practice by advanced computational optimization techniques and identify specific small molecules as starting material in the *hit-to-lead* process. In chemical genomics, we seek ultimately to develop methods to generate selective compounds from a *structural genomics* understanding of the proteome. Furthermore, it should be possible to predict potential lack of selectivity of current drugs through *protein-binding promiscuity*.

Pharmacogenomics

Pharmacogenomics offers a way to study the effect of small changes in gene sequence on drug action; *pharmacogenetics* studies the effect of these changes on human populations. There are strong relationships between the two *disciplines*.

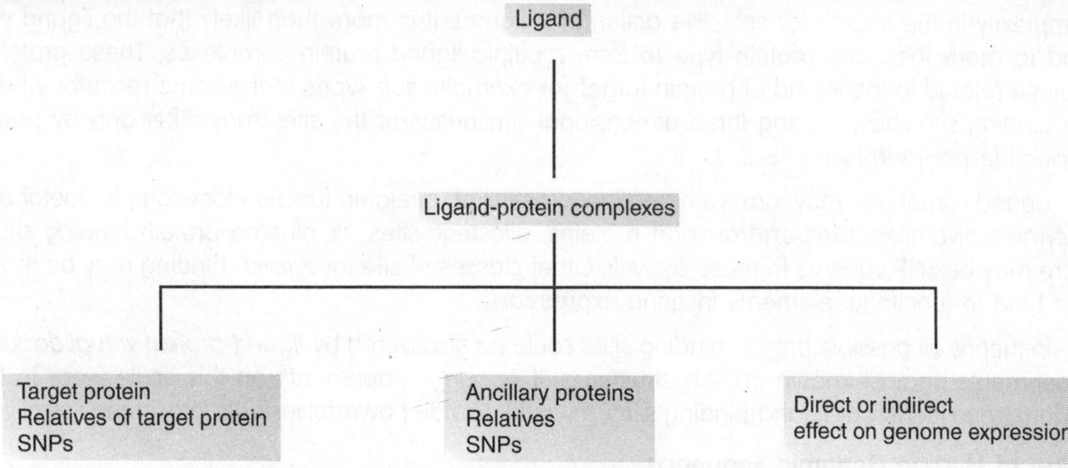

Figure 17.1: Types of ligand-protein complex classed by function.

The International Union of Basic and Clinical Pharmacology (IUPHAR) has recognized the importance of this link in the establishment of a section on the field of pharmacogenomics and *pharmacogenetics*. The general points for pharmacogenomic influences on drug action are at the steps of:

- Absorption
- Distribution
- Site of action
- Metabolism
- Elimination

At each of these general steps, there are many molecular mechanisms that can be involved in drug handling that have a significant bearing on chemical genomics. For example, absorption should not be thought of as a passive process but as a dynamic flux, forwards and backwards, across a set of molecular barriers from the site of entry into the body to the target cell.

Any change in the molecular structure of the barriers, pump mechanisms, or the rate of production of barrier molecules will have significant effects on the concentration of the drug at the therapeutic molecular target. Furthermore, the target cell may have induction mechanisms that can be triggered by the drug to start up a cycle of expulsion of the drug from the cell, leading to drug resistance.

Historically, pharmacogenetics has focused on mutational changes in drug-metabolizing enzymes, the reason being that the enzymes have been simple to identify, isolate, and compare with the wild type. The determination of the human genome sequence has potentially provided a much wider spectrum of points for interference in drug action.

In principle, these genomic differences can be identified, and therefore diagnostic kits could be developed to identify problems in specific drug handling by individuals. A goal has thus been set in chemical genomics for personalizing medical treatment. Figure 17.1 gives an overview of the complexity in the search for selective action of a ligand. It is more than likely that the ligand will bind to more than one protein type to form multiple ligand-protein complexes. These proteins may be related to the intended protein target-for example, sub-types of the same receptor where the binding site shows strong three-dimensional similarities-or the sites may differ only by single nucleotide polymorphisms (SNPs).

Ligand complexes may form with ancillary proteins of unrelated function for example, metabolic enzyme active sites, sites on transport proteins, allosteric sites, or plasma protein binding sites; there may be SNP variants to these as well. Other classes of site for *ligand*- binding may be those that bind to functional elements in gene expression.

In theory, all possible protein binding sites could be discovered by *ligand-protein virtual docking* experiments against known protein structures. If *docking* experiments on this scale were to be performed to identify all ligand-binding sites, it would provide powerful insights into ligand selectivity.

Value of Human Genomic Sequence

Clearly, the completion of the human genome sequence offers medicine the *possibility* of huge steps forward in the conquest of disease. Probably most scientists in drug discovery have thought through the implications of this new *knowledge* for our own specific research area and can now envision a *rational* path towards possible solutions.

The problems we face in converting vision into reality are largely problems of scale; hence the drive to industrialize the drug-discovery process using large *computational* efforts, *micro-miniaturization* of experiments, and robotic *procedures* to enhance the throughput of *experiments*. It is well-known that drugs can have an effect on the up- and *down-regulation* of genes.

The question that arises is: how do we relate gene expression in a tissue(s) to the molecular structure of drugs acting there? If there are differences in responses between individual *patients*, how do we handle genetic polymorphisms to small drug structures? Would this provide scientists with a handle for approaching the *problem* of *personalized medicine* through advances in chemical *genomics*?

STRATEGIES

Pharmacogenomic Regulation

The chemical structure of the active drug molecules has to be related to their activity on all expressed genes, or gene products, within the body, whether that activity is viewed as a positive effect or one with drawbacks. Although in totality this is an *exceptionally* complex problem, it can be broken down into logical components for a more structured analysis.

The general problem can be expressed by considering a class of drug molecules, A, composed of a set of structurally similar members $\{A_1, A_2, A_3, A_k .. . A_m\}$. Each drug molecule, A_k, of the set can be specified by a set of atomic coordinates for each atom a $\{x_{1a}, x_{2a}, x_{3a}\}$ and the molecule is composed of a set of molecular features F $\{F_1, F_2, F_3, F_k ... Fn\}$. Within a set of active compounds, there are likely to be substructural similarities; these similarities can be quantified so that similar molecular *substructures* in the set A can be found.

A simple *quantification* system for molecular similarity is provided by the *Tanimoto* index of a pair of compounds. The index works by identifying common substructures within the molecules. Conversely, structural differences can also be identified. Correspondences in substructures and features can be determined by experiment and subsets of features can be related to activity.

Consider a set of organ tissues, O $\{O_1, O_2, O_3, O_k...O_p\}$, on which gene expression can be measured, then for each organ tissue there is a subset of genes G $\{G_1, G_2, G_3, G_k ... G_q\}$ that may exhibit modified expression. This expression of gene, G_k, may show a dynamic pattern of expression, E_{Gk}, where E_{Gk} is a time-dependent relationship over the period of stimulus, t, so

$$E_{Gk} = f(t)$$

This formal representation helps to provide a framework for investigating two questions:

1. Can the structural differences in a set of molecules, A, be correlated with changes in the genes, G, expressed in a particular organ, O_k?
2. If we consider a superset of drug molecules composed of M structurally different classes, M $\{A_{1...n}\ B_{1...n}, C_{1...n}, ... Z_{1...n}\}$, can a particular subset of molecular features, F_k, be correlated with changes in the genes, $G_{1...n}$ expressed in organ, O_k? If the answer to this question is yes, we would discover a general set of sub-structure and feature markers that regulate a gene expression pattern, EGk.

Note that this formal representation says nothing about the mechanism of expression. It is still a superficial representation that is useful only for a top-level classification. What is needed is a further layer of sophistication that relates structural molecular features in the set of drug molecules with 3D interactions with their binding site. There is also the strong possibility that a structural molecular feature may also interact with, more than one binding site, as is frequently the case with closely related gene families.

A further layer of complexity needs to be added to consider gene expression networks with feedback to give a more realistic representation of the chemical genomics problem, but that is beyond the scope of this chapter. Useful reviews of the theory are given by Zhou et al. and Hartemink et al..

Modified Chemical Genome

Polymorphisms commonly affect chemical genomics in two ways: modifications to the active site for the drug or modifications to downstream handling of the drug molecule, e.g., metabolism or cell permeability. In many cases these differences are caused by SNPs.

Polymorphism in an Active Site

Consider a polymorphic site, S, for which there are a set of positions P $\{P_1, P_2, P_3, P_k, ... P_r\}$ where protein polymorphisms are possible, at each position, P_k, there is a set of possible residues R $\{R_1, R_2, R_3, R_k, ... Rs\}$; if a particular residue, R_k, at position, P_k, is absent or replaced within a patient and that residue is believed to contribute significantly to the activity of the current drug molecule, then that patient will show a marked difference in affinity for the drug.

In a crude sense, the interaction energy between a drug and its binding site can be expressed as a summation of all pair-wise atomic interactions between the binding components. Replacement of atoms in a *polymorphism* therefore has a significant effect on *affinity*. The drug designer then has two choices for discovering active compounds: one choice would be to design to the subset of site points that are common, with the hope that all patients could be treated effectively by the same drug; the other would be to design a different drug for each *subpopulation* of individuals.

The latter option is the personalized medicine approach. In many respects, this problem is identical to that encountered in practice with designing either promiscuity or selectivity into the *drug-receptor interaction*.

A good example of SNP problems in practice for drug designers is the case of *treatment* of *chronic myeloid leukemia* through the enzyme Abelson tyrosine kinase (ABL). Gleevec binds to ABL in its inactive form and keeps it inactive. However, many SNP mutations have arisen that prevent Gleevec binding. The promising new compound BMS-354825 is able to bind to the mutated ABL.

Physiological Polymorphism

Polymorphisms in drug metabolism have a profound effect in chemical genomics and could also be considered formally. Drug metabolism is governed by the recognition of sub-structural molecular fragments by specific *drug-metabolizing* enzymes. The problem can be formally expressed: drug *molecule* A is composed of a set of sub-structural fragments F $\{F_1, F_2, F_3, F_k, \dots F_r\}$; each fragment, F_k, is recognized by an enzyme Z_k, where Z_k is a member of the set, Z, of all drug-metabolizing enzymes.

Recognition of the fragment by Z_k leads to the metabolism of A to metabolite M_k, where M_k is a member of the set of metabolites M. If the drug-metabolizing enzyme Z_k, for a particular fragment F_k on drug A, shows a set of polymorphisms where there are a set of positions P $\{P_1, P_2, P_3, P_k, \dots P_r\}$ on the enzyme, and at each position P_k, there is a set of possible residues R $\{R_1, R_2, R_3, R_k, \dots R_s\}$ for the site S, then different individuals will metabolize the drug less predictably. In turn, the drug metabolites may then create different gene-expression patterns. The questions to be addressed are:

1. What are the predicted routes of *metabolism* for the drug, given the set of *polymorphisms*?
2. How do we redesign the drug to eliminate the polymorphism problem?

Comparative Studies

The evolution of protein structures shows conservation in recognition features for common binding sites. For example, the adenosine triphosphate (ATP)/ guanosine triphosphate (GTP) binding site contains an amino acid motif A (P-loop) of [AG]-x(4)-G-K-[ST]. Within this motif, ample variation is possible, with consequent differences in 3D structure in the binding sites.

These differences will affect the binding of purine analogs to the binding sites, and a representation of this effect is quite similar to that given under Subheading modified chemical genome for polymorphisms.

The problem for chemical genomics is twofold:

1. Many compounds will cross-react by binding to variants of the site.
2. Different target proteins may well contain the same binding motif even though the function of the protein may be different; consequently, the drug will bind to different proteins and may interfere in different signaling pathways. Although in this case, the compound would

be perfectly selective for the site, its action would be nonspecific, and that target should be avoided for drug design.

Progress in chemical genomics has been made towards linking ligand structural motifs to protein structural motifs by Deng, Chuaqui, and Singh; they used a method for fingerprinting the ligand and obtaining the complementary fingerprint of the binding site.

This formal representation of problems in chemical genomics identifies the areas where there will be considerable impact on drug design in the future. Currently, progress has been limited to very few examples in practice. However, chemical genomics is a rapidly expanding field, and these problems are ripe for exploration. The rest of this chapter focuses on the application of computational methods used in drug design to genomically derived information.

VALIDATION

Therapeutic target identification and validation is a key problem for drug discovery. Protein interactions and pathway analysis provide clues for identifying proteins that may be key control points within a pathology. Where protein-protein interactions are involved, there is potentially an all-against-all problem to be resolved to pinpoint the interacting proteins.

If we are to industrialize chemical genomics and drug discovery efficiently, we need the 3D structures of all proteins, both for man and pathogens. This is the current challenge for structural genomics. Although there have been huge advances in the application of X-ray crystallography to protein structure determination using many thousands of crystallization attempts per day, the number of structures determined is still small: the total number of structures in the Protein Databank was 26,500 in July 2004.

The structural genomics initiatives in North America, Europe, and Japan are tasked with providing high-resolution structures, either by X-ray crystallography or by nuclear magnetic resonance. However, experimental progress is slow, and in parallel to this large-scale computational structure prediction, methods are constantly being developed to automate the generation of new structures rapidly.

Technique for Homology

The raw data for protein-structure generation is the protein sequence. In principle this data string encodes the 3D structure of the protein. How the protein folds is still unclear. Nevertheless, certain folding patterns are associated with specific sequence patterns, so it is possible to develop expert systems to provide 3D models of new folded proteins. Commercial software for homology modeling is widely available.

Automated homology methods can be used to model significant numbers of proteins on a genome-wide scale. However, the important issue is how closely the model represents structural reality. The Structural Genomics initiative has selected representatives from around 20,000 groups of proteins that could be used in a large-scale homology modeling process for most protein sequences.

It is estimated that this represents 150,000 proteins. Active sites are quite closely conserved in sequence and structure, so it may be possible to use crystal data to represent the core elements of the site while modeling with molecular dynamics the more peripheral elements of the site.

DESIGNING MICRO MOLECULES

If genome-wide drug-design methods are to be developed, they have to be fast and automated so that the algorithm can design a good selection of molecular structures that can be synthesized readily. Synthetic tractability has always been a problem for *de novo* design methods, because by their nature the algorithms are geared to creating novel structures.

However, if the molecules are to be built from small molecular fragments, the fragment assembly process can follow a limited set of rules that try to ensure synthetic tractability. There are two general approaches to *de novo* drug design.

First, where 3D coordinate data for the site are available, e.g., X-ray crystal data, nuclear magnetic resonance data, or good homology model data. Second, where this type of site data is missing, *de novo* design can proceed from the use of active ligands to build pharmacophore models of the site.

Structural Variability

Consider a set of atoms, S, in the site, together with a set of tolerances, T, for the fit of a ligand(s), and a set of bounds, B, that limits the design region within the site; these constraints govern the local fit between the designed ligand and the site constraints.

If the *de novo* structures are to be built from a complete set of molecular fragments, F, then in principle, all possible molecular structures could be considered by the generator, although not all would be acceptable.

In the extreme case where the tolerances are set to zero, only perfectly fitting structures could be designed. These structures would be few if the bounds for design are restricted; they would be essentially isosteres. As the tolerances are relaxed by increasing T, different local molecular substructures will be generated that satisfy the set of constraints imposed by the site. Hence, molecular-scaffold diversity will be introduced into *de novo* ligand design.

For example, suppose there is a local steric constraint within a binding pocket for a six-membered aromatic ring; if the tolerance criteria are relaxed, then other ring systems may be accommodated. If the set of tolerances is uniformly relaxed, then by analogy, a larger structural variation could be accommodated across the site, but within the limits of B.

This generation problem has been studied in some detail for the methotrexate binding site of dihydrofolate reductase, where 10,000 runs were performed; 26 solutions had the same connectivity as methotrexate but with slightly different orientations within the site; the principal known binding mode was mimicked. Each solution had a slightly different pose in the site.

If the site bounds are increased, ligand structural variation will be increased. To explore how much variation could be quantified, consider a simple case. Let there be five site points, with required features, that are needed for producing activity, and say *n* structures could be generated within a particular constraint set.

If the limits of B are expanded twofold to include 10 possible site points, then the number of combinations of 5 from 10 is 252. It is more than likely that *>252n* structures are possible that fit the design criteria.

General Strategies

In this case, detailed site coordinate data are used to control *de novo* structure generation. The quality of the data is important; crystal-structure data with $R < 2.1$ Å are ideal. Furthermore, it is often found that bound water molecules are associated with hydrogen bonding to the site and ligand; these clearly have an effect on ligand binding and consequently on drug-design strategies.

Two general strategies are often used. Interacting fragments are positioned in the site and a design mechanism tries to link the fragments together by other linker fragments. Alternatively, a single fragment is selected and positioned n the site, and structural elaboration ensues from that fragment, with other fragments being pieced together stepwise.

Automated *de novo* design methods seek to explore different molecular structures that could fit and bind to the site. They have general features in common:

1. A description of the site, usually the coordinates of the protein with a bounding box for the site; design is confined within the bounding box.
2. A description of the strategy to be adopted for design, i.e., which site points to use to force certain interactions to be utilized.
3. A file of chemical fragments from which the ligand structures are to be built.
4. Rules for assembly of fragments within the site to ensure that appropriate chemistries are used in linking fragments together.
5. A penalty function, which enables a suggested structure to be assessed in the site for goodness of fit.
6. An optimization procedure, which controls the evolution of novel structures.
7. A stochastic process, which enables the algorithm to provide alternative solutions. Two fundamental questions have to be asked about *de novo* design algorithms:
    1. Can they reproduce known chemotypes for a range of targets without the algorithm having any knowledge of the known solutions?
    2. How many novel structures can be grown for a target?

Table 17.1: Comparison Between the Numbers of Known Chemotypes for Four Targets With the Number of Chemotypes Reproduced by *De Novo* Structure Generation Using SkelGen Within 600 Generations Per Target.

Target protein	Known chemotypes/motifs	Chemotypes/motifs reproduced by SkelGen
COX2	5 chemotype classes	5 chernotype classes
Estrogen receptor	10 chemotype classes	6 chemotype classes
CDK2	9 binding motifs	9 binding motifs
MMP3	5 binding motifs	5 binding motifs

Stahl et al. have attempted to answer the first question using four widely different targets: cyclo-oxygenase-2 (COX2), estrogen receptor (ER), cyclin-dependant kinase-2 (CDK2), and matrix metalloproteinase-3 (MMP3). In each case, the algorithm was run 600 times, and the output of each generated structure was compared with the known active chemotypes.

To make this comparison more stringent, the output structures were compared with the known structures using a combined two-dimensional and 3D metric. The 3D metric took the known structure in the site as a reference structure and compared this with the coordinates of the output structure generated in the site. In this way, it is possible to superpose the two structures as bound in the site and generated in the site. Table 17.1 summarizes the results for the four target sites.

In each target, close structural chemotypes were found to the known active compounds. For COX2, CDK2, and MMP3, all known chemotypes were found. With ER, 6 of the 10 chemotypes were found; known cis-stilbene scaffolds could not be modeled into ER (1 err Protein Databank structure) in a strain-free manner, so it is not surprising that the chemotype could not be generated.

Many different alternative chemotypes were also grown. Compute farms are ideally suited to molecular-scaffold generation in de novo design, because the process can be split into complete runs that can be distributed onto different processors. It scales linearly and is embarrassingly parallel.

Ligand-Based Technique

In the absence of a 3D structure for the target protein, and where there is little confidence in a homology model for the binding site, it may be possible to use compounds discovered to be active as the basis for further de novo structure generation. The compounds may be derived from either a high-throughput screen or, in the worst case, from a single active compound. This procedure uses de novo structure generation for a scaffold-hopping exercise.

Because the information derived by this process is less precise, the method can be used to generate focused libraries that could be of value in screening with a novel target protein within a related gene family. In practical terms, this approach could be very useful, since most novel targets do not have 3D structural data. Lloyd et al. have shown that at the estrogen receptor binding site, coordinate data of the site can be ignored and active ligands used as the training set to create a negative-image model of the site in which de novo structure generation can be performed.

The structures generated can then be compared with known chemotypes. Once again, many known active chemotypes were generated for the ER, but this time without structural site information. Even with only one active ligand, a similar procedure can be used to create an image of the site and hence perform de novo structure generation. The big commercial value of this approach is in a scaffold-hopping exercise to explore intellectual property that is unencumbered by structures from the original training set or high-throughput screening hit(s).

Micro-Molecular Selectivity

One aim of chemical genomics is to design potential drug structures that are selective enough for individual patient therapy. Suppose we are confronted with a target for which there are documented SNP variations within the binding site together with a 3D structure of the wild type;

assume that design to the wild type generates an active compound, but let us say that the compound is inactive on the SNP variants. What approaches should we use to help ensure that the designed ligand is active in all known pharmacogenetic cases?

The first step would be to build all models of the SNP variations compared with the structurally known wild type. In the tailored therapy approach, for individual patients, different effective compounds could be designed and made. However, this is not likely to be a commercially viable strategy. A more attractive approach would be to establish a single design strategy that avoids interference by the SNP regions within the site.

The altered SNP residues become a further constraint set that has to be avoided as *de novo* design proceeds. Alternatively, a less elegant solution would be to design ligands separately to all variant site structures within the SNP set. Each structure could then be cross-checked for binding to each of the other sites.

Promiscuity of Micro-molecules

The opposite of selectivity in drug design is *promiscuity*; in this situation, a single compound acts on a number of sites. In chemical genomics, this could be useful as a tool for inhibiting all members of a gene family. It would provide *pharmacologists* with a chemical toolkit to *ascertain*

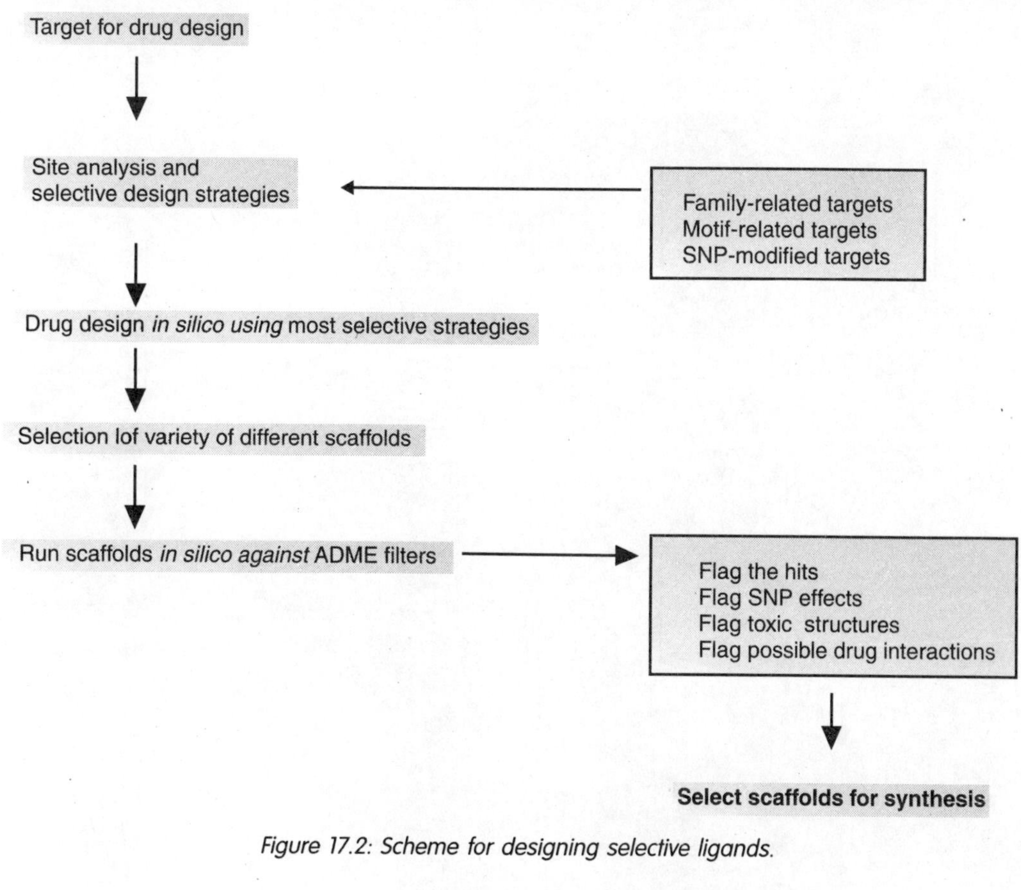

Figure 17.2: Scheme for designing selective ligands.

whether a particular gene family could be important in an experimental or *pathological scenario*. The design strategy here would be very similar to the case of tailoring design to create active structures against a SNP set.

Metabolome

Many problems in drug toxicity are related to how the drug is degraded. The inclusion of early-stage absorption, distribution, metabolism, and *excretion* properties in drug design has long been established as an important goal in drug discovery. The output of *de novo* structure generators is so varied that different scaffolds and substituent groups can be generated easily.

Thus, the number of choices available for drug discovery is massive, and *in silico* predictors of metabolic pathways could be linked to the output of *de novo* generators. Post-processing the output could be performed such that structures that could be *metabolized* by undesired metabolic pathways could be eliminated from the generated set for consideration.

Chapter 18

Silencing of Specific Genes

The silencing of specific genes by small interfering RNA (termed RNA interference or RNAi) has provided a revolutionary approach to studying gene function. While technologies such as homologous recombination and antisense targeting have been used successfully in reverse genetics, these are expensive and laborious. RNAi, in contrast, is comparatively rapid and, due to its remarkable potency, cost effective.

The phenomenon of RNAi was first described in 1998 by Nobel laureates Andrew Fire and Craig Mello, who developed a method of gene silencing using specific double-stranded RNA in *Caenorhabditis elegans.* Following on previous work indicating that both sense and antisense RNA were equally effective in suppressing the expression of specific genes in *C. elegans,* Fire and Mello discovered that sense and antisense RNAs had a synergistic effect on gene expression, with the dsRNA (double-stranded RNA) showing far greater potency than either RNA alone.

Indeed, only a small amount of dsRNA was required to dramatically reduce the expression of the target gene. This elegant study shed mechanistic light on previous work in plants showing that exogenously added DNA could affect endogenous target gene expression and plants combated RNA viruses by destroying their RNA. The technique of RNAi was quickly adopted by investigators studying the genetics of *C. elegans,* with considerable success.

The subsequent identification of short interfering RNAs (siRNAs) led to the adoption of RNAi to study gene function in mammals, revolutionizing this field of research. The widespread adoption of the technique has led to an increasing commercialization of both dsRNA production and methods of delivery; indeed, prevalidated siRNAs are now widely available to study the function of an increasingly large array of genes.

Since the discovery of RNAi, remarkable progress has been made in understanding the underlying mechanisms. One of the first clues as to the mechanism of RNAi came from studies in *Drosophila* embryo extracts in which the ability of the lysates to synthesize luciferase from an exogenous mRNA was diminished following incubation with dsRNA. This raised the possibility that the silencing effect was mediated by a nuclease targeted by the dsRNA to degrade homologous transcripts.

Subsequently, this nuclease was identified in *Drosophila* as a complex known as RISC (RNA-induced silencing complex,). It has since become clear that this complex is able to recognize target molecules following the conversion of the dsRNAs (including miRNAs and replicating viruses) to short interfering RNAs (siRNAs) produced by Dicer, a member of the RNase III family. siRNAs are 21-23 nucleotides with 3' overhangs and 5' phosphate and 3' hydroxyl groups which are incorporated into the RISC complex.

Following incorporation, the duplex siRNA is unwound and the antisense strand targets the

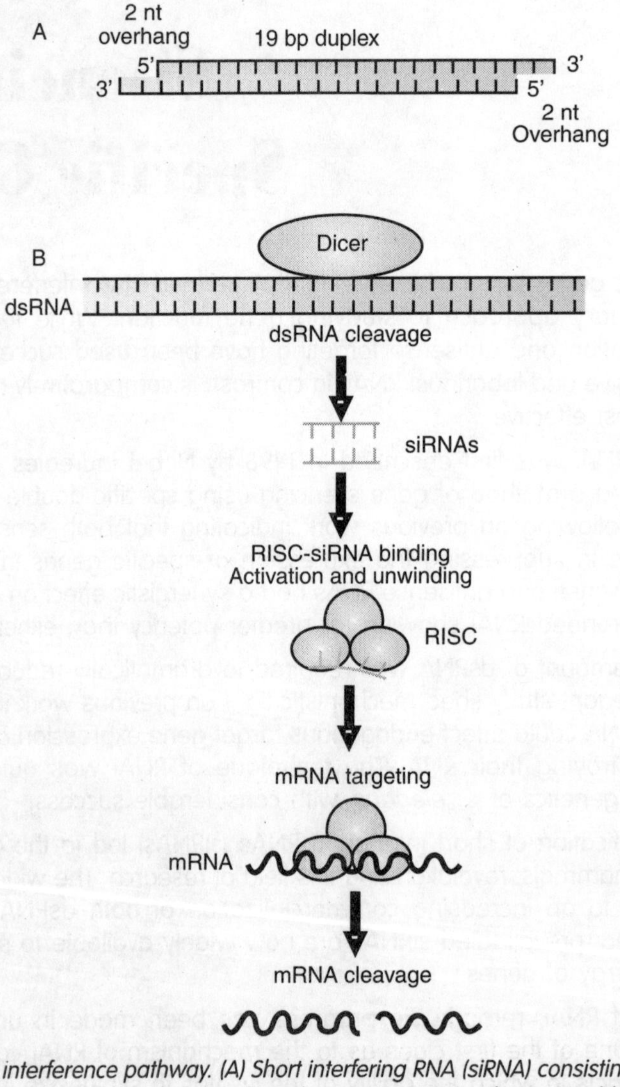

Figure 18.1: The RNA interference pathway. (A) Short interfering RNA (siRNA) consisting of 21 nucleotides (nt), a 19bp duplex with 3' dinucleotide unphosphorylated overhangs. (B) RNAi pathway. Double-stranded RNA is cleaved into short interfering (si)RNA by Dicer, a member of the RNase III family member. The siRNA is incorporated into the RNAi silencing complex (RISC). RISC is activated and the siRNA unwinds. RISC is targeted to the complementary sequence in the target mRNA and mediates mRNA cleavage and gene silencing.

complex to homologous sequences, mediating mRNA cleavage and gene silencing. The elucidation of this pathway, and in particular the identification of siRNAs, paved the way for the utilization of RNAi in mammalian cells. While the introduction of dsRNA into *C. elegans,* plants, and other species was extremely effective, this technique was not suitable for mammalian cells due to their antiviral response mechanisms to foreign dsRNA.

The demonstration that siRNAs could be utilized to evoke RNAi without triggering nonspecific antiviral effects opened the door for the use of RNAi-mediated gene silencing in mammalian cells. The ease of delivery of siRNAs into mammalian cells using standard transfection reagents and the potency and specificity of gene silencing has quickly led to RNAi becoming a routine technique for studying gene function in mammalian cells.

The ability to efficiently and selectively silence target gene expression has numerous applications. In a basic science setting, RNAi has been utilized successfully to study the function of single genes, both in vitro and in vivo.

In addition, the identification of microRNAs (miRNAs), endogenously expressed small RNAs that evoke RNAi, led to the development of short hairpin RNAs (shRNAs), which mimic pre miRNA hairpins and can be transcribed from DNA vectors, allowing their stable expression and the use of retrovirus or adenovirus as vectors.

This allows the construction of expression libraries of shRNAs targeting a range of genes, allowing RNAi to be used on a large, genomewide scale. Such large-scale studies are useful in studying biochemical pathways and in identifying and validating drug targets. In the clinic, efforts are underway to utilize RNAi as a therapeutic tool to target a number of diseases ranging from HIV to cancer.

While promising results have been achieved using RNAi in mice, it remains to be seen whether it can be successfully utilized in humans, due to the difficulties in developing delivery strategies and the danger of off-target effects. Despite these limitations, however, it seems likely that RNAi will evolve from an extremely useful tool to study gene function in the laboratory into a therapeutic strategy in the future.

REQUIREMENTS

Oligonucleotides

1. siGLO™ RISC-free siRNA (Dharmacon, D-001600-01-20; Thermo Fischer Scientific UK). This is a nontargeting siRNA with a fluorescent label and impaired ability for RISC interaction. It is useful as both a negative control and a cotransfection marker.
2. siCONTROL™ Lamin A/C siRNA (Dharmacon, D-001050-01-20; Thermo Fischer Scientific UK). This is a positive silencing control for guaranteed silencing of Lamin A/C in human, mouse, and rat cells.
3. siGENOME™ *SMART*pool reagent targeting endothelin-converting enzyme-1 (Dharmacon; Thermo Fischer Scientific UK). This is a mixture of four SMARTselection" designed siRNAs.
4. Custom synthesized individual siRNA duplex targeting ECE-1 using the siDESIGNTM Centre on the Dharmcon website).

Transfection of Oligonucleotides

1. Mouse embryonic fibroblasts (STO; kindly provided by Prof. N. J. Maitland, Yorkshire Cancer Research Unit, University of York, UK). They are available from both the American Tissue Culture Collection (ATCC, CRL-1503) and the European Collection of Cell Cultures (ECACC, 86032003).
2. Dulbecco's Modified Eagles Medium (DMEM; BioWhittaker: BE12-604F; Wokingham, UK).
3. Trypsin/EDTA. (BioWhittaker, Wokingham, UK).
4. Phosphate buffered saline (PBS; BioWhittaker, Wokingham, UK).
5. Six-well plates (Nuns ", Denmark).
6. Oligofectamime™ (Invitrogen;, 12252-011; Paisley, UK).
7. Opti-MEM reduced serum medium (GIBCO, 31985; Paisley, UK).
8. Fetal bovine serum (FBS; BioWhittaker, Wokingham, UK).
9. Cell culture facility.

Technique

1. RNeasy Kit (Qiagen, UK).
2. iScipt cDNA Synthesis Kit.
3. iCycler thermal cycler (Bio-rad).
4. SYBR-Green Supermix (Bio-rad: 170-8880; U.K.): the kit contains KCl, Tris-HCl pH8.4, 0.4 mM of each dNTP, Taq DNA polymerase, MgCl2, SYBR-Green, and fluoroscein.
5. 0.2-mL thin-wall PCR plates (AbGene, AB-0624; Surrey, UK).
6. Oligonucleotide primer pairs at 10pmol (MWG Biotech, Ebersberg, Germany).

Technique of Western Blotting

Cell Lysates

1. PBS (Oxoid, Basingstoke, UK).
2. Triple detergent lysis buffer: 50mMTris-HCl pH7.4, 150mMNaCl, 0.5% (w/v) sodium deoxycholate, 0.1% (w/v) SDS, 0.02% (w/v) sodium azide, 1 % (v/v) Nonidet P-40, 0.l mM Na_3VO_4, EDTA-free protease inhibitor cocktail (lX; Roche) (pH 8.0) for the complete inhibition of serine and cysteine proteases.
3. Benzonase nuclease (Novagen).
4. 23G syringe needles (Tyco Healthcare, UK).
5. 1-mL syringes (Terumo, Belgium).
6. Bovine serum albumin (BSA; Sigma, Poole, UK).
7. Bicinchoninic acid (BCA; Sigma, Poole, UK).
8. 4% (w/v) $CuSO_4$ solution (Sigma, Poole, UK).

9. Microtiter plate reader (Anthos Labtec Instruments, Salzburg, Austria).

SDS-PAGE Method

1. SDS gel equipment (Invitrogen, Paisley, UK).
2. 3-8% Tris-acetate precast gels (Invitrogen, EA0375BOX; Paisley, UK).
3. 1X LDS sample buffer (Invitrogen, NP0007; Paisley, UK).
4. 0.5M DTT (Invitrogen, Paisley, UK).
5. SeeBlue Plus 2 prestained standards (Invitrogen, LC5625; Paisley, UK).
6. 1X NuPAGE Tris-acetate SDS running buffer (Invitrogen, LA0041; Paisley, UK).
7. NuPAGE antioxidant (Invitrogen, NP0005; Paisley, UK).
8. Blotting pads (Invitrogen, Paisley, UK).
9. Nitrocellulose membrane (Invitrogen, Paisley, UK).
10. Methanol (B.D.H. Chemicals Ltd, UK).
11. 1X NuPAGE transfer buffer (Invitrogen, NP0006-1, Paisley, UK).
12. Blotting module (Invitrogen, Paisley, UK).
13. Filter paper.
14. Ponceau red stain: 1% (w/v) in 30% (v/v) trichloroacetic acid and 30% (v/v) sulfosalicyclic acid (Sigma, Poole, UK).
15. TB ST: 0.1 % Tween-20 in 10 mM Tris-HCl, pH 7.4.
16. Blocking solution: TBST with 5% (w/v) milk powder and 1 % (w/v) BSA.
17. Reference primary antibody (ECE-1 monoclonal antibody 1:100; kindly provided by Dr K. Tanzawa, Sankyo Research Laboratories, Tokyo).
18. Antimouse horseradish peroxidase-conjugated secondary antibody (Amersham Ltd, UK).
19. Enhanced chemiluminesence (ECL; Amersham Ltd., UK).
20. Saran wrap.
21. Cassette.
22. Autoradiography film (Kodak, UK).

Staining Method

1. Coverslips (B.D.H. Chemicals Ltd, UK).
2. PBS (Oxoid, Basingstoke, UK).
3. Methanol/acetone (1:1 ratio: B.D.H. Chemicals Ltd, UK).
4. Blocking buffer: TBS, 1% (v/v) normal goat serum and 0.2 % (w/v) gelatin.
5. Reference primary antibody: ECE-1 monoclonal antibody (1:100; kindly provided by Dr. K. Tanzawa, Sankyo Research Laboratories, Tokyo).
6. Control IgG1 subclass antibody (Sigma, Poole, UK).

7. FITC conjugated antimouse IgG secondary antibody (Jackson ImmunoResearch Laboratories, USA).
8. 6-Diamidino-2-phenylindole (DAPI): 1:1000 in PBS from 0.5 mg/mL stock in DMSO (Sigma; Poole, UK).
9. Vectashield (Vector Laboratories, USA).
10. Nail varnish.
11. Olympus IX70 inverted wide-field fluorescence microscope.

Invasion Technique

1. Matrigel basement membrane (Becton Dickinson, 354234). BD Matrigel™ matrix is a solublized basement membrane preparation extracted from Engelbreth-Holm-Swarm mouse tumor mouse sarcoma.
2. BD Falcon™ cell culture inserts for a 24-well plate, 8.0-μm translucent PET membrane (Becton Dickinson, 353097).
3. BD Falcon™ cell culture insert companion plate, 24-well (Becton Dickinson, 353504).
4. DMEM containing 0.1 % bovine serum albumin (BSA).
5. PC-3 prostate epithelial cells were kindly provided by Prof. D. Nanus, Cornell University, New York. They also are available from both the ATCC, CRL-1435, and the ECACC, 90112714.
6. 0.1 % crystal violet (Sigma, Poole, UK).

PROCEDURES

Oligonucleotides of Mammalian RNA

Many websites are available for functional siRNA search and design for mammalian RNAi studies. We designed siRNA duplexes to the ECE-1 sequence using the siDESIGN™ Centre on the Dharmacon website. The conventional design strategy does the following:

1. Identifies a region that begins 75-100 bases downstream of the start codon to avoid nucleotide sequences occupied by regulatory or translational proteins and exonexon junctions.
2. Identifies the first AA(N19)TT motif, where N 19 represents the 19 nucleotide core duplex and AA and TT define the overhang sequences of the duplex.
3. Evaluates the GC content of the potential siRNA sequence. Ideally, the G:C ratio should be near to 50% or 30-70% according to Tuschl's guidelines.
4. Compares the candidate siRNA with known expressed sequence tags (ESTs) in publicly available databases to ensure specificity of targeting (e.g., www.nebi. nlm.nih.gov/BLAST).

Biosynthesis of Oligonucleotides

Chemical synthesis provides the best material of choice for introduction of siRNAs in vitro. Although the natural intermediates that siRNAs mimic carry a 5' phosphate, the 5' phosphate is omitted in chemical synthesis as a cellular kinase rapidly phosphorylates the siRNA once it is inside the cell. Dharmacon has developed siRNAs suitable for in vivo protocols, which include 2' deprotection, desalting, and duplex annealing.

These modifications enhance the persistence of the siRNA duplex inside the cell. Other modifications to consider are delivery, concentration, and dosage. For a complete review, see Nature Methods' "Focus on RNA Interference-A User's Guide".

Regulatory Method

To maximize the specificity and validity of any RNAi experiment, the researcher should

1. Include untreated cells to serve as a baseline standard.
2. Titrate the siRNA duplex being transfected to its lowest effective level; siRNA concentrations greater than 100nM produce nonspecific off-target effects at a much higher frequency than concentrations of 20-100nM.
3. Confirm that the target mRNA (northern analysis/real-time PCR) or protein (Western analysis) is specifically reduced by comparing to a nontargeting, non-functional siRNA control that will reveal any off-target effects.
4. Include a positive silencing control such as high-efficiency siRNAs targeted to known genes, such as lamin A/C, cyclophilin B.
5. Confirm that the observed experimental phenotype is not the result of off-target effects by using multiple siRNAs targeted to distinct regions of the same gene. Dharmacon's readily available siRNA SMARTpool may be an option.
6. If possible, rescue expression of the target gene by expressing it in a form resistant to the siRNA being used.

Procedures of siRNA Delivery

siRNAs can be delivered using many protocols, including lipofection, electroporation, microinjection, and infection. For a lipid mediated oligonucleotide transfection into adherent mammalian cells the following protocol is used for one well of a six-well plate:

1. Culture STO cells routinely in 75-cm^2 flasks in lOmL of DMEM medium supplemented with 10 % (v/v) FBS.
2. To seed the cells, remove the culture medium and wash twice with 10 mL PBS. Add 1 mL of 1X trpsin/EDTA and incubate at 37°C until the cells detach from the flask. Stop the trypsin reaction by adding 8 mL of DMEM.
3. Count the cells and seed them into a six-well plate at 50% confluency 24 h prior to transfection.
4. For one well of a six-well plate, add 4µL of OligofectamimeTM to a sterile tube containing 26 µL of Opti-MEM.
5. In *a* separate tube, dilute 4 µL of a 20µM stock of the siRNA oligonucleotide duplex in 66 µL of Opti-MEM to give a 100 nM oligo concentration (based on 400 µL total volume).
6. Incubate the individual mixes for 10 min at room temperature.
7. Add the diluted OligofectarnimeTM reagent to the diluted siRNA duplexes. mix gently, and

incubate the pooled mixture at room temperature for 20 min.

8. For each six-well transfection add 300µL of Opti-MEM to the mix, giving a final volume of 400 µL per well.
9. Wash the cells once with Opti-MEM.
10. Overlay 400µL of the complex into each well of cells.
11. Incubate the cells for 4h at 37°C in a CO2 incubator (5% CO_2).
12. Following incubation, add 2mL of normal growth media containing three times the normal amount of FBS to the cells without removing the transfection mixture.
13. Assay cell extracts for gene activity 12-72h post transfection.

Procedures for siRNA Gene Silencing

Analyzing mRNA

For a real-time (quantitative) PCR analysis,

1. Isolate total RNA using the RNeasy Kit following the manufacturer's instructions.
2. Quantify the RNA using a spectrophotometer at 260 nm. The purity of RNA can be estimated from the ratio of absorbance readings at 260 nm and 280 nm.
3. Perform first-strand synthesis using the iScript cDNA Synthesis Kit (Bio-rad). For a single reaction, pipette lXx iScript reaction mix, 1 µg of total RNA template, and 1 µL of script reverse transcriptase into a PCR tube and add ddH2O to give a total volume of 20µL for each reaction.
4. The reaction protocol is as follows: 25°C for 5 min, 42°C for 30 min, and 85°C for 5 min.
5. Assay the resulting cDNA using real- time PCR. Real- time PCR was performed in 0.2-mL thin-wall PCR plates using the iCycler thermal cycler and carried out with SYBR-Green Supermix.
6. In a single reaction, include 1X SYBR-Green Supermix, 10 pmol of each primer and 10 ng of template in a total reaction volume of 20µL. Prepare a master mix of SYBR Green Supermix containing the primers and probe. Add 18 µL to each well containing 2 µL of cDNA.
7. Heat the mixture initially to 95°C for 3 min followed by 35 cycles with denaturation at 95°C for 30 s, annealing at 57°C for 30 s, and extension at 72°C for 30 s. Perform each assay for unknown samples simultaneously with standard samples. Standard samples should consist of serial dilutions of candidate DNA (0.01 pg-100 pg). A reference gene such as GAPDH should be used.

Analyzing Protein

Western Blotting Analysis

For preparation of the cell lysates,

1. Wash cells three times in PBS and collect using a rubber policeman.

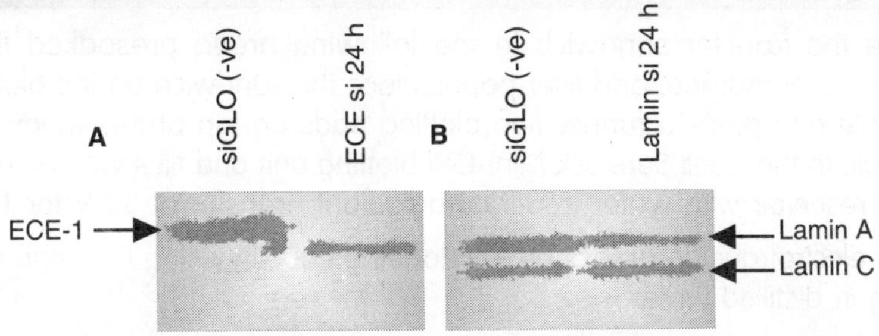

Figure18.2: Western blot analysis showing knock-down of ECE-1 expression in mouse stromal (STO) cells. (A) STO cells were transfected with 200nM ECE-I siRNA (B) and 200nM Lamin siRNA (B) using Oligofectamine then incubated for 24h. The siGLO siRNA was used as a negative control. Samples were resolved using SDS PAGE and transferred to a nitrocellulose membrane. Membranes were incubated in ECE- 1 C (1:200) and Lamin A/C (1:500) primary antibody. Protein expression was reduced following siRNA treatment.

2. Centrifuge the cells at 190g for 10min.
3. Resuspend the resulting pellet in 200µL of ice-cold triple detergent lysis buffer containing EDTA-free protease inhibitor cocktail and pass five times through a 23G needle.
4. To reduce sample viscosity, add 2 µL of benzonase nuclease and incubate the samples for 20 min on ice.
5. Centrifuge the samples at 8,500g for 2 min and collect the supernatant.
6. To determine protein concentration using the the bicinchoninic acid (BCA) protein assay, first, generate a standard curve using bovine serum albumin (BSA). Add a $CuSO_4$/BCA solution to the sample, standard or blank, in a 96-well plate and incubate at 37°C for 30min. Measure the absorbance at 570nm in a colorimetric plate reader.

For SDS-PAGE and Western Nlotting,

1. Dilute the protein isolated from whole cell lysates (10-40 µg) to 1.1 X the desired concentration in 4X LDS sample buffer and ultrapure water.
2. Mix the resulting sample solution with reducing agent in a 9:1 ratio and incubate the samples at 70°C for 10 min on a heating block.
3. Resolve the prepared samples by SDS-PAGE using Tris-acetate 3-8% (v/v) acrylamide gels. Remove all packaging from the gel and rinse the wells three times with IX NuPAGE Tris-acetate SDS running buffer. Fill the wells with running buffer.
4. Place the gels in an electrophoresis unit with the well side facing inward then load 5 µL of prestained standard and 20 µL of each of the samples onto the gel.
5. Fill the inner chamber with 1X running buffer containing 0.5% (v/v) NuPAGE antioxidant. Fill the outer chamber with running buffer and resolve the proteins in the sample by electrophoresis at 150V for 1 h.
6. Following electrophoresis soak the blotting pads, nitrocellulose membrane, and two pieces of filter paper in IX NuPAGE transfer buffer containing 10% (v/v) methanol per gel and 0.5% (v/v) antioxidant. Position two blotting pads in the cathode core of the blotting module. Remove the Tris-acetate gels from their cassettes and excise the stacking gel.

Assemble the transfer sandwich in the following order: presoaked filter paper, gel, nitrocellulose membrane, and filter paper. Place the sandwich on the blotting pads in the blot module and place a further two blotting pads on top of the sandwich. Position the blot module in the XCell Surelock Mini-Cell blotting unit and fill it with IX transfer buffer. Fill the outer reservoir with water to act as a coolant. Transfer at 30 V for 1 h.

7. Following electrotransfer, assess protein-loading equality using Ponceau red dye. Destain by rinsing in distilled water.
8. Block nonspecific binding sites with TBST containing 5% (w/v) milk powder and 2% (w/v) BSA. Incubate the membranes for 1 h with agitation.
9. Add primary antibody for 2h at room temperature with gentle rotation at an appropriate dilution in TBST containing 2% milk powder. ECE-1 (1:500).
10. Wash the membranes three times over 10min in TBST.
11. Incubate with horseradish peroxidase-conjugated secondary antibody for 1 h at room temperature.
12. Wash the membranes three times over 10min with TBST.
13. Visualize immunoreactive bands using enhanced chemiluminescence (ECL). Incubate the membrane in 6 mL of a 1:1 mix of detection solutions A and B provided for 1 min at room temperature.
14. Wrap the membrane in Saran wrap and expose it to autoradiography film.

Staining Methods

1. Sterilize the glass coverslips by dipping them in 70% ethanol and then flaming while loosely holding between forceps. Place one sterile coverslip in each well of a six-well plate.
2. Plate 2mL of the cell suspension into the wells containing the coverslips at 60% confluency and allow the cells to adhere over 24 h.
3. Wash the cells twice with 2 mL PBS and fix and permeabilize for 10 min in 2 mL of freshly prepared methanol/acetone (1:1 ratio) at room temperature. Set up a humidified chamber consisting of a petri dish with a lid containing water-saturated filter paper on which a layer of parafilm is placed.
4. Place the coverslips cell face down onto 50μL of blocking buffer, which is pipetted directly onto the parafilm in the humidified chamber to block nonspecific binding. Incubate the coverslips for 30min at room temperature in the humidified chamber.
5. Incubate the coverslips in 50μL of primary antibody at an appropriate dilution in blocking buffer for 2 h at room temperature in the humidified chamber. For the control antibody, see Note elsewhere in this chapter.
6. Wash the cells three times in PBS over 5 min on a shaker.
7. Incubate the coverslips for 30min at room temperature with 50μL of labeled secondary antibody (FITC conjugated antimouse IgG, 1:1000) diluted in blocking buffer.

8. Wash the cells once over 5 min in PBS and then add 1 mL of DAPI at 1.1000 dilution in PBS from 0.5 mg/mL stock in DMSO to the coverslips in the wells of a six-well plate and incubate for 1 min at room temperature.
9. Remove the DAPI immediately and wash the coverslips three times in PBS over 5 min on a shaker.
10. Invert and mount the coverslips onto glass slides using Vectashield and seal the edge of the coverslips using nail varnish.
11. Examine the cells using a fluorescent microscope. Images were captured using Delta Vision from Applied Precision.

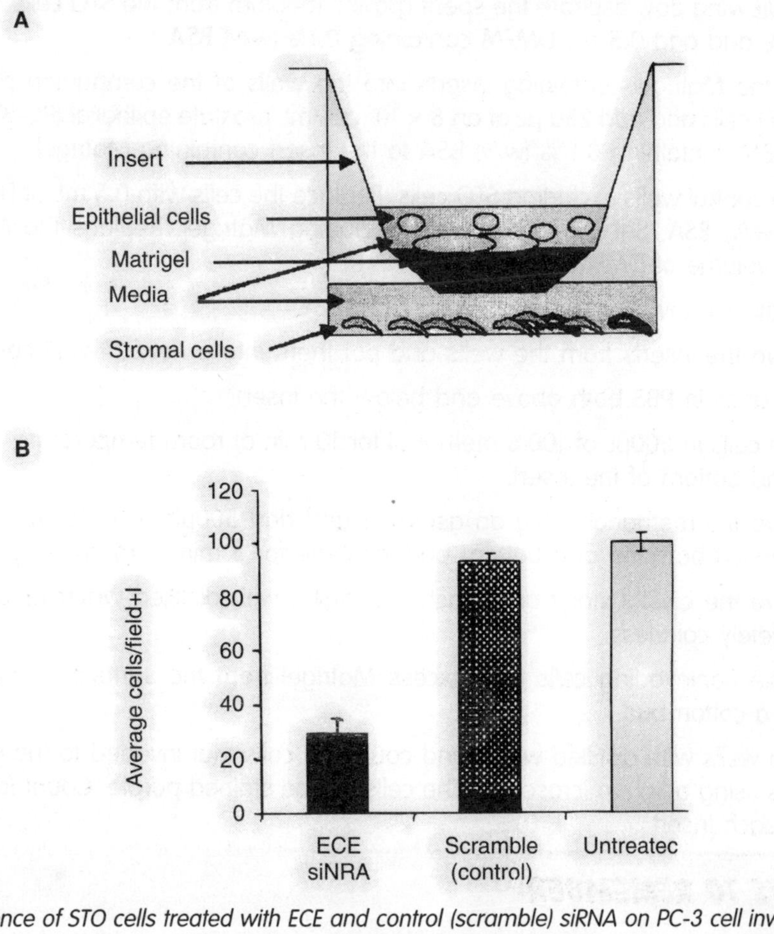

Figure 18.3: Influence of STO cells treated with ECE and control (scramble) siRNA on PC-3 cell invasion. (A) Stromal cells are plated onto the bottom of the well and Matrigel is layered over the 8-µm pore insert. Both well and insert are incubated separately at 37 °C for 24h. Next day, the insert is united with the well and epithelial cells are seeded over the Matrigel. Supplements are added to both well and insert and the assay is incubated overnight. (B) STO (stromal) cells transfected with ECE and scramble (control) siRNA were harvested 48h post transfection. Treated STO cells and prostate cancer (PC-3) epithelial cells were plated as in (A). PC-3 cells invading through the Matrigel were counted after 24h incubation. Each bar represents the mean value of eight fields counted.

Biochemical Assays

1. Defrost the Matrigel overnight at 4 °C on ice. Matrigel will gel rapidly at 22-35 °C (Matrigel may even gel at slightly elevated temperatures in a refrigerator). Dilute the Matrigel to 250μg/mL (v/v) with DMEM using precooled pipettes, plates, and tubes.
2. Place the plastic inserts into the wells of a 24-well companion plate using sterile forceps.
3. Layer 200μL of Matrigel onto each cell culture insert housed in a 24-well companion plate and incubate overnight at 37 °C.
4. In parallel, seed 1 mL of STO cells at 100% confluency into the wells of a 24-well companion plate and incubate overnight at 37'C.
5. The following day, aspirate the spent growth medium from the STO cells, wash once with DMEM, and add 0.5 mL DMEM containing 0.1% (w/v) BSA.
6. Place the Matrigel-containing inserts into the wells of the companion plates containing the STO cells and add 250 μL of an 8×10^5 cell/mL prostate epithelial (PC-3) cell suspension in DMEM containing 0.1 % (w/v) BSA to the insert containing Matrigel.
7. Set up control wells excluding STO cells. Replace the cells with 0.5 mL of DMEM containing 0.1% (w/v) BSA. Set up control wells excluding Matrigel. Replace the Matrigel with an equal volume of DMEM (250μL).
8. Incubate the invasion assay overnight at 37 °C.
9. Remove the inserts from the wells and put them into a new 24-well companion plate.
10. Wash once in PBS both above and below the insert.
11. Fix the cells in 500μL of 100% methanol for 10 min at room temperature. Add to both the top and bottom of the insert.
12. Remove the methanol using an aspirator and add 500μL of 0.1 % (w/v) crystal violet to each insert both top and bottom and incubate for 5 min at room temperature.
13. Remove the crystal violet and wash four times with distilled water or until the wash is completely colorless.
14. Remove noninvading cells and excess Matrigel from the surface of the Matrigel layer using a cotton bud.
15. Fill the wells with distilled water and count the cells that invaded to the underside of the inserts using a light microscope. The cells will be stained purple. Count four fields of view from each insert.

POINTS TO REMEMBER

1. A cDNA synthesis kit should be purchased with respect to the thermal cycler being used for real-time PCR.
2. At room temperature, BD Matrigel matrix polymerizes to produce biologically active matrix material resembling the mammalian cellular basement membrane. It is stable for at least 3 months when kept frozen at -20°C or 12 d at 37 °C. Avoid multiple freeze-thaws. Do not

store in a frost-free freezer.

3. This protocol can be adapted for other invasive cell line models.
4. Dharmacon's conventional design method (www.dharmacon.com) selects 19-21 nucleotide targets followed by a nucleotide (BLAST) search. Due to the variable success rate of this method and to rule out off-target effects, investigators are advised to complement custom siRNA sequence studies with a Dharmacon commercial SMARTpool targeted to the same candidate gene.
5. If no suitable sequence can be found, the search can be relaxed to identify sequences with the motif NA(N21).
6. Optimal cell density differs between cell lines and should be determined empirically for each cell line. Transfection of oligonucleotides is very sensitive to cell density, and so it is important to maintain a standard seeding protocol to ensure reproducibility.
7. We compared a number of transfection reagents, such as Ribojuice, Dharmafect, and Fugene-6, but maximum transfection efficiency was observed with Oligofectamine.
8. The melt curve should be checked to ascertain specificity of primer by identifying a single peak. Multiple peaks imply poor primer specificity for target sequence.
9. Due to the sensitivity of quantitative PCR, pipetting errors can easily affect results. To avoid this prepare a master mix of SYBR Green Supermix containing both the primers and probe. For optimal reproducibility of replicate samples, the template DNA sample could be added to aliquots of the master mix.
10. Choose the reference gene carefully; for example, GAPDH expression fluctuates with hormone treatment.
11. Equivalent loading also can be checked by probing the membrane with an antibody against a housekeeping protein such as b*-actin. Levels of housekeeping proteins remain constant.
12. Alternative fixing solutions can be employed, as some antibodies cannot tolerate acetone. Paraformaldehyde or formalin is less harsh on the membrane and often the method of choice for studying membrane proteins.
13. For monoclonal antibodies, the control antibody should be of the same IgG subclass, such as IgG1. For polyclonal antibodies, incubate with serum specific to the animal in which the polyclonal was raised.
14. A separate coverslip incubated with secondary antibody alone is a useful reference point for identifying background nonspecific antibody binding.
15. DAPI is a nuclear stain and useful for identifying the presence of cells when the reference protein is not visible.
16. Removing noninvading cells from the surface of the Matrigel layer prevents them being accidentally counted along with cells that invaded to the underside. After counting, wipe away invading cells to ensure no more cells are visible.

Chapter 19

Chromosomes of Parasitic Protozoans

In contrast to the chromosomes of metazoan parasites, the chromosomes of many protozoan parasites cannot be visualized using a microscope because they are too small or do not condense sufficiently. The development of pulsed-field gel electrophoresis (PFGE), combined with the release *in situ* of unbroken chromosomes from cells embedded in agarose plugs, allowed definition of the karyotypes of many of these parasites for the first time.

PFGE also revealed the considerable size heterogeneity between apparently homologous chromosomes in different strains and closely related species. Indeed, many analyses have used that size variability to their advantage to determine the number of chromosomes, as we illustrate in Figure elsewhere in this chaptere.

One problem of PFGE, compared to visualization of chromosome spreads, is that non-homologous chromosomes may be of similar size and, therefore, not resolved in the gel. However, hybridization of single-locus markers to a range of strains with size-variable chromosomes allows the researcher to determine syntenic groups and, hence, the total number of chromosomes.

The separation of individual chromosomes in gels also allows their further manipulation. Early chromosome maps were based on the excision of individual homologs and the use of rarecutting enzymes to digest the intact chromosome into fragments, again *in situ* in agarose to prevent random breakage, and these methods remain useful for the analysis of sections of individual chromosomes or, indeed, artificial chromosomes or fragmented chromosomes.

The ability to elute individual chromosomes from a gel was also exploited by several sequencing teams working with protozoan parasite genomes. The malaria genome is remarkably AT-rich. Large fragments of this genome cloned into vectors that replicate in bacteria (e.g., bacterial artificial chromosomes or cosmids) were found to be highly unstable, enforcing a reliance on yeast artificial chromosomes.

To circumvent the many problems inherent in this approach, the sequencing teams developed whole chromosome shotgun , whereby a chromosome is eluted from the gel and shotgun cloned to produce a chromosome-enriched library (usually containing approx 10-13% contamination from

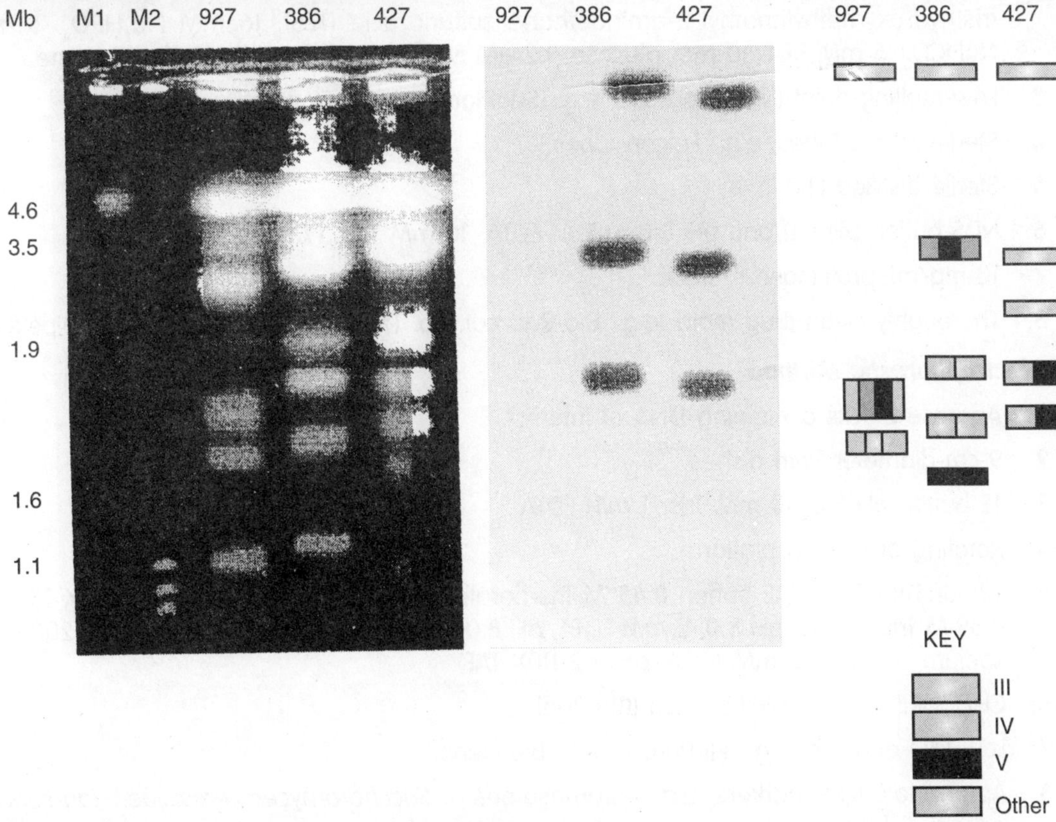

Figure 19.1: Resolution of chromosomes III, IV, and V in the T. brucei *karyotype using PFGE, hybridization of a chromosome-specific marker, and comparison of three karyo-typically variable stocks. Note that hybridization of markers specific to these chromosomes to stock 927 only does not determine chromosomal location. CHEF DRIII conditions: 3000-2400 s, 72 h, 1.5 V/cm, 106°, 12°C, followed by 850—450 s, 72 h, 2.4 V/cm, 106°, 12°C. 1% agarose gel in 0.5x TBE. M1,* S. pombe; *M2,* S. cerevisiae; *Mb, Megabases.*

other chromosomes). This method has now been used in several sequencing projects. Here, we provide basic protocols for the manipulation of chromosome-sized DNA molecules in agarose; their resolution, whole or digested, by PFGE; and their excision and cloning for the production of small-insert chromosomal libraries in plasmid vectors. We base these protocols on our work with African trypanosomes and *Leishmania* species and include additional notes on how to adapt the protocols to work with other protozoan parasites.

REQUIREMENTS

Isolation Chromosomes for Trypanosoma

Intact Chromosomes in Agarose Plugs

1. Parasites: e.g., *Trypanosoma brucei* procyclics in semi-defined medium SDM-79.
2. Trypanosome dilution buffer (TDB), pH 8.0: 118 mM NaCl, 1.2 mM KH_2PO_4, 30 mM N-

tris[hydroxymethyl]methyl-2-aminoethane sulfuric acid (TES), 16 mM Na_2HPO_4, 5 mM $NaHCO_3$, 5 mM KC1, 10 mM glucose, 1.2 μM $MgSO_4$ or phosphate-buffered saline.

3. Low-melting-point (LMP) agarose: e.g., SeaPlaque"GTG", Flowen.
4. Sterile 50-mL tubes, e.g., Falcon tubes.
5. Sterile distilled H_2O.
6. NDS buffer, pH 9.0 and pH 8.0: 0.5 *M* EDTA, 10 mM Tris, 1% lauroylsarcosine.
7. 10-mg/mL proteinase K stock.
8. Thoroughly clean plug mold (e.g., Bio-Rad cat. no. 170-3713) 9. Sterile Pasteur pipets.

Gel Electrophoretic Method

1. Agarose blocks containing DNA of interest.
2. 9-cm-diameter Petri dishes.
3. TE buffer, pH 8.0: 10 mM Tris. 1 *mM* EDTA.
4. Rotating or rocking platform.
5. TBE or TB(0.1)E or TAE buffer: 0.45 *M* Tris-borate, pH 8.0, 10 mM EDTA, pH 8.0(5X TBE) or 0.89 M Tris-borate, pH 8.0, 2 *mM* EDTA, pH 8.0 (I OX TB [0.1]E) or 0.4 M Tris-HCl, 200 mM sodium acetate, 10 mM EDTA, pH 8.2 (10X TAF).
6. CHEF DRII or CHEF DRIII system (Bio-Rad).
7. Low EEO agarose: e.g., Hi-Pure Ultra"', Bio Gene.
8. Appropriate size markers, e.g., chromosomes of *Saccharomyces cerevisiae, Han-senula wingei, Schizosaccharomyces ponibe* in agarose plugs.
9. 1-L conical flask.
10. 50-mL conical flask.
11. Sterile distilled H_2O.
12. Sterile Pasteur pipets.
13. 10 mg/mL ethidium bromide stock solution.
14. UV transilluminator.
15. Camera/gel imaging system.

Extraction of a Single Chromosome

From a Pulsed Field Gel

1. Glass plate.
2. Ruler. 3. Ethanol.
4. Black paper/dark surface.
5. Sterile scalpel blades.
6. Containers and electrophoresis buffer from gel tank.

7. 10 mg/mL ethidium bromide stock solution.
8. UV transilluminator.
9. Camera/gel imaging system.
10. Foil/sterile needles.
11. 15-mL Falcon tubes.
12. 0.5 M EDTA, pH 8.0.

Removal of a Chromosome in Agarose

1. Excised chromosomal DNA in molecular biology-grade agarose.
2. Sterile scalpel blade.
3. Petri dishes.
4. TE buffer, pH 8.0: 10 mM Tris-HCl, 1 mM EDTA.
5. Enzymes needed for digestion plus appropriate buffers.
6. Eppendorf tubes.
7. Water bath.
8. Appropriate size markers, e.g., 0.1-200-kb Pulse Marker (Sigma).

Methods of Southern Blotting

1. UV transilluminator.
2. Denaturing solution: 0.5 *M* NaOH, 1 *M* NaCl.
3. Neutralization solution: 0.5 M Tris-HCl, 1.5 M NaCl.
4. Nylon membrane (e.g., Hybond N+, Amersham Biosciences).
5. 3MM paper (Whatman) and paper towels.
6. 10X SSC, pH 7.0: 1.5 *M* NaCl, 0.15 *M* Na-citrate.
7. Large tray and glass plates.
8. Labeling and detection reagents (e.g., AlkPhos Direct, Amersham Biosciences).
9. Hybridization oven and bottles.
10. X-ray film (e.g., Kodak X-OMAT AR) and cassette.

Plasmid Library of Chromosomal DNA

Isolation of Chromosome DNA

1. β-Agarase and buffer (NEB).
2. Ultrapure phenol (Invitrogen).
3. Butanol (Sigma).
4. 1 M *NaCl.*
5. 1 mg/mL glycogen (Sigma) in deionized H_2O.

6. Ethanol (absolute and 70%).
7. TE buffer: 10 mM Tris-HCl, pH 8.0, 1 mM EDTA.

Blunt-End Fragment Preparation

1. 10X MBB buffer: 50% (v/v) glycerol, 10 mM $ZnCl_2$, 0.5 *M* NaCl, 0.3 *M* sodium acetate, pH 5.0.
2. Sonicator fitted with a cup horn probe (e.g., model CL4, Misonix, Inc., Farmingdale, NY).
3. DNA molecular-weight markers. e.g.. a mixture of lambda digested with *Hind* III and pBR322 digested with *BstNI* (NEB).
4. Agarose gel electrophoresis equipment and ethidium bromide (EtBr). 5. Mung bean nuclease (Amersham, 156 U/µL).
6. 1 M NaCl.
7. Pellet paint (Novagen),
8. Absolute ethanol.
9. 1X TAE buffer: 40 mM Tris, 20 mM sodium acetate, 1 mM EDTA, pH 8.2
10. Long-wavelength UV transilluminator.
11. TE buffer: 10 mM Tris-HCl, pH 8.0, 1 mM EDTA.
12. Dye solution: 10% (w/v) ficoll 400 (Sigma), 0.1% w/v bromophenol blue in 1X TBE. 13. AgarAce (Promega).

Shotgun Libraries

1. T4 DNA ligase (Roche, 5 U/µL) and buffer.
2. pUC 18 SmaI-CIP (40 ng/µL), Qbiogene.
3. 14 mg/mL proteinase K stock (Roche).
4. Electroporation apparatus.
5. Electrocompetent cells, e.g., Electromax DH I 0B (Invitrogen).

PROCEDURES

Isolation of Chromosomes from Leishmania

When preparing agarose plugs containing chromosomes from *Trypanosoma* or *Leishmania* species, we begin with a parasite density of approx 1×10^8/mL of culture medium.

Intact Chromosomes in Agarose Plugs

1. Look at organisms under the microscope and determine health and number. Work out the concentration of the cell culture by your usual method (we use a hemocytometer) and the resuspension volume required to achieve the correct concentration. Prepare an appropriate amount of 1.4% LMP agarose solution and hold at 37°C until required.
2. Spin down culture at 3000-4000g, at 4-8°C, for 5-10 min only (you do not want to lose too

many, but you do not want them packed down such that you damage them when resuspending). Work fast and try to keep them all alive.

3. Resuspend in cool (not too cold) TDB with sterile Pasteur pipet (tip cut off to widen aperture). Resuspend fully but without damaging the cells. This is a wash, so they must resuspend.
4. Spin down again. Wash again. Do not leave them sitting around.
5. Resuspend cells in TDB at twice the concentration you require in the plugs. (We aim for 5 $\times 10^1$ cells/mL final concentration, i.e., 5 × 107 per 100 µL plug; therefore, we resuspend to 1×10^9 cells/mL).
6. Warm cells briefly in 37°C for 1-2 min, swirling constantly. Add equal amount of 1.4% LMP agarose solution to give 0.7% agarose plugs. Mix quickly, but thoroughly.
7. Fill plug mold quickly (the pipet tip should not have too narrow a bore, we use a sterile 3-mL plastic Pasteur pipet). Leave to solidify for a few minutes, then place carefully in a dish large enough to accommodate the plug mold but not so large as to waste buffer. Add enough NDS, pH 9.0, to the dish to cover the plug mold and place in the fridge for 10-15 min. This allows the plugs to set completely without sticking to the sides of the plug mold or becoming distorted.
8. Push plugs out of mold with minimal distortion and drop them into NDS, pH 9.0, with 1 mg/mL proteinase K (minimum volume, as the enzyme is expensive). Leave at 50°C for 24 h.
9. Cool plugs at 4°C (to make them easier to handle and less likely to distort, which leads to breakage of large DNA molecules) and change buffer to NDS, pH 8.0, 1-mg/mL proteinase K. Incubate again for 24 h at 50°C to allow the enzyme to destroy all cellular proteins released into the agarose.
10. Cool plugs. Transfer to fresh NDS, pH 8.0 (sufficient to cover), and store at 4°C. They will keep for many years if they are well prepared and not damaged.
11. Dialyze plugs thoroughly against TE or TBE (electrophoresis buffer) before setting up gel to remove all detergent and reduce EDTA concentration (either overnight, or you may shorten the required time by using a very large volume, movement, and changing buffer several times).

Gel Electrophoresis

1. Place required number of plugs into 9 cm-diameter Petri dishes and add enough TE buffer, pH 8.0, to cover the plugs. Place on a slowly rotating platform and leave for 30 min-1 h. Replace with fresh TE buffer at least twice more to ensure thorough dialysis.
2. Prepare 2 L of the required buffer by diluting the stock solution with sterile distilled H_2O.
3. Place the buffer into the gel tank, switch on the pump and cooling module, and set the cooler to the required temperature. Allow the buffer to equilibrate to the required temperature.
4. Remove 120 mL buffer from gel tank. Weigh required amount of agarose and place in a 1-L conical flask. Add the buffer, weigh the flask plus contents and make a note. Heat

carefully to boiling point and check that all the agarose has dissolved. Weigh flask plus contents, and add sterile H_2O to make back up to original weight. Allow to cool to approx 55°C.

5. Cover the teeth of the comb with autoclave tape and place the comb horizontally on a level surface. Place the plugs in the required order onto the comb. Using a sterile Pasteur pipet, place a few drops of agarose solution onto and between each plug and allow to set for 2-3 min.
6. Position the comb into the mold carefully and pour in the agarose solution, retaining approx 5 mL for use later. Allow the gel to set for 30 min to 1 h.
7. Carefully remove the comb. Melt the retained agarose and use to fill the hole left by the comb. Allow to set for 2-3 min.
8. Transfer the gel to the gel tank, positioning carefully within the support. Ensure that the pump and cooler are switched on and appropriate conditions set, then start run. Check that bubbles are visible from the electrodes and make a note of the starting current.
9. When the run is completed, remove the gel from the tank and place into a box containing enough running buffer to cover the gel. Add EtBr to 0.5 μg/mL and allow the gel to stain for approx 1 h (sometimes overnight staining in a lower concentration of EtBr gives a superior gel picture).
10. Place the gel onto a UV transilluminator to visualize the DNA and photograph. It may be necessary to stain the gel for a longer period of time or to destain the gel depending on the concentration of DNA in the gel. Avoid exposing the DNA to UV for long periods of time, as this will break the DNA and allow it to diffuse out of the gel.

Restriction Mapping of Individual Chromosomes

It is absolutely vital that the chromosomal DNA is never allowed to come into contact with ethidium bromide or is exposed to UV light.

Chromosome From a PFG

1. Clean a glass plate and a ruler with ethanol and place on a dark surface. e.g., a piece of black paper. This will make it easier to see where to cut.
2. Using the ruler as a guide, cut a strip off the right and left ends of the gel ensuring that a few millimeters, or a lane, of your sample and/or markers are included in the cut area.
3. Place these gel strips into a container and add enough buffer from the gel tank to cover the gel plus 0.5 μg/mL ethidium bromide. Allow to stain for approx 1 h, Place the remaining gel into a container with electrophoresis buffer only; *Do not add EtBr.*
4. Place the gel strips onto a UV transilluminator to visualize the DNA and photograph. Mark the position of the band(s)/region of interest using small cuts, above and below the band, with a sterile scalpel blade. If desired, small pieces of foil or sterile needles can be placed in the cuts to make them more visible and aid alignment.
5. Reassemble the unstained gel plus gel strips on the glass plate and locate the region of interest.

Table 19.1: Examples of PFGE Separation Parameters for *T. brucei* and *L. major* Chromosomes.

Size range	Switch time (s)	V/cm	Angle	Run time (h)	Temperature (°C)	Agarose (%)	Buffer
0-200 Kb	3-8	6.0	120	16	13	1.0 1	X TB (0.1)E
0.1 Mb-1 Mb	175	2.5	120	120	12	1.0	1X TB(0.1)E
0.5 Mb-2 Mb	400-200	2.5	120	144	12	1.0	1X TB (0.1)E
1 Mb-1.5 Mb	450	2.5	120	96	14	1.0	1X TB(0.1)E
1Mb-2 Mb	850-450	2.5	120	144	14	1.0	1X TB(0.1)E
1 Mb-3 Mb	1400-700	2.5	120	144	14	1.2	1X TB(0.1)E
3 Mb-6 Mb	3600	1.2	120	168	14	0.8	0.5X TBE
1 Mb-3 Mb	1100-600	2.4	106	72	14	1.0	0.5X TBE
3 Mb-6 Mb	3600-3000	1.5	106	96	14	0.8	1X TB(0.1)E
1Mb-6 Mb	(a) 3000-2400	1.5	106	72	14	1.0	0.5X TBE
	(b) 850-450	2.4	106	72			

6. Carefully excise the band using a sterile scalpel blade and using the ruler and cut marks as a guide. Place the excised chromosome into a 15-mL Falcon tube and store in 0.5 M EDTA, pH 8.0, at 4°C if not used immediately.
7. Stain the remaining gel pieces in running buffer containing 0.5 μg/mL ethidium bromide, then check that the correct region has been removed by reassembling the remaining gel pieces on the UV transilluminator.
8. If you intend to extract the chromosomal DNA for preparation of a shotgun clone library, then reset the chromosome slice into a gel prepared with LMP agarose as described under Subheading Gel Electrophorenis (This can be difficult to handle and costly, in which case it is possible to produce a composite gel by preparing a standard gel exactly as described above, then excising an area below the chromosomal band and refilling with LMP agarose.)

Electrophorese using the same parameters as previously, for approx half the time, then excise the chromosome again as described in steps 1-7.

Removal of a Chromosome in Agarose

This method may be used to digest individual chromosomes, or large-insert clones such yeast artificial chromosomes (YACs) excised from a PFG. Do not assume that restriction enzyme sites that are rare in other genomes, such as mammals, will be rare in the parasite genome also. For example, *NotI* cuts the *T. brucei* genome much more frequently because of the higher GC content of this genome.

After electrophoresis of the restricted fragments, you will not necessarily see the fragments following EtBr staining. It is not always best to increase the concentration of the DNA in the plugs, as this leads to increased contamination of the excised chromosome, but you should be able to detect the fragment of interest on hybridization of a suitable marker.

1. Cut the gel fragments so that they are approx 6 mm long, 3 mm wide, and 6 mm deep and dialyze them in TE buffer, pH 8.0, for several hours as described in Subheading Gel Electrophoresis.
2. Place the fragments in 1.5-mL Eppendorf tubes and add sufficient 1 X restriction buffer to cover fragment. Incubate for I h at room temperature.
3. Remove buffer and replace with 1 X restriction buffer containing 50 U enzyme. Place at required temperature for at least 24 h.
4. Dialyze fragments in TE buffer, pH 8.0, for 2 h before setting into an appropriate agarose gel.
5. Electrophorese using conditions suitable for expected size.

Method of Southern Blotting

Because of the large size of the chromosomal DNA in the PFG, it is necessary to nick the DNA before transferring onto a nylon membrane. This can be achieved by depurination or by UV irradiation. We routinely nick the DNA with UV light at 0.12 J/cm^2, then transfer the DNA onto nylon membrane (Hybond N+, Amersham) using capillary transfer as described by Sambrook. We extend

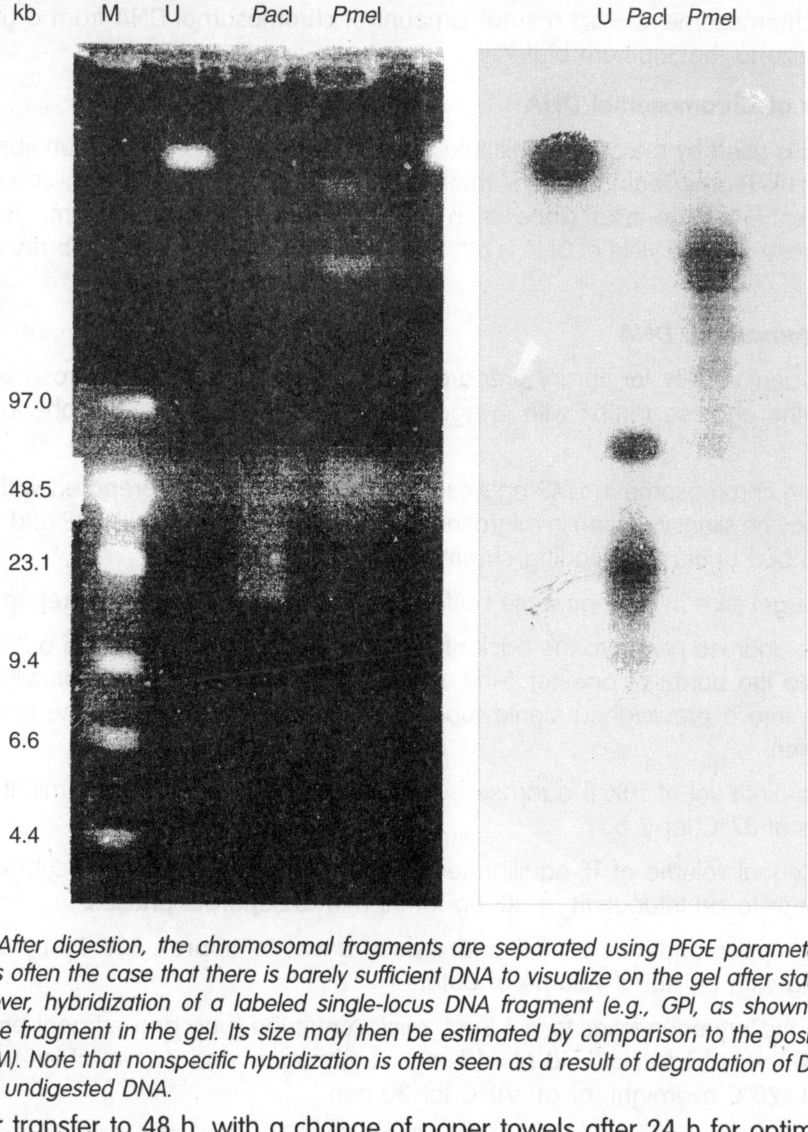

Figure 19.2 : After digestion, the chromosomal fragments are separated using PFGE parameters appropriate to their size. It is often the case that there is barely sufficient DNA to visualize on the gel after staining with EtBr (left panel); however, hybridization of a labeled single-locus DNA fragment (e.g., GPI, as shown here) reveals the position of the fragment in the gel. Its size may then be estimated by comparison to the position of markers of known size (M). Note that nonspecific hybridization is often seen as a result of degradation of DNA fragments. kb, kilobases; U, undigested DNA.

the time for transfer to 48 h, with a change of paper towels after 24 h for optimal results. After allowing the membrane to dry, cross-link the DNA to the membrane by exposing it to 0.12 J/cm2 UV light.

Hybridization of DNA probes may be carried out using standard radioactive labeling methods or nonradioactive methods. We routinely use the AlkPhos direct labeling and detection system (Amersham Biosciences), following the manufacturer's instructions, although we increase the temperature of hybridization and washing steps to 65°C instead of 55°C.

You may be able to determine the order of the fragments derived from the chromosome by hybridization of entire BAC or cosmid clones to Southern blots, if such clones do not contain repeated sequence found in multiple fragments. Also, to visualize all fragments derived from restriction of a chromosome, extract a small amount of chromosomal DNA from a gel slice and label and hybridize to the Southern blot.

Plasmid Library of Chromosomal DNA

This method is used by the Sanger Institute to produce chromosome shotgun libraries (http://www.sanger.ac.uk/Teams/Team53/psub/ methods.shtml), but it is equally applicable to a large restriction fragment or a large-insert clone, such as a bacterial artificial chromosome (BAC), excised from a LMP agarose gel. The yield of DNA is often quite low, 100-500 ng, and so library construction can be difficult.

Isolation of Chromosome DNA

DNA of sufficient quality for library preparation can be isolated from agarose gel slices by digesting away the agarose matrix with (3-agarase and then extracting with phenol to remove contaminants.

1. Excise the chromosome in LMP agarose, preferably from a gel prepared with TAE buffer (there may be slight variation in migration between gels prepared with TAE and TBE buffers), as described under Subheading chromosome from PFG.
2. Soak the gel slice in 1X β-agarase buffer for 30 min on ice. Drain and repeat this wash.
3. Load the agarose plug into the back of a 5-rnL syringe barrel fitted with a 23-ga needle. Force into the barrel of another 5-mL syringe. Repeat four times, eventually forcing the agarose into a preweighed sterile tube. Reweigh and calculate volume of agarose by subtraction.
4. Add one-ninth vol of 10X β-agarase buffer and 8 U of β-agarase per milliliter. Mix and incubate at 37°C for 2 h.
5. Add an equal volume of TE-equilibrated phenol and vortex for 1min. Incubate on ice for 5 min prior to centrifugation at 3000g for 10 min to separate phases.
6. Take the upper aqueous phase and reduce down to approx 300 PL by successively extracting with an equal volume of butanol.
7. Remove the aqueous layer to a 1.5-mL microcentrifuge tube and ethanol precipitate by adding 0.1 vol of 1 M NaCl, 10 µL of 1-mg/mL glycogen, and 2.5 vol of ice-cold ethanol. Leave at -20°C overnight, or at -70°C for 30 min.
8. Pellet DNA by centrifugation, wash with 70% ethanol and dry. Resuspend pellet in 21 µL TE, then ascertain yield by running 1 µL of prep, alongside markers of known DNA concentration (we use 50 ng of lambda *Hind*III digest), through a standard agarose minigel (e.g., 0.8% agarose in TBE).

Blunt-End Fragment Preparation

1. Pool sufficient material gained from several extractions until you have around 10 µg of extracted DNA, and pipet into a sterile 1.5-mL microcentrifuge tube. Add 6 µL of 10X MBB buffer and water to give a final volume of 60 L. Place on ice.
2. Microfuge briefly to settle the tube contents and fragment the DNA by sonication, keeping

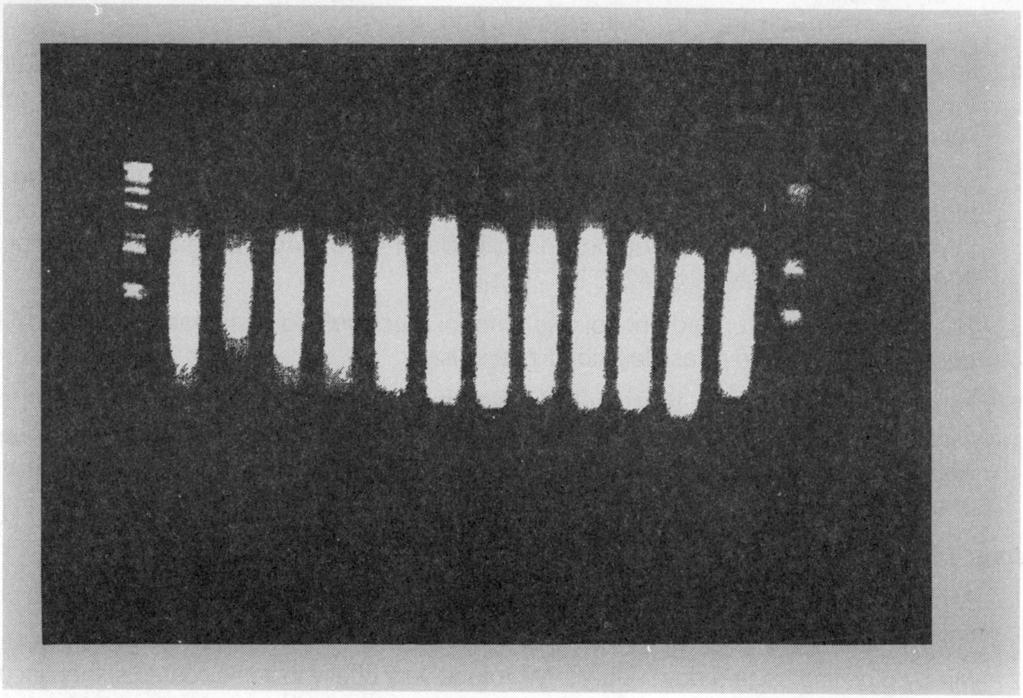

Figure 19.3: After sonication of the eluted chromosomal DNA, the size of the resulting fragments is assessed on an agarose gel. You should see a smear of DNA fragments around 500 by to 4 kb, but very little remaining HMW DNA. This gel picture shows ideally sonicated DNA samples, flanked by DNA molecular-weight markers (a mixture of λ digested with HindIII and pBR322 digested with BstNI).

the sample cold throughout to prevent denaturation of DNA. We use a sonicator fitted with a cup-horn probe on top of which the sample is suspended in a 4°C water bath. Cooling is maintained by siting of the whole apparatus in a cold room. Sonicate at 15% of full power for 2 × 10 s, at a distance of 1 mm from the probe surface. Incubate on ice for 1. min and briefly microfuge to settle contents between sonications.

3. Run a 1-μL aliquot of each sample alongside a suitable marker on a 0.8% agarose minigel. There are three possible outcomes:

   a. Complete sonication (no sign of high-molecular-weight [HMW DNA, smear between > 4 kb and 500 bp).

   b. Near-complete sonication (smear and faint HMW DNA).

   c. Unsonicated (faint smear and substantial HMW DNA).

   In the case of b., resonicate for 5 s, and check again on a minigel if desired. In the case of c., resonicate again for 2 ×10 s and check again on a minigel before proceeding.

4. Add 0.3 μL of mung bean nuclease to each tube, mix gently, briefly microfuge to settle contents, and incubate at 30°C for 10 min.

5. Add 141 μL of water, 20 μL of 1 M NaCl, 1 μL of pellet paint, and 550 μL cold ethanol. Leave at -20°C overnight or at -70°C for 30 min. Pellet DNA by centrifugation at 13,000g, 4°C for

30 min, wash with 1 mL ice-cold ethanol, and dry.

6. Resuspend DNA pellet in 6.25 µL TE, 0.75 µL 10X TAE, and 2 µL dye solution. Run out through a 0.8% LMP agarose, IX TAE gel (with 2 µL ethidium bromide per 50 mL gel volume). We run a 15 × 10 cm gel at 50 V, 70 mA, for 2 h.
7. On a long-wave UV transilluminator, cut out 0.6-1 kb, 1-1.4 kb, 1.4-2 kb, and 2-4 kb agarose blocks and place in fresh 1.5-mL microcentrifuge tubes. Melt at 65°C for 5 min or pass through a 26-ga needle to fragment. Equilibrate at 42°C, add 1 U AgarAce (Promega) per 200 µL vol and incubate at 42°C for 20 min.
8. Extract with TE-equilibrated phenol and ethanol precipitate, adding one-tenth vol 1 M NaCl and 1 µL of pellet paint as described previously.
9. Pellet DNA fragments to be cloned by centrifugation at 13,000g, 4°C for 30 min, wash with 1 mL ice-cold ethanol, and dry. Resuspend in 3 µL of T0.1E if you started the procedure with < 1 µg, 5 µL of T0.1 E if you started with 1-5 µg, and 10 µL if you started with >5 µg of extracted DNA.

Shotgun Libraries

1. Set up ligations as follows, in 0.5-mL microcentrifuge tubes: 3 µL insert DNA, 0.3 µL pUC18 SmaI-CIP, 0.4 µL lox ligase buffer, 0.3 µL ligase.
2. Mix by pipetting up and down slowly. Microfuge very briefly to settle contents, then incubate overnight at 12-14°C.
3. Next morning dilute to 50 µL with sterile distilled H_2O and add 1 µL of 14-mg/mL proteinase K solution and incubate at 50°C for 1 h to denature the ligase. Thereafter, keep ligations cold (preferably frozen), because there is no buffering in the solution.
4. Test libraries for quality and titer, by electroporation of a 0.2µL aliquot into high-efficiency electrocompetent cells.

POINTS TO REMEMBER

1. Some people use PBS instead of our TDB. We find that TDB works best, probably because it contains ingredients that keep the trypanosomes alive and in good condition until the last moment.
2. Many companies produce PFG-specific agarose, optimized for use with large DNA fragments. These are generally more expensive, and we do not use them routinely.
3. If using parasites that grow intracellularly, e.g., *Toxoplasma,* or parasites grown in laboratory rodents, then it is necessary to completely separate the intact parasites from the host cells before preparing the plugs. The parasites must remain intact until the moment they are set into agarose.

   When deciding how many parasites per plug you require, take into account variation in genome size-aim for approximately the amount of DNA suggested here, as increasing the amount too much leads to problems in removal of proteins, etc.
4. We have found that estimates of cell numbers by this method vary among laboratory

workers. It is best to make plugs at several concentrations in the first instance, e.g., half and/or twice the given concentration also. If there are too few cells in the plugs, then this can make analysis difficult, as you cannot compensate for this subsequently. If there are too many, then you can compensate when setting up gels by cutting up the plugs.

However, if cell density is too great, the proteinase K may not remove all proteins, reducing quality and preventing DNA migration into the gel.

5. The cells must be alive until they die slowly *in situ* after addition of agarose. If they die earlier than this, then the cell wall can degrade and the DNA is broken and/or degraded enzymatically. Ideally, the cell wall will be lysed on addition of the plugs to the detergent in NDS.

   At this point, at least while learning, you could look at them under the microscope and see if they are still wriggling. But don't take long! Just smear a drop onto a slide or a Petri dish and look under any microscope available.

6. The detergent in NDS buffer lyses the cell walls and releases the chromosomes *in situ*. *Trypanosoma* and *Leishmania* parasites have cell walls that are easily and gently disrupted in this way. If the parasites of interest do not lyse so easily, nevertheless a method must be found to lyse them *in situ* in the agarose. If they lyse prior to embedding in agarose-while still in liquid medium-the chromosomes will break
7. Dialysis removes the EDTA and detergent from the storage buffer. The length of time to allow for dialysis is dependent on the number of plugs in each Petri dish; the greater the time and numbers of changes of TE buffer.
8. Use autoclaved deioinized H_20 to avoid introducing fungi and other contaminants.
9. Volumes given are for gels cast using the 14 × 12.7 cm casting stand supplied with the CHEF-DRII or CHEF-DRIII system and will result in a 6-mm-thick gel.
10. Loading plugs onto a comb covered in autoclave tape rather than pushing them into the well prevents distortion of the plugs-and it is easier.
11. It is important to cut a strip from either side of the gel in case the gel has not run straight. Cut enough to get a clear idea of where the bands are. For a wide gel, it may be beneficial to remove a narrow strip from the center of the gel as well.
12. The final size of cut gel fragments is largely dependent on the initial size of the chromosomal band. However, if the gel fragments are too large, the efficiency of digestion may be reduced and a larger volume of enzyme may be needed, increasing the cost.
13. In this instance, we avoid any contact with EtBr during preparation of the chromosomal DNA because this increases cloning efficiency, in contrast to Subheading chromosome from PFG where we avoided EtBr and UV to avoid breakage of the large DNA molecules.
14. This step is designed to allow the b*-agarase enzyme access to the agarose matrix. The gel plug may be dissipated either as described here or by incubating at 65°C for 10 min to melt the agarose. In our experience, melting at 65°C is more efficient but can lead to denaturation of A/T-rich DNA sequences and the underrepresentation of such sequences in the resulting library.

15. Note that butanol forms the upper layer in this mixture and that you must retain the lower aqueous layer. If too much butanol is added, all the aqueous layer will be absorbed into it. In such an event, gradually add TE until an aqueous phase can be observed. The DNA should segregate back into the aqueous layer.
16. 10 µg will give the best results, though with care as little as 100 ng for plasmid libraries will be sufficient.
17. An alternative fragmentation method such as nebulization may be used, in which case the fragmented DNA may need to be concentrated by ethanol precipitation before proceeding with the remainder of the protocol. Details of nebulization can be found at the Web site of Bruce Roe's group at the University of Oklahoma: http:// www.genome.ou.edu/ protocol book/protocol parti1.html#II, and descriptions of other alternate DNA fragmentation techniques.
18. The desired size range of fragments varies depending on your experiment. However, in general, smaller fragments clone more efficiently. Therefore, usually it is best to clone DNA from several blocks and check the actual size range of the cloned inserts in the resulting libraries.

Chapter 20

Bioinformatic Methods

In the postgenomic era, attention is turning towards the systematic assignment of function to proteins encoded by genomes. Bioinformatics methods are now used ubiquitously in initial efforts to assign function to predicted open reading frames. However, while such methods can give helpful insights into possible function, there are now many examples of proteins that have closely related sequences and/or structures but prove to have quite different functions.

There is thus an emerging need for high-throughput methods that are suitable for the experimental determination/verification of protein function. At the forefront of this monumental task, the field of proteomics is now segregating into discovery- and systems-oriented proteomics. Discovery-oriented proteomics is mainly concerned with documenting the abundance and localization of individual proteins as well as building a picture of protein-protein interaction networks.

This is the realm of two-hybrid screens, two-dimensional (2D)-gel electrophoresis, and more direct mass spectrometry-based methods; the latter two methods in particular are commonly used to understand the way in which expression profiles change in response to different stimuli by comparing, for example, diseased and healthy cell extracts. However, these discovery-oriented proteomics methods tell us little about the function of individual proteins or protein complexes.

Systems-oriented proteomics takes a different approach; rather than re-discovering each protein in each new experiment, the focus is on a predefined set of proteins, enabling more precise questions to be asked regarding the functionality of each member of that set. However, obtaining quantitative and genuinely comparative functional data across large sets of proteins with any degree of accuracy is technically difficult, requiring isolation of each individual protein in an assayable format.

We and others have chosen to focus on protein function microarray-based methods because the parallel, high-throughput nature of microarray experiments is attractive for analyzing large numbers of protein interactions, while the uniform intra-array conditions both simplify and increase accuracy of assays. Additionally, the small volumes of ligand or reaction solution required to perform assays, typically tens to hundreds of microliters, can provide economic advantages-for example, when using expensive recombinant proteins or labeled compounds.

The key element to such microarray experiments is that the arrayed, immobilized proteins retain their folded structure such that meaningful functional interrogation can then be carried out. There are a number of approaches to this problem, which differ fundamentally according to whether the proteins are immobilized through nonspecific, poorly defined interactions or through a specific set of known interactions.

The former approach is attractive in its simplicity and is compatible with purified proteins derived from native or recombinant sources, but suffers from a number of risks-most notably that the uncontrolled nature of the interactions between each protein and the surface might give rise to a heterogeneous population of proteins or destroy activity all together.

In practice, an intermediate situation frequently occurs, where a fraction of the immobilized proteins either have undergone conformational change as a result of the nonspecific interactions or have their binding/active sites occluded by surface attachment; these effects will effectively reduce the specific activity of the immobilized protein and therefore decrease the signal-to-noise ratio in any subsequent assay.

The advantages of controlling the precise mode of surface attachment are that, providing the chosen point of attachment does not directly interfere with activity, the immobilized proteins will have a homogeneous orientation, resulting in a higher specific activity and signal-to-noise ratio in assays, with less interference from nonspecific interactions. This may be of particular advantage when studying protein-small-molecule interactions in an array format.

The disadvantages of this approach, though, are that it is really compatible only with recombinant proteins or with families of proteins, such as antibodies, that have a common structural element through which they can be immobilized.

However, in a systems-oriented approach, the disadvantage of working with recombinant proteins is currently largely outweighed by the problems encountered in individually purifying large numbers of active proteins from native sources.

In this chapter we therefore describe the expression, array fabrication, and assay of a set of recombinant proteins in which the mode of surface attachment is tightly controlled. Furthermore, we show how laborious pre-purification of the recombinant proteins prior to array fabrication can be avoided through use of a suitable affinity tag, thus greatly simplifying array fabrication.

We illustrate this approach to array fabrication with respect to a set of disease-associated variants of the human tumor suppressor protein p53, expressed as fusions to a polypeptide tag that becomes biotinylated in vivo. In addition, we show representative data from a number of different assays carried out on protein function arrays made in this way.

REQUIREMENTS

1. pQE-80L expression system (Qiagen).
2. *Escherichia coli accB* gene.
3. Human p53 cDNA.
4. *E. coli* strain XL1-blue.
5. Luria-Bertani (LB) media.

6. Biotin.
7. Ampicillin.
8. Isopropyl-b*-D-thio-galactopyranoside (IPTG).
9. Phosphate-buffered saline (PBS): 1.5 mM KH_2PO_4, 4.3 mM Na_2HPO_4, 137 mM NaCl, 3 mMKCl (pH 7.3).
10. PBST: PBS, 0.1% (v/v) Tween-20.
11. Marvel/PBST (PBS, 100 mg/mL Marvel, 0.1% [v/v] Tween-20).
12. p53 buffer: 25 mMHEPES (pH 7.6), 50 mMKCl, 20% (v/v) glycerol, 1 mMdithio- threitol (DTT), 1 mg/mL bovine serum albumin (BSA), 0.1% (v/v) Triton X-100. 13. p53 freeze mix (250 mL p53 buffer plus 250 mL glycerol; keep at 4°C). 14. Lysozyme (4 mg/mL in water).
15. SAM2 membrane (Promega).
16. Anti-His antibody (Sigma).
17. Cy3 Mono-Reactive Dye labeling kit (Amersham Biosciences).
18. GADD45 duplex oligo.
19. Genetix QArray robot.
20. DNA microarray scanner.

PROCEDURE

The methods described below outline (1) the construction of the expression plasmids, (2) the induction of protein expression, (3) the extraction of the protein from E. *coli,* (4) the printing of a protein microarray, and (5) the assay of the protein microarray for DNA binding function.

Expression Plasmid

The *E. coli* expression system used is based on the pQE-80L expression system. Sequences inserted into the multiple cloning site can be expressed as native proteins bearing an N-terminal hexahistidine tag upon IPTG induction of the T5 promoter in suitably transformed E. *coli* cells.

Cloning of the Wild-Type p53 Gene

All DNA manipulations were carried out using standard recombinant DNA methods to construct the expression plasmids, and are accordingly not described here in detail. The gene encoding the E. *coli* biotin carboxyl carrier protein (BCCP) domain (amino acids 74-156 of the *E. coli accB* gene) was amplified by polymerase chain reaction (PCR) from an E. *coli* genomic DNA preparation using primers that added a *5'-Barn* HI site, removed the natural stop codon of BCCP, and added a 3' multiple cloning site including *Sac I, Not I,* and *HindIII* restriction sites.

This PCR product (approx 280 bp) was cloned into pQE-80L downstream of and in frame with sequences encoding six histidine residues as *a Bam HI/ Hin* dIII fragment to generate an intermediate vector, pQE-80L H6-BCCP. The full-length, wild-type *p53* gene was amplified by PCR from a HeLa cell eDNA library (Clontech) using primers that added a 5' *Sac I* site and a 3' *Not I* site.

Figure 20.1: Crystal structure of the biotin carboxyl carrier protein domain of Escherichia coli. *The N- and C-termini and, 50Å away, the single lysine residue that is biotinylated in vivo can all clearly be seen. Figure prepared from PDB file 1 BDO using Swiss PDB Viewer.*

The PCR product (approx 1190 bp) was cloned into the intermediate vector downstream of and in frame with the BCCP domain as *a Sac I/Not I* fragment, such that the resulting construct (pQE-80L H6-BCCP-p53), encoded a His6-BCCP-p53 fusion protein. The *p53* gene was sequence verified and confirmed as matching SWISS-PROT entry P04637.

Preparing p53 Variants

Forty-eight variants of p53 were generated by inverse PCR using the wild-type p53 expression vector, pQE-80L H6-BCCP-p53, as template. Phosphorylated forward primers bore the sequence variation at the 5'-terminus followed by 20-24 nucleotides of *p53* sequence.

Unphosphorylated reverse primers were complementary to the 20-24 nucleotides immediately before the mutated position. PCR was performed using *Pwo* polymerase (Roche Molecular Biochemicals) and generated blunt-ended products corresponding to the entire p53-containing vector.

PCR products were gel purified and ligated, and methylated parental template DNA was digested with *Dpn I* (New England BioLabs). E. *coli* XL1 Blue cells were transformed to ampicillin resistance and mutant *p53* genes verified by DNA sequencing.

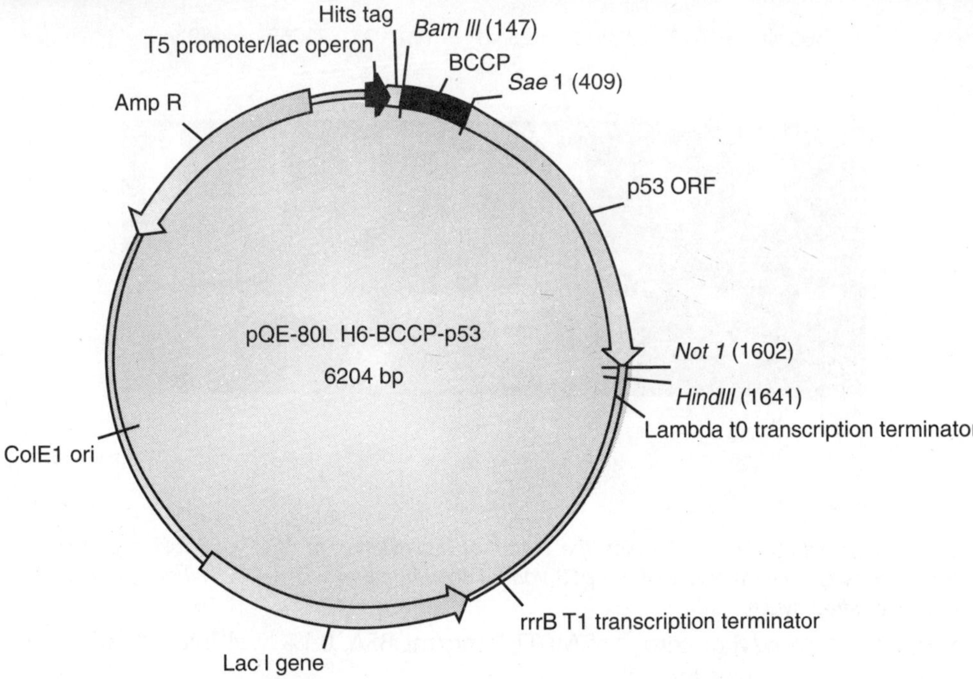

Figure 20.2: Map of plasmid vector used to express BCCP-p53 fusion proteins.

Transformation of the His6-BCCP-p53 Fusion Proteins

1. Transform E. *coli* XL1-blue cells with the pQE-80L H6-BCCP-p53 plasmid DNA (or the plasmids encoding the individual mutants) using standard methods. Plate cells on LB agar plates supplemented with ampicillin at 100 µg/mL and incubate overnight at 37°C.
2. Select single colonies and grow overnight at 250 rpm, 37°C in 5 mL Luria-Bertani medium supplemented with ampicillin at 100 µg/mL (LBamp).
3. Inoculate separate 200-mL quantities of LBamp with 4 mL of the individual overnight cultures and incubate at 250 rpm, 37°C until the cultures reach OD_{600} approx 0.4.
4. Reduce the temperature to 30°C and induce protein expression by addition of IPTG (100 µM) and free biotin (50 µM) for 4 h. It should be noted that the induction time can be extended to increase expression, but for the BCCP-p53 fusion proteins this usually results in a decrease in protein quality.
5. Harvest cells by centrifugation (4500g, 20 min).
6. Resuspend the cells in 100 mL PBS and re-centrifuge to remove excess free biotin.
7. Aliquot cells and store as cell pellets, each the equivalent of 2 mL of expression culture, at -80°C.

Protein Extraction

1. Thaw individual cell pellets on ice, add 80 µL of p53 buffer (25 mMHEPES [pH 7.6], 50

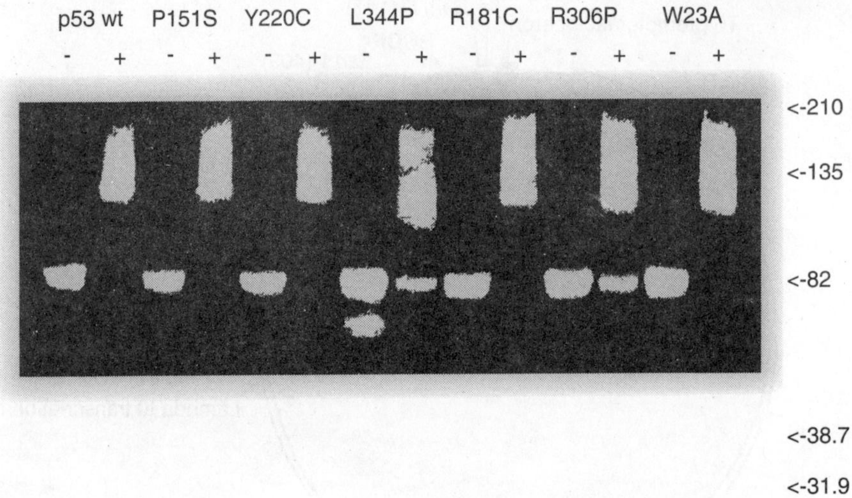

Figure 20.3: Supershift assay to determine the extent of biotinylation of BCCP-p53 fusion proteins. Each pair of lanes corresponds to a crude lysate of the p53 variant shown, either with (+) or without (-) preincubation with a molar excess of streptavidin.

mMKCl, 20% [v/v] glycerol, l m.MDTT, 1 mg/mL BSA, 0.1% [v/v] Triton X-100) and vortex the tubes to resuspend the cells.

2. Inspect the tubes visually to confirm that the cells have become efficiently dispersed; if not, continue vortexing until no clumps can be seen.
3. Add 20 µL of a 4 mg/mL lysozyme solution to each resuspended cell aliquot, mix, and incubate at 4°C with gentle agitation for 30 min.
4. Pellet cell debris by centrifugation at 10,000g for 10 min at 4°C.
5. Transfer the supernatants from each tube into individual wells of a 384-well V- bottom plate (Genetix) and keep at 4°C. Take care not to mix or confuse the sample identities at this stage. This is the source plate for the print runs.
6. Centrifuge the 384-well plate at 2200g for 5 min at 4°C to pellet any cell debris that has carried over. Store plate on ice prior to print run.
7. Determine the protein concentration of the soluble protein extracts by Bradford assay to confirm that effective cell lysis has occurred.
8. Determine the approximate expression level of soluble BCCP fusion by sodium dodecyl sulfate (SDS)-polyacrylamide gel electrophoresis (PAGE) together with western blot analyses using an anti-His primary antibody.
9. To determine the extent of biotinylation of the BCCP fusion protein, carry out a supershift Western blot assay (again with an anti-His primary antibody) in which equivalent crude lysate samples are preincubated with or without streptavidin (0.1 g/mL).

Fabrication

In the procedures described below, we do not employ a prepurification step prior to array fabrication, but instead rely on a rapid, single-step immobilization and purification procedure to create arrays of biotinylated BCCP fusion proteins.

Printing Technique

1. Stick one small piece of double-sided tape onto each end of a 7.5 H 2.5 cm glass microscope slide.
2. Remove SAM2 membrane from -20°C and warm to room temp.
3. Peel the backing off the tape and place the slide face-down onto SAM2 membrane.
4. Cut around the slide with a scalpel blade.

Fabrication Technique

In general, any microarray printer could be used to print the arrays. The printing procedures can be carried out at room temperature providing the source plate is kept at 4°C and the atmosphere in the print chamber is humidified. However, preferably the printing device itself should also be cooled.

Here we describe one specific set of parameters that work well on SAM2 membranes using a Genetix QArray, cooled to 4°C and equipped with 8 H 300 µm tipped solid pins.

1. Load the 384-well source plate into the QArray.
2. Load the SAM2 slides onto the print bed of the QArray.
3. Print the arrays using the QArray settings in Table elsewhere in this chapter.

Mechanism of Arrays After Printing

1. Remove slides from microarraying robot.
2. Cut across membranes at top and bottom using scalpel blade to separate the membranes from the tape and remove the arrays from slide by holding edge with tweezers.
3. Make a notch at top right of each array to indicate orientation.

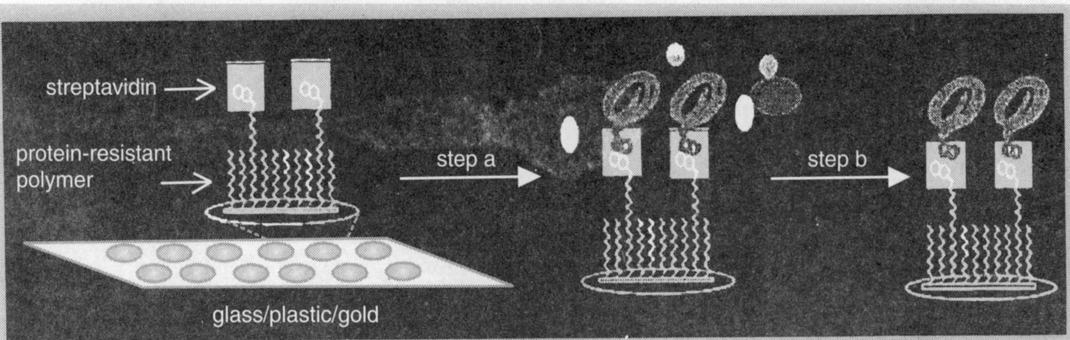

*Figure 20.4: Schematic of a single-step array fabrication process for in vivo biotinylated proteins.*Step a: *A crude lysate containing the desired biotinylated recombinant protein is printed onto a streptavidin-coated surface coderivatized with a polymer that resists nonspecific protein absorption.* Step b: *Unbound proteins are washed away to leave the purified recombinant protein, specifically immobilized and oriented on the array surface via the biotin moiety on the BCCP tag.*

Table 20.1: Genetix QArray Parameters for SAM2® Membranes.

Container = "STACKER SOURCE PLATE HOLDER"

Plate = "GENETIX PLATE 384 WELL"

Use Test plate = blank

Use stacker = blank

Source order by = Columns

Run print test = blank

Inking Time (ms) = 2

Max plates = 9,999

Substrate = "3 H 1' SLIDE (8-PINS/7-FIELDS)"

Holder = "SLIDE HOLDER"

Microarraying pattern = "4 H 4 1000 micron"

Use duplicate pattern = Yes (ticked)

1	1	2	2
3	3	4	4
5	5	6	6
7	7	8	8

(note: above pattern set up with 1000 micron spacing)

Field order = "p53 run —fields 1 and 2 shown below are used.

7	6	5	4	2	1	3

No. of substrates = "X"-dependent on the desired print run size

No. of fields in substrate = "2"

Max stamps per ink = "1"

No. identical fields = "2"

No. stamps per spot = "3"

Stamp time = "2 ms"

Microarraying by = Spot position

Printing depth = "250 microns"

Head = "8 PIN MICROARRAYING HEAD"

Water washes = 2000 ms wash, 1000 ms dry

Ethanol wash = 2000 ms wash, 7000 ms dry

4. Submerge the arrays in p53 buffer at room temp until all arrays are removed and notched.
5. Now transfer the arrays into a vessel containing 50 mL Marvel/p53 buffer (5% Marvel in p53 buffer). Rock the vessel at room temperature for 30 min, making sure that the arrays do not stick together.
6. Wash all arrays three times for 5 min in 50 mL p53 buffer.
7. Place in p53 freeze mix in a polypropylene tube and store arrays at -20°C.

Binding DNA

In order to assess the quality of the p53 protein microarrays printed on SAM2 membranes, two different assays are carried out on replica arrays. The antibody-binding assay comprises binding a Cy3-labeled anti-His antibody to the arrayed p53 proteins. This assay is independent of protein activity but provides a measurement of the relative amounts of protein immobilized in each spot.

This assay can also be used to determine intra- and inter-array variabilities. p53 is a tumor suppressor protein and, as part of its function as a transcription factor, it has a specific DNA-binding domain. The DNA-binding assay therefore comprises binding a Cy3-labeled duplex DNA containing a p53 recognition site to the active immobilized p53 proteins, and provides a measurement of activity of the proteins immobilized in each spot.

Fluorophore Technique

1. Following recommended protocols, buffer exchange an anti-His antibody into 0.1 Msodium carbonate buffer (pH 9.3) using a PD-10 desalting column (Amersham Biosciences) to give an antibody concentration of 1 mg/mL.
2. Add the protein solution (1mL) to one vial of Cy3 Mono-Reactive Dye (Amersham Biosciences) and mix thoroughly.
3. Incubate at room temperature for 30 min with occasional agitation.
4. Separate the labeled antibody from the free dye-labeled protein using a PD-l0 desalt ing column (Amersham Biosciences), eluting protein in 2 mL PBS to give a final antibody concentration of 0.5 mg/mL.
5. Estimate the molar dye/protein ratio by measuring absorbance at 280 nm and 552 run, using approximate extinction coefficients of 170,000 $M^{-1}cm^{-1}$ for protein and 150,000 $M^{-1}cm^{-1}$ for Cy3.

Antibody-Binding Method

1. Dilute 10 μL of the Cy3-labeled anti-His antibody in 1 mL Marvel/PBST to give an antibody concentration of 5 μg/mL.
2. Remove the array from freezing buffer and equilibrate in PBST at room temperature for 5 min.
3. Drain away the PBST, add the antibody solution to the array, and incubate with gentle agitation at room temperature for 30 min.
4. Wash the array three times for 5 min with 1 mL of PBST.

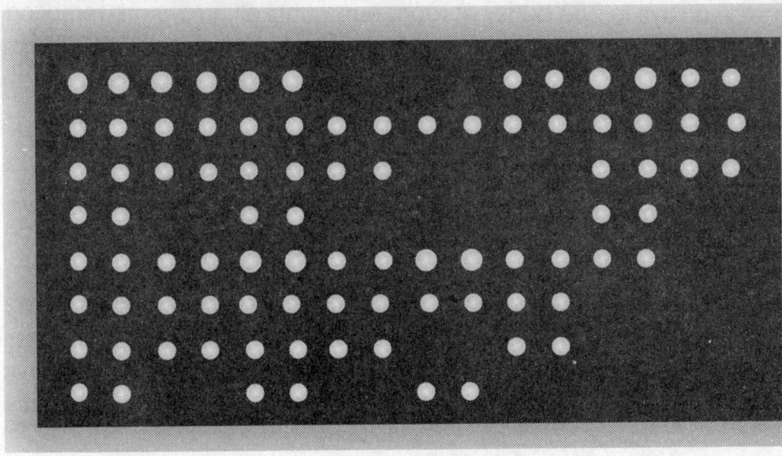

Figure 20.5: Measurement of the relative amount of protein immobilized in each spot in the array. A p53 protein function microarray probed with a Cy3-labeled anti-His antibody. Quantification of the signal intensity from each spot reveals the relative amounts of recombinant protein present in each spot in the array.

5. Dry the array by blotting it between two sheets of Whatman paper.
6. Mount the array onto a glass microscope slide using double-sided tape, trimming the edges with a scalpel blade if required.
7. Scan the array at 550 nm using a DNA microarray scanner and process the data using a DNA microarray data analysis software package.

DNA-Binding Assay

1. 5'-label complementary GADD45 promoter oligos (5'-gta cag aac atg tct aag cat get ggg gac-3' and 5'-gte ccc age atg ett aga cat gtt ctg tac-3') with Cy3 during synthesis.
2. Anneal the oligos together in p53 buffer and store at 4°C.
3. Dilute the 2Cy3-GADD45 duplex DNA in p53 buffer to a concentration of 250 nM. 4. Remove array from freezing buffer and equilibrate in p53 buffer at room temperature for 5 min.
5. Drain away the p53 buffer, add 1 mL of the DNA solution, and incubate at room temperature for 30 min with gentle agitation.
6. Wash the array three times for 5 min with 1 mL of p53 buffer.
7. Dry the array by blotting it between two sheets of Whatman paper.
8. Mount the array onto a glass microscope slide using double-sided tape, trimming the edges with a scalpel blade if required.
9. Scan the array at 550 nm using a DNA microarray scanner and process the data using a DNA microarray data analysis software package.

Other Techniques

We have also carried out other protein-protein interaction assays using p53 protein function microarrays fabricated according to the protocols described above. In such assays, we probe the

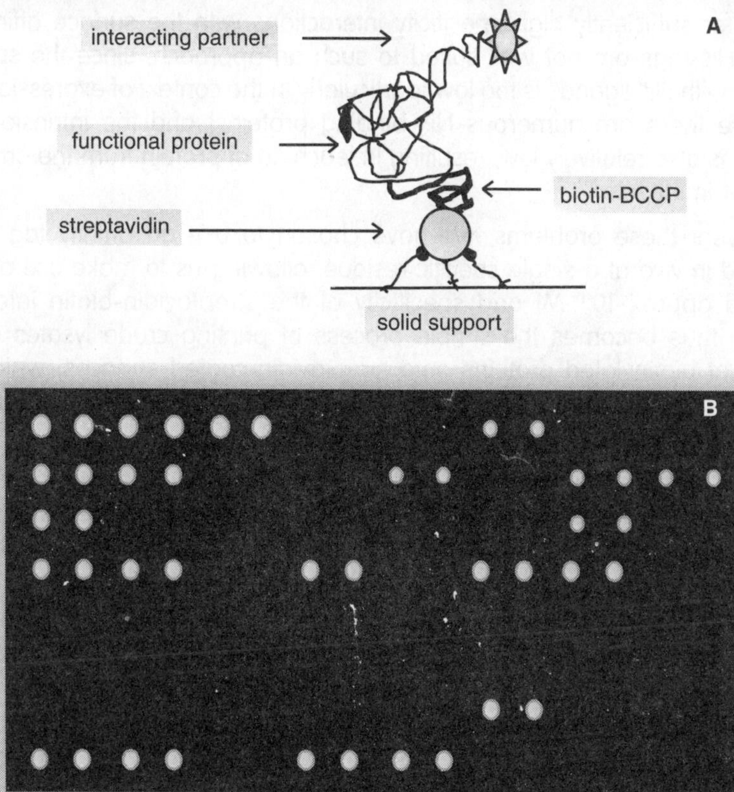

Figure 20.6: Measurement of the relative amount of ligand bound to each protein in the array. (A) Schematic of on-chip binding assay in which a fluorescently labeled interaction partner binds to the functional, arrayed protein immobilized to the streptavidincoated surface via the biotinylated BCCP tag. (B) p53 protein function microarray probed with Cy3-labeled GADD45 duplex oligo. Quantification of the signal intensity from each spot allows the effect of polymorphic and functional variation on the DNA binding function of p53 to be determined.

array with a Cy3-labeled solution-phase partner (for example, Cy3-MDM2) in a manner essentially identical to that described above for the antibody-binding assay.

In other work, using a range of different protein function microarrays, each fabricated in the form of BCCP fusion proteins as described here, we have been able to monitor a wide range of protein activities, including protein-protein interactions and protein-small molecule interactions, as well as carrying out on-chip phosphorylation assays.

POINTS TO REMEMBER

1. In order to avoid laborious prepurification of each recombinant protein prior to array fabrication, we have developed a procedure in which we combine the immobilization and purification stages into a single step. For this approach to work, the array surface itself must have a low capacity for nonspecific binding of proteins, yet must have a high specificity and high binding affinity for the proteins to be arrayed.

   In principle, a range of different affinity tags could be used here, but in practice, few actually offer sufficiently high specificity interactions with the surface affinity matrix. For example, His-tags are not well suited to such an approach, since the specificity of the interaction with Nit' ligands is too low (particularly in the context of expression in eukaryotic cells where there are numerous $Ni^{2'}$-binding proteins) and the intrinsic affinity of the interaction is also relatively low, resulting in leaching of protein from the array, even in the absence of imidazole.

To circumvent these problems, we have chosen to use an affinity tag that becomes biotinylated in vivo at a single specific residue, allowing us to make use of the very high affinity (Kd approx 10^{-15} M) and specificity of the streptavidin-biotin interaction. Array fabrication thus becomes the simple process of printing crude lysates containing the recombinant biotinylated proteins onto streptavidin-coated surfaces, which have a low nonspecific protein-binding capacity, followed by washing to remove all non-biotinylated proteins from the array surface.

One such commercially available surface that works well is streptavidin-derivatized phospho-cellulose membrane (SAM2), although we have also used streptavidin-coated glass or plastic surfaces. In the latter cases, we have found that organic polymer coatings, such as those based on dextran or polyethylene glycol, are superior to proteinaceous blocking agents such as bovine serum albumin or powdered milk in reducing the nonspecific binding background, and a number of such surfaces are now also commercially available.

We have observed that using the streptavidin-biotin interaction as the basis for array fabrication confers one further major advantage: the very high affinity of the streptavidinbiotin interaction means that we quickly start to saturate the available biotin-binding sites on the surface so a crude normalization of protein loading can be achieved without pre-adjusting the concentrations of the crude lysates to compensate for differences in the individual expression levels of the different recombinant proteins.

Obviously, host cell proteins that are endogenously biotinylated will be expected to compete with the biotinylated recombinant proteins for the available streptavidin-binding sites on the surface. E. *coli* has one such endogenous protein, the approx 25 kDa *accB* gene product, but under native conditions we have observed that this protein does not compete efficiently with biotinylated recombinant proteins for binding to streptavidin.

This perhaps reflects low natural expression levels and the fact that the endogenous, fulllength E. *coli* biotin carboxyl carrier protein naturally forms part of a heteromultimeric complex in which the biotin may not be readily accessible to streptavidin. As with many affinity tags, the biotinylated affinity tag can be positioned at either the N- or C-terminus of the protein to be arrayed, dependent on the structural and functional characteristics of the protein.

"Biochemical" biotinylation is greatly preferable to "chemical" biotinylation, since the latter offers little control over the site of biotinylation and still requires prepurification of each protein. There are three main alternatives for biochemical introduction of a biotin moiety into a recombinant protein: two involve affinity tags that can be biotinylated in vivo or in vitro, while the third involves an intein-mediated introduction of biotinylated cysteine.

The AviTag (Avidity, Denver, CO) is an in vitro evolved 15-residue peptide that is specifically biotinylated exclusively by the E. *coli* biotin ligase. We have chosen, however, to use a compact, folded, biotinylated, approx 80-residue domain derived from the E. *coli* biotin carboxyl carrier protein, because this affords two significant advantages over the AviTag.

Firstly, the BCCP domain is cross-recognized by eukaryotic biotin ligases, enabling it to be biotinylated efficiently in yeast, insect, and mammalian cells without the need to coexpress the E. *coli* biotin ligase. Secondly, the N- and C-termini of BCCP are physically separated from the site of biotinylation by approx 50 Å so the BCCP domain can be thought of as a stalk that presents the recombinant proteins away from the surface, thus minimizing any deleterious effects due to immobilization.

Recently, a third route to biochemical biotinylation has been described, in which a single biotinylated cysteine moiety is added to the C-terminus of recombinant proteins during intein-mediated protein splicing of fusion proteins. However, it is not clear at this time what advantages this route offers over use of the AviTag.

2. Vectors for expressing proteins as fusions to a BCCP domain derived from *Bleb-siella pneumoniae* are now available from Invitrogen (pET 104 DEST Biocase). This *K. pneumoniae* domain is highly homologous to the *E. coli* BCCP protein and confers the same properties.
3. Although the growth temperatures and induction times were specifically designed for BCCP-p53 fusions, they should be generally applicable to any other BCCP fusions as well. However, for different recombinant proteins, various combinations of growth temperature and induction time should be explored to find the optimal conditions.
4. We observed that addition of free biotin to the growth medium increases the extent of biotinylation of the recombinant BCCP fusion protein. Strains overexpressing the E. *coli* biotin ligase are also available from Avidity (www.avidity.com) if desired, although we have not found this necessary when using the BCCP tag.

   The wash step prior to cell lysis is needed to remove free biotin before array fabrication; if the protein is purified before array fabrication, this is not necessary.

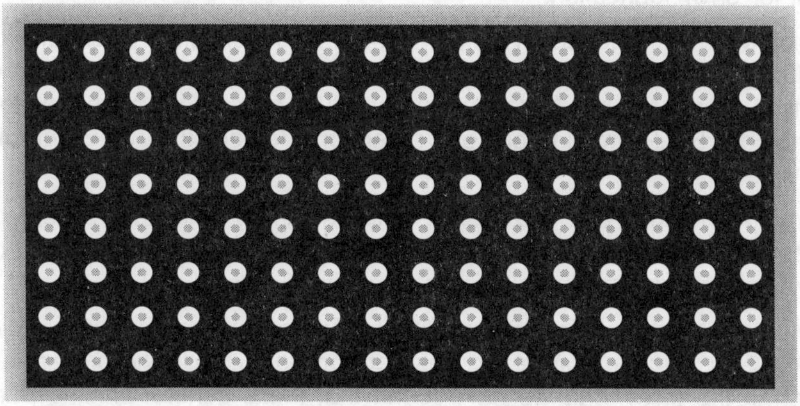

Figure 20.7: Determination of coefficients of variation for protein microarrays. Replica spotting of Cy3-labeled, biotinylated bovine serum albumin allows the intra- and interarray spot-to-spot reproducibility to be determined.

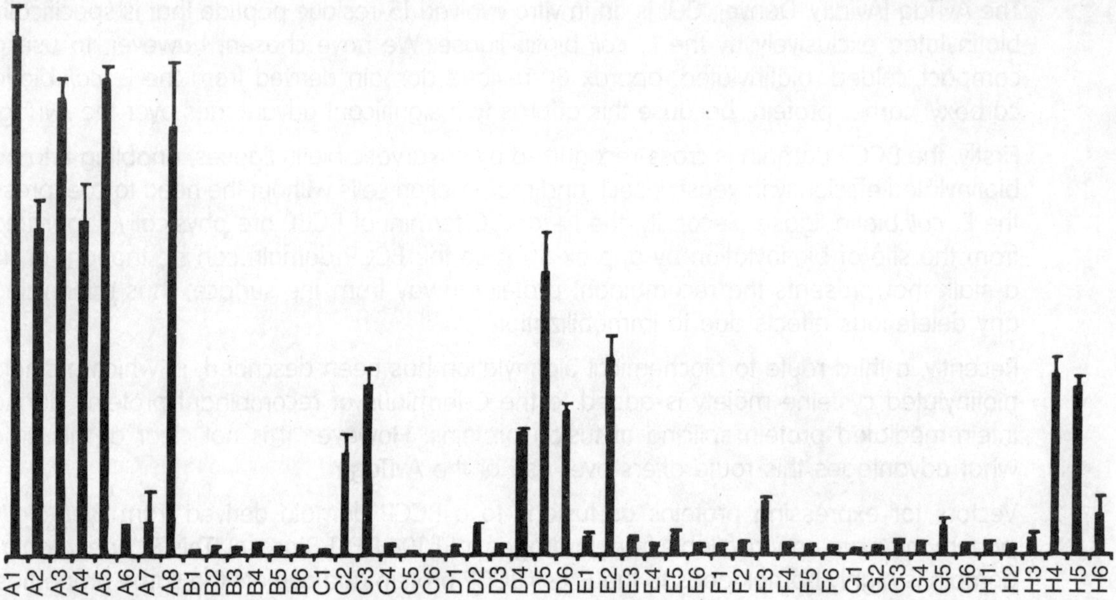

Figure 20.8: Determination of the relative amount of ligand bound per unit protein. The signal intensities from the DNA binding assay have been divided by the relevant anti-His tag antibody binding signal to show the relative amount of DNA bound per unit protein for each p53 variant in the array.

5. For expression of BCCP-p53 fusion proteins, we typically found the protein concentration in crude lysates to be 5 mg/mL, and we estimated that BCCP-p53 was present at approx 1 % of total soluble protein. When expressing a number of clones in parallel for array fabrication, the Bradford assay can conveniently be done on all clones in parallel using a microtiter plate format.

   However, it would be laborious to carry out SDS-PAGE analysis on all clones, so typically we assess only a selection of clones in this way, since the absolute expression level is not critical for array fabrication.

6. The biotinylated component of the sample will be supershifted by streptavidin even under denaturing conditions, enabling a simple side-by-side comparison to be carried out. To save time, this assay need not be done on all clones; we have typically found that if the recombinant BCCP proteins are expressed and folded, the BCCP domain is efficiently biotinylated.

7. We have found solid pins the easiest to clean rapidly, and have observed no carryover between samples using such pins. In addition, we have found that solid pins also work well when printing directly onto glass or plastic surfaces.

8. Each spotting event delivered approx 50 nL liquid containing approx 10 finoles biotinylated p53. We typically use multiple stamps per spot to increase the protein loading at each position in the array, and we have found that these general printing parameters work well with other surfaces, including glass microscope slides, although care must be taken in calibrating the z-height on the robot when using fragile surfaces.

In addition, by using the same printing parameters under the same conditions of printing (i.e., in a glycerol-containing buffer to reduce evaporation rates), prepurified proteins can be spotted onto nonselective surfaces that bind proteins by chemical cross-linking (e.g., epoxide- or aldehyde-coated glass or by noncovalent adsorption (e.g., supported nitrocellulose, agarose, or polyacrylamide).

9. The blocking and wash steps should remove all nonbiotinylated proteins from the array surface, while the biotin in the milk powder blocks any remaining biotinbinding sites on the streptavidin surface, such that any biotinylated proteins that do dissociate cannot rebind to the array. Arrays stored in p53 freeze mix at -20°C have been shown to be stable for > 1 y.
10. By printing, processing, and assaying arrays using these protocols, we have been able to achieve spot-to-spot CVs of 4-5%.
11. Fluorophores other than Cy3—for example, the Alexa dyes (Molecular Probes) or Oyster dyes (Denovo Biolabels)—also work well in these applications. However, when using any amine-reactive dye to label the protein, buffers containing primary amino groups such as Tris and glycine should be avoided, because they will inhibit the conjugation reaction. The presence of low concentrations (<2%) of biocides such as sodium azide do not affect protein labeling.
12. In principle, any microarray scanner can be used to read the protein microarrays, since the differences lie mainly in the sensitivity. However, models that read in a transmission mode might not work with membrane-based arrays.
13. The oligonucleotides can also be radiolabeled- for example, with ^{32}P, ^{33}P, or ^{35}S-in which

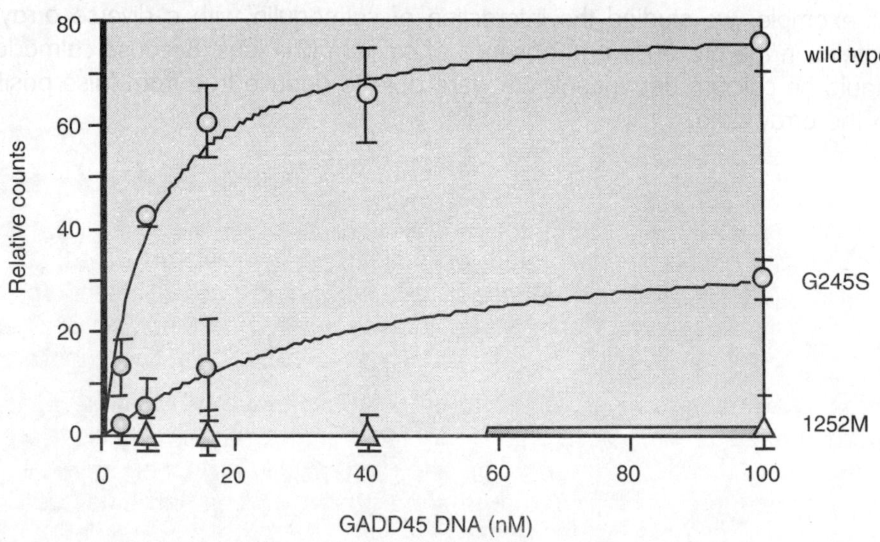

Figure 20.9: Quantification off ligand binding to arrayed proteins. Binding curves generated by measuring the relative amount of bound GADD45 DNA at increasing DNA concentrations allow binding affinities and specific activities to be calculated for each p53 variant in the array.

case the arrays can be read either using a phosphorimager or by autoradiography.

14. After initial data acquisition and processing, further data analysis can be carried out. For example, the raw data can be normalized for the relative amount of protein immobilized in each position in the array, as measured by an anti-His antibody-binding assay, by simply dividing each measured DNA-binding signal intensity by the relevant measured antibody-binding signal intensity.

    This gives a measure of the relative specific activity of each protein in the array. Alternatively, a number off replica arrays can be probed with different concentrations of ligand in solution (or potentially even by using increasing ligand concentrations on the same array, taking a reading at each concentration).

    The relative amount of bound ligand can then be plotted as a function of ligand concentration in solution for each protein in the array and fitted to a simple hyperbolic concentration-response curve according to

$$R = BmaxL/(Kd + L)$$

    where R is the response in relative counts and L is the ligand concentration. From this, the ligand binding constants and maximum ligand-binding capacities for each arrayed protein can be determined.

15. In the phosphorylation assay, we used ^{33}P-ATP to radiolabel those elements of the array that were substrates for the relevant kinase; the arrays were then read using a phosphorimager. We have also shown that by manipulating the experimental conditions in array-based protein-protein interaction assays, we can readily distinguish true from false positives.

    For example, we studied the interaction of calmodulin with a diverse array of human proteins in the presence and absence of calcium (2+) ions. Because calmodulin binding should be calcium dependent, we were able to deduce true from false positives based on the array data.

Chapter 21

Cytology of Terminal Deletion

Deletion of the most distal, telomeric band of human chromosomes can result in a variety of mental retardation and multiple congenital anomaly syndromes. These terminal deletions are some of the most commonly observed structural chromosome abnormalities detected by routine cytogenetic analysis.

Terminal deletions of 1p36 occur in approx 1 in 5000 live births, making it the most frequently observed terminal deletion and one of the most commonly observed mental retardation syndromes in humans. Molecular characterization of subjects with monosomy 1p36 indicates that, like other terminal deletions, 1p36 deletions have breakpoints occurring in multiple locations over several megabases and are comprised of terminal truncations, interstitial deletions, complex rearrangements, and derivative chromosomes.

In addition, cryptic interrupted inverted duplications have been observed at the end of terminally deleted chromosomes, suggesting premeiotic breakage-fusion-bridge (BFB) cycles can be intermediate steps in the process of generating and stabilizing terminal deletions of 1p36. Overall, these observations are identical to those made in yeast and other model systems in which a double-strand break (DSB) near a telomere can be repaired by a variety of mechanisms to stabilize the end of a broken chromosome.

Furthermore, sequence analysis and fluorescent *in situ* hybridization (FISH) mapping of the terminal 10.5 Mb of 1p36 including a variety of terminal deletion breakpoint junctions indicate that segmental duplications, low-copy repeats (LCRs), and short repetitive DNA sequence elements may mediate the generation and stabilization of terminal deletions of 1p36.

We hypothesize that nonallelic homologous recombination (NAHR) between palindromic or inverted LCRs in the subtelomeric region of 1p36 could generate a dicentric chromosome that is broken at a random location during the subsequent anaphase as the centromeres move to opposite poles.

This model suggests that the molecular basis of terminal deletions may be directly linked to genomic architectural features in the subtelomeric regions that generate the initial, variable-sized terminally deleted chromosome, and that stabilization of the broken chromosome occurs by one of a variety of competing DSB repair pathways.

LINEAR CHROMOSOMES

The distal ends of all linear chromosomes are capped by a specialized protein-DNA complex known as the telomere. The term "telomere" was originally put forth by Muller in 1938 to describe the "terminal gene" required for "sealing the end" of a broken chromosome. This work was subsequently expanded upon by McClintock, who demonstrated that broken chromosomes that have lost a telomere form end-to-end fusions resulting in dicentric chromosomes that are inherently unstable.

In addition to preventing chromosome fusions, telomeres have now been shown to play critical roles in a number of cellular processes, most notably in the replication of linear chromosome ends. Furthermore, telomere associations during meiosis suggest that they may be involved in homolog pairing and recombination and that they are necessary for faithful segregation of the chromosomes.

Telomeric DNA consists of tandem repeats of simple G-rich sequences that show remarkable conservation throughout eukaryotic evolution. All human chromosomes terminate with approx 2-20 kb of the simple tandem repeat $(TTAGGG)_n$. Proximal to this telomeric repeat tract is a structurally complex region of subtelomeric DNA that can extend several hundred kilobases from the end of most chromosomes and has been shown to be highly polymorphic.

These subtelomeric regions are enriched with segmental duplications interspersed with interstitial $(TTAGGG)_n$-like sequences and other subtelomeric repeat sequence blocks that can be shared by a number of different chromosome ends. In addition, early chromosome banding (G-banding) studies indicated that the telomeric regions of most human chromosomes stain G-negative and contain GC-rich isochores and are, therefore, thought to be regions that contain abundant genes.

Thus, telomeric deletions and/or rearrangements have been predicted to result in more significant and serious phenotypic abnormalities than similar-sized deletions and/or rearrangements at other chromosomal locations. Analysis of the draft sequence of the Human Genome assembly has recently confirmed that telomeric regions are gene-rich, although not to the extent originally estimated.

Cytogenetic Abnormality

Mental retardation affects approx 3% of the world's population. Although chromosomal and genetic disorders are thought to account for 30-40% of all cases with an additional 10-30% accounted for by environmental factors, the cause of mental retardation in the vast majority of subjects is still unknown. It has been estimated that approx 20% of mental retardation is caused by a cytogenetic abnormality, with aneuploidy, mainly trisomy 21, being the most common.

Although estimating the second most common cytogenetic cause may have some biases, in recent years telomeric abnormalities have emerged as a major yet previously under-recognized cause of mental retardation in the population. In the past decade, numerous submicroscopic chromosomal abnormalities in subjects with apparently normal karyotypes and mental retardation or multiple congenital anomalies have been identified by molecular methods such as DNA polymorphism analysis, FISH, and microarray-based comparative genomic hybridization (array CGH).

Although it is difficult to compare directly these various studies because of differences in patient selection criteria and variations in the subtelomeric screening methodologies, the combined data suggest that approx 5 % of subjects with idiopathic mental retardation or dysmorphic features and apparently normal karyotypes may have submicroscopic deletions and/or cryptic rearrangements at the telomeres.

First Identified Deletion

Telomeric regions of the genome are of particular interest in clinical cytogenetics because, owing to their light G-negative staining on G-banded chromosomes, rearrangements of these regions are difficult to identify by conventional cytogenetic methodology. Nonetheless, terminal deletions are one of the most commonly observed structural chromosomal abnormalities detected by routine chromosome analysis.

The first deletion identified in humans was the terminal deletion of 5p associated with Cridu-chat syndrome (5p-). Shortly thereafter, terminal deletions of 4p were identified and the resulting syndrome, Wolf-Hirschhorn, was described. With the advent of chromosome banding techniques and the more recent molecular cytogenetic technology, visible or subtle terminal deletions have been identified for every human chromosome.

This has allowed the delineation of a number of mental retardation syndromes associated with deletion of the telomeric bands of the chromosomes including Miller-Dieker (17p-) and monosomy 1p36 (1p-). However, the molecular mechanism(s) that generates and stabilizes terminal deletions is only beginning to be elucidated.

Terminal deletions of p36 occur in approx 1 in 5000 live births, making it the most frequently observed terminal deletion and one of the most commonly occurring mental retardation syndromes in humans. Monosomy p36 is characterized by distinct facial features including deep-set eyes, flat nasal bridge, asymmetric ears, and pointed chin. Other features include mild to severe mental retardation, seizures, hearing impairment, vision disorders, growth failure, hypothyroidism, oropharyngeal dysphagia, orofacial clefting, and cardiac defects.

More than 100 subjects with deletions of 1p36 have now been systematically analyzed (and unpublished data). Although the majority of cases of monosomy p36 have been identified through cytogenetic analysis, this deletion is difficult to visualize by routine chromosome analysis at the 400-550 band resolution. However, the majority (98%) of terminal deletions of 1p36 can be visualized retrospectively using good quality chromosomes at a resolution of 550 bands or greater, with careful attention to chromosome band p36.

Nevertheless, nearly half of patients studied had at least one chromosome analysis, sometimes two, interpreted as normal. Furthermore, the size and complexity of the deletion cannot be distinguished at the resolution of the light microscope. A variety of molecular cytogenetic techniques have revealed that deletion sizes in monosomy 1p36 vary widely with no single common breakpoint.

Additionally, terminal deletions, interstitial deletions, complex rearrangements, and derivative chromosomes have been identified among the study cell lines. Given that p36 deletions are relatively common and, like other terminal deletions, comprise a heterogeneous group of submicroscopic rearrangements with breakpoints occurring in multiple locations over several megabases, monosomy 1p36 has proven to be a valuable model for investigating the molecular basis of terminal deletions.

TELOMERIC DEGENERATION

Telomeres provide structural stability to the ends of all linear chromosomes. Loss of a telomere, by either a DNA DSB or by improper telomere maintenance can, if not properly repaired, lead to apoptosis, cell senescence, or severe genomic instability characterized by gene amplifications and deletions as a result of perpetual BFB cycles.

It is now clear that there are at least two general repair pathways that exist within the cell whereby a terminally deleted chromosome can acquire a new telomeric cap. The first pathway is known as "telomere healing" and results in the stabilization of terminal deletions by the direct addition of telomeric repeats onto the end of a broken chromosome. The results of telomere healing events are pure terminal truncations.

Although true telomere healing events may occur by telomerase-mediated, *de novo* addition of telomeric repeats on the end of a chromosome, pure terminal truncations can also result from telomerase-independent, recombination-based telomere capture mechanisms from a pre-existing telomere.

The second general pathway has been termed "telomere capture" and refers to the process through which a terminally deleted chromosome acquires a new telomeric sequence from another chromosomal location. Telomere capture events could potentially occur between sister chromatids, homologs, or nonhomologous chromosome ends and result in derivative chromosomes. Although a number of examples of what appear to be telomere healing and telomere capture events have now been described in a variety of model organisms, relatively few cases have been reported for constitutional chromosome abnormalities. in humans.

Moreover, the precise mechanisms that mediate these pathways in humans are not understood. Furthermore, it is not clear why terminal deletions are so common and which mechanism of repair and stabilization is used most frequently. Molecular studies on a large cohort of subjects with cytogenetically defined terminal deletions of 1p36 using microsatellite analysis, FISH, and array CGH indicate that deletion sizes range from approx 1 Mb to more than 10.5 Mb with the majority (40%) of breakpoints clustering between 3 and 5 Mb from the telomere.

Furthermore, telomere FISH studies have identified interstitial deletions (6%), derivative chromosomes (17%), complex rearrangements (5%), and apparently pure terminal deletions (72%). Ten rearrangements of lp36 have been examined at the DNA sequence level and an additional two were examined by microarray analysis and FISH.

Based on the previous molecular analyses, seven were apparently pure terminal deletions, three were known derivative chromosomes 1 [der(1)t(1;1)(p36;q44)], and two were apparent duplications of 1p36. Analysis of the breakpoint junctions in these subjects uncovered evidence for BFB cycles, confirmed a telomere healing event, and clarified the mechanisms of telomere capture involved in terminal deletions of 1p36.

In the next sections we discuss in detail the various mechanisms by which terminal deletions can be generated and stabilized, most of which have already been observed in terminal deletions of 1p36, and suggest that terminal deletions of 1p36 are a good model for understanding the breakage and stabilization of human chromosomes.

Monosomic Terminal Deletion

Breakpoint junction sequences from two subjects with monosomy 1p36 identified terminal deletions associated with cryptic interrupted inverted duplications at the end of the chromosome.

These breakpoint sequences are identical in structure to those found in embryonic stem (ES) cell lines and tumor cell lines that have gone through BFB cycles as a result of uncapped sister chromatids being fused by nonhomologous end joining.

The analysis of two additional subjects with apparent duplications of l p36 showed terminal deletions, duplications, and triplications. The mechanisms postulated to result in these complex rearrangements involve BFB cycles. These data represent the first evidence of a BFB cycle generating a constitutional chromosome rearrangement resulting in a human genetic disease.

Furthermore, this suggests that the observed variability in size of terminal deletions of 1p36 (and other terminal deletions) may be the result of random breaks generated when a fused, dicentric chromosome is broken during anaphase separation of the centromeres.

Perhaps most significant is that the identification of BFB cycles suggests that at least some terminal deletions of 1p36 are not generated and stabilized in a single step but pass through an intermediate phase as a broken chromosome that must then be stabilized by some other cellular mechanism.

Terminal Truncation

Pure terminal truncations have been reported in at least 12 terminal deletion cases from a variety of human chromosome ends. Recently, sequence analysis of the precise breakpoint junctions in subjects with monosomy l p36 confirmed at least one case to be a pure terminal truncation stabilized by telomeric repeat sequences. Telomere healing has also been identified in human tumor cell lines, mouse ES cells, and a number of other model organisms.

Although it is presumed that pure terminal truncations were stabilized by *de novo* telomere healing by telomerase, broken chromosomes stabilized by acquisition of telomeric sequences from another chromosome end could appear virtually identical to ones stabilized by *de novo* telomere synthesis.

However, the reported findings of pure $(TTAGGG)_n$ telomeric repeats at the sites of healing, as opposed to the variant telomeric repeats that are commonly found in the proximal telomeric region of existing telomeres, support *de novo* synthesis of telomere repeats by telomerase. Even though the lack of variant repeats is consistent with *de novo* addition, it is not definitive because greater than half of an existing telomere's 2-20 kb sequence consists of pure telomeric repeats.

In support of *de novo* synthesis, using an in vitro assay system, human telomerase was shown to preferentially recognize the sequence at the breakpoint of one 16p terminal truncation and to add $(TTAGGG)_n$ telomeric repeat sequences.

Telomeric Aberration

In the past few years, a variety of cryptic telomeric aberrations have been identified. making it clear that many apparently pure terminal deletions, viewed at the resolution of the light microscope, are in fact terminal deletions/broken chromosomes stabilized by telomere capture events. At least four models have been proposed for the molecular basis of telomere capture. First is the mechanism of NAHR between LCRs in nonhomologous telomeric regions, which would predict the simultaneous generation and stabilization of a terminal deletion as a derivative chromosome.

Analysis of der(1)t(1;1)(p36;q44) chromosomes from three subjects with monosomy 1p36 revealed variability in the breakpoint locations without any evidence for homologous sequences or LCRs between the p36 deletion and l q translocation breakpoints. This suggests that NAHR did not play a role in mediating these particular rearrangements.

However, this does not preclude LCRs within the subtelomeric region of 1p36 from being involved in generating and/or stabilizing terminal deletions by other mechanisms such as subtelomeric pairing with proximal recombination. The second major mechanism for telomere capture, which has been termed "subtelomeric pairing with proximal recombination" (SPPR), suggests that LCRs located within the complex subtelomeric regions can facilitate mispairing of nonhomologous chromosome ends.

A more proximal recombination event between the mispaired chromosomes at one of the more abundant and much shorter repetitive DNA sequence elements could generate a terminal deletion and stabilize it as a derivative chromosome in a single step. Although data from the three der(1)t(l;1)(p36;q44) chromosomes is consistent with a SPPR model in that no LCRs were identified at the breakpoints, microsatellite data from these subjects suggest that these telomere capture events were mediated by intrachromosomal rearrangements.

Therefore, to remain consistent with a SPPR model, in these three der(1) cases lqter sequences must have been translocated to 1pter from a sister chromatid. A third mechanism has recently been proposed for generating terminally deleted chromosomes that are stabilized by telomeric sequences originating from the opposite end of the chromosome, as seen in the three der(1)t(1;1)(p36;q44) chromosomes.

This model proposes that a premeiotic exchange between chromosome 1 homologs could result in the "transposition" of 1 pter and 1 qter sequences. If normal homolog pairing in meiosis occurred followed by a recombination event in 1p36 proximal to the site of interchromosomal "transposition," then the result could be both a deleted 1p, duplicated 1q chromosome and a duplicated 1 p, deleted 1 q chromosome.

Although technically possible, this mechanism does not account for the wide range of derivative chromosomes seen in monosomy 1p36 and in terminal rearrangements of other chromosomes. The fourth, and most likely, model for telomere capture suggests that telomeric sequences from another chromosome end could be copied directly onto the end of a broken chromosome through a process known as break-induced replication (BIR). BIR has been studied in yeast systems and it has been recently shown to play a significant role in DSB repair and preventing gross chromosome instability by maintaining telomere integrity. BIR is a very appealing mechanism in that a broken chromosome could theoretically be healed by copying telomeric DNA from any chromosome end.

This could account for the wide range of derivative chromosomes seen in monosomy p36, as well as those in other terminal deletion syndromes, in addition to the lack of LCRs at or near the deletion/translocation breakpoints.

Molecular Analysis of Deletion

As a whole, the molecular analysis of terminal deletions of 1p36 is consistent with a number of recent reports that suggest a chromosome broken near the telomere can be stabilized by a variety of competing DSB repair pathways to provide the necessary telomeric cap. In particular, telomere dysfunction because of improper telomere maintenance, chromosome breakage, or mutations in various DSB repair and/or S-phase checkpoint pathways results in severe genomic instability typical of cancer.

This genomic instability is characterized by chromosome fusions and BFB cycles that generate inverted duplications, terminal deletions, and nonreciprocal translocations that are stabilized by several telomere healing and telomere capture mechanisms. The recent studies by Kolodner et al. of yeast are of particular interest because they suggest extensive and redundant DNA repair pathways function to keep genomic rearrangements at very low levels. Their data indicate that disruption of the S-phase checkpoints, compared to disruption of the G_1, G_2, or mitotic spindle checkpoints, significantly increases the rate of genome rearrangements. These results indicate that replication errors play an important role in generating genomic instability.

In addition, all of the genome rearrangements observed following inactivation of the S-phase checkpoint were terminal deletions stabilized by telomeric repeat sequences. Kolodner et al. have also shown that mutations in yeast that disrupt the BIR repair pathway, and not the classical DSB repair by homologous recombination, result in a significant increase in genome instability.

Furthermore, telomerase-defective yeast show a dramatic increase in the number of BIR-induced nonreciprocal translocations and nonhomologous end-joining chromosome fusions. We propose that the diversity in rearrangements observed in terminal deletions of 1p36 illustrates a similar set of circumstances in human terminal deletions in that they reflect the various DSB repair pathways that were taken to stabilize a broken chromosome. Figure 1 shows a schematic representation of some of the known mechanistic pathways whereby a broken chromosome can be stabilized.

Processing a broken chromosome through any one of these pathways can result in a terminal deletion coupled with a subtle telomeric rearrangement-all of which have been observed in monosomy 1p36 subjects. Furthermore, these observations on terminal deletions of 1p36 suggest that the same variety of subtle telomeric rearrangements observed in other terminal deletion syndromes are also likely to be the result of various DNA repair pathways stabilizing broken chromosomes.

DOCUMENTED SEQUENCE ELEMENTS

Sequence Elements

In recent years, numerous examples in the literature have documented repetitive DNA sequence elements as promoters of aberrant homologous recombination and DNA DSBs. This is evidenced by the vast number of cases in which repetitive DNA sequence elements are found at the

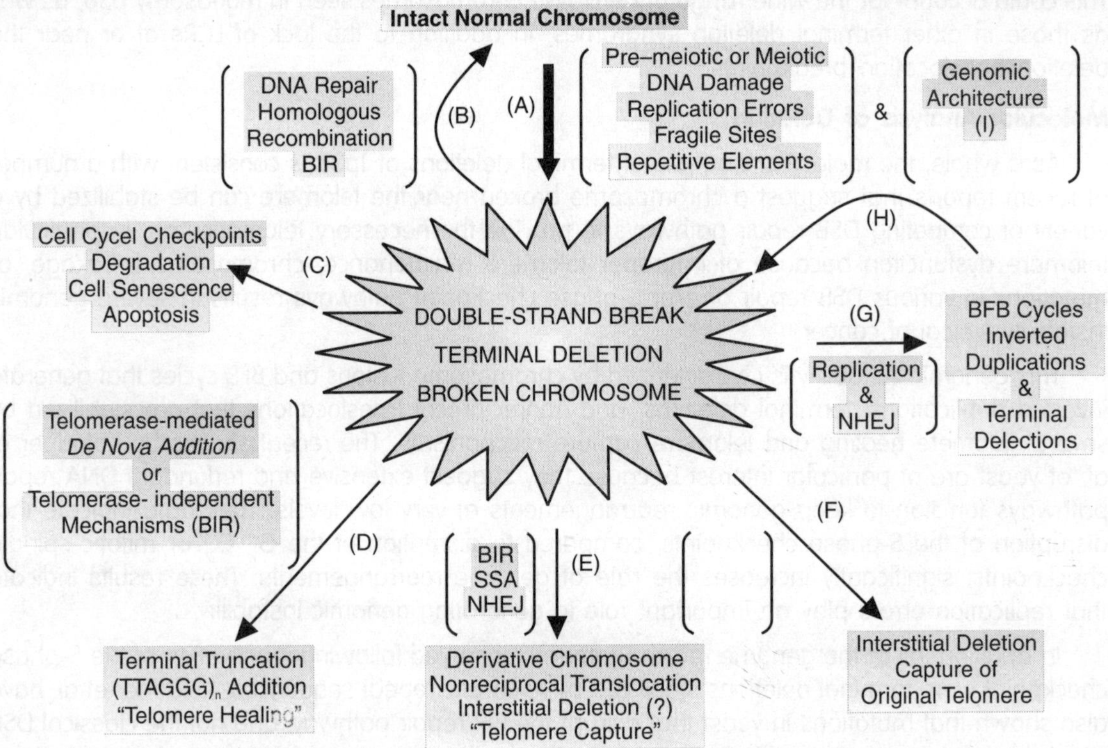

Figure 21.1: Showing the mechanisms for stabilizing a broken chromosome.

breakpoints in constitutional chromosomal aberrations that cause genetic disease and in sporadic chromosomal abnormalities that cause cancer. Presumably, aberrant homologous recombination between closely related repetitive DNA sequence elements (such as Alit and L1 sequences) can mediate these rearrangements without a prolonged period as a broken chromosome. In four subjects with apparently terminal deletions of 1p36, the breakpoints occurred in repetitive DNA sequence elements.

This suggests that aberrant homologous recombination may have been involved in generating and/or stabilizing these terminal deletions. Aberrant homologous recombination between Alit elements has also been attributed to the generation of two subtle interstitial deletions of I 6p that were initially thought to be simple chromosomal truncations. A similar situation has also been reported for a terminal deletion of 18q in which satellite III DNA sequences were located at the breakpoint. Some repetitive DNA sequence elements have also been shown to be susceptible to genomic rearrangements including deletions. Microsatellite, mini satellite. and other short repetitive sequences are particularly susceptible to replication errors and can generate palindromic, hairpin structures that are prone to cleavage, resulting in DSBs and other chromosomal rearrangements.

Because the breakpoints of four subjects with terminal deletions of1p36 were shown to be located within repetitive DNA sequence elements, it is reasonable to suggest that some terminal deletions of 1p36 may have been generated by DNA sequences that were susceptible to DNA DSBs.

These DSBs generate broken chromosomes that must then become stabilized by telomere healing or capture mechanisms to avoid potential BFB cycles. However, repetitive DNA sequence elements found at the end of a broken chromosome may also constitute a favorable site for stabilization by DNA repair mechanisms.

Genomic Instability

What generates the initial broken chromosome and why terminal deletions of 1p36 are ascertained so frequently is still unknown. This could reflect higher instability in this particular region of the genome or increased survival of terminal deletions of 1p36 over other telomeric abnormalities. Interestingly, one method uses an inverted duplication of the short arm.

This was how McClintock originally formed a dicentric chromosome to generate broken chromosome sand initiate BFB cycles using a maize chromosome 9. A crossover between the normal homolog and the inverted portion of the duplicated chromosome, or between duplicated regions of sister chromatids, resulted in a dicentric chromosome that formed a bridge in anaphase with subsequent breakage.

Large-scale sequence analysis of the terminal 10.5 Mb of 1p36 indicates that segmental duplications along with palindromic and inverted LCRs may be present in the distal subtelomeric regions. The region surrounding these LCRs forms a natural division between the interstitial deletion breakpoints of 1p36 and nearly all of the terminal deletion breakpoints (Ballif, Gajecka, and Shaffer, unpublished data).

It is tempting to speculate that NAHR between palindromic or inverted LCRs in the subtelomeric region of 1p36 could be responsible for generating a dicentric chromosome that is broken at a

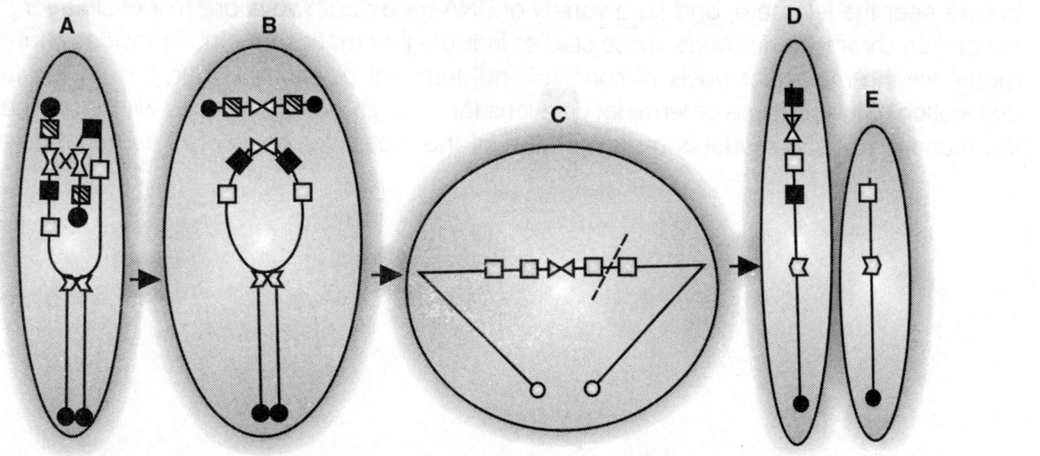

Figure 21.2: Model for nonallelic homologous recombination (NAHR) between palindromi low-copy repeats (LCRs) in the subtelomeric region of 1p36. Palindromic LCRs are shown as arrowheads. (A) Palindromic LCRs can promote misalignment of homologous chromosomes. (B) NAHR between misaligned palindromic LCRs results in the formation of a dicentric chromosome. (C) In anaphase, when the centromeres move to opposite poles, the resulting dicentric chromosome forms a bridge that will eventually break at a random location. (D) Random breakage results in one chromosome with an inverted duplication and a distal deletion and a second chromosome (E) with a terminal deletion.

random location during the subsequent anaphase as the centromeres move to opposite poles. Similar models have been proposed for generating inverted duplication/deficiency chromosomes, although those models were based on subtelomeric LCRs located in inverted orientations and positioned several megabases apart.

The palindromic repeats proposed in this model for terminal deletions of 1p36 would provide a single site for NAHR with subsequent breakage of the newly formed dicentric chromosome generating variablesized terminally deleted and inverted duplication/deficiency chromosomes. Stabilization of the broken chromosomes could then occur by a variety of DSB repair pathways.

Although duplication/deletion chromosomes would also be expected to occur, it is likely that these reciprocal products have not been ascertained in studies of monosomy 1p36 because they would be predicted to have much smaller imbalances and may present with a mild phenotype that is not typical of the syndrome.

Because the genomic sequence of 1p36 is not in its finished form, the precise size and orientation of these putative LCRs is still under investigation. However, regions of the genome that contain large duplications are notoriously difficult to map and sequence. A closer examination of the subtelomeric regions of all chromosomes may be useful for understanding the molecular basis of other terminally deleted chromosomes.

CONCLUDING REMARKS

Combined, the data presented suggest that: (1) genomic architectural features may ultimately play a role in the generation of terminal deletions of 1p36 by generating the doublestrand DNA breaks near the telomere, and (2) a variety of DNA repair pathways are responsible for stabilizing the broken chromosomes ends, these studies indicate that monosomy 1p36 provides a fundamental model for the molecular basis of constitutional terminal deletions in humans. In the future, the delineation of the structure of terminal deletions for other chromosome ends will determine whether the monosomy 1p36 model is representative of the molecular basis of all terminal deletions.

Chapter 22

Neoplastics Disorders

Neoplastic disorders are characterized by recurrent *somatically* acquired chromosomal *aberrations* that alter the structure and/or expression of a large number of genes. Most "cancer genes" discovered to date in human neoplasms have been identified through isolation of genes at the breakpoints of balanced chromosomal translocations.

Although functional studies of such cancer-causing genes have demonstrated their causal role in tumorigenesis, the mechanisms underlying the formation of recurrent chromosomal changes in cancer remain enigmatic. Low-copy repeats (LCRs) are important mediators of *erroneous meiotic* recombination, resulting in constitutional chromosomal rearrangements.

Recently, LCRs have been implicated in the formation of the frequent and characteristic neoplasia-associated chromosomal aberrations t(9;22)(q34;q11) and i(l7q), suggesting that similar genome architecture features may play an important role in generating also other somatic chromosomal rearrangements.

Neoplasia originates in a single cell through acquired somatic genetic changes. To date, approx 300 "cancer genes" (i.e., genes causally implicated in oncogenesis), have been identified. These cancer-causing genes encode proteins that normally regulate cellular proliferation, differentiation, and/or cell death. A probably too simplistic, but generally accepted, way of classifying cancer genes is based on the way they act at the cellular level.

Socalled oncogenes act dominantly (i.e., a single-mutated allele is sufficient to contribute to oncogenesis), whereas so-called tumor suppressor genes (TSGs) are recessive (i.e., both genes need to be mutated). It has also become clear over the last few years that, haploinsufficiency, which is loss of only one allele, also may be an important factor in tumor development and progression.

Several types of mutations in cancer genes have been identified, including point mutations within coding and non-coding regions, small insertions/deletions altering the open reading frame, as well as gene copy-number changes and rearrangements resulting from chromosomal abnormalities. The latter group of aberrations has been studied extensively, with more than 300 recurrent balanced chromosomal abnormalities reported.

Most cancer genes discovered to date have been identified at the breakpoints of balanced chromosomal translocations that result in either chimeric fusion genes or deregulation of genes at the breakpoints by juxtaposition of regulatory elements to another gene.

Most likely, however, there is a heavy bias towards fusion genes in the current estimation of the total number of cancer genes not only because of *methodological* limitations in the past to detect subtle changes at the DNA level on a *genome*-wide basis (e.g., point mutations or small deletions), but also because of difficulties in identifying single cancer genes in neoplasias *characterized* by complex and/or *unbalanced* chromosomal aberrations affecting multiple genes.

Although functional studies of individual cancer genes using various transformation assays or *transgenic* animal models have demonstrated their causal role in tumorigenesis, the mechanisms underlying the formation of recurrent chromosomal changes in neoplasia remain elusive. Most of our knowledge is based on cloning individual fusion genes and looking for repeat elements or putative recombinogenic sequence motifs at the vicinity of the breakpoints.

In many instances, however, it has been difficult to determine whether the presence of a certain motif at a chromosomal breakpoint represents a chance occurrence or a bona fide association. This lack of knowledge seems to contrast with the rapid progress made recently in the field of constitutional genomic abnormalities, exemplified by the increasing number of microdeletion syndromes being identified and the involvement of LCRs in generating such disorders.

However, LCRs have been implicated recently in the formation of the most common isochromosome, i(17q), in human neoplastic disorders and also in the balanced translocation, t(9;22)(q34;q11), the cytogenetic hallmark of chronic myeloid leukemia (CML). This chapter focuses on mechanisms involved in the generation of somatically acquired structural chromosomal aberrations with an emphasis on recent advances made in elucidating how i(17q is formed. Whenever applicable, conceptual and mechanistic differences between somatically and constitutionally acquired genetic changes are also discussed.

NON RECIPROCAL TRANSLOCATIONS

At the chromosomal level, genomic changes in tumor cells are visible either as balanced abnormalities (reciprocal translocations, inversions, and insertions) or unbalanced changes, including nonreciprocal translocations, deletions, duplications, numerical aberrations, and amplifications (visible as double minutes, or homogenously staining regions).

Individual cancer genes become altered through several different mechanisms; oncogenes are typically activated by point mutations, structural rearrangements, and amplifications, whereas TSGs become inactivated by mutations, deletions, or silencing through methylation.

The overwhelming majority of oncogenes in human malignancies have been identified by isolating genes located at the breakpoints of balanced chromosomal translocations that result in tumor-specific fusion genes. To date, 271 fusion genes have been described and because many genes recombine with shared partner genes (e.g., *ETV6, MLL, EWSR1),* the number of rearranged genes (275 unique genes) is smaller than expected from the number of fusion genes.

Fusion genes are particularly common in hematologic malignancies and in mesenchymal tumors and are considered key mediators of malignant transformation in these tumor entities. In contrast, tumors of epithelial origin, the most common tumor type in man, are generally believed to originate through loss of TSGs or activation of oncogenes by genomic amplification.

This dichotomy, suggesting a fundamental tissue-specific difference in the genetic mechanisms by which neoplasia is initiated, has been challenged recently, however, and it has been suggested that fusion genes in epithelial tumors may be more frequent than previously anticipated.

What are the mechanisms underlying the formation of balanced chromosomal translocations resulting in fusion genes? In hematologic malignancies, two qualitatively different types of rearrangements have been identified, each probably arising through different mechanisms. In a subset of malignancies of B- or T-cell origin, illegitimate recombinations result in the juxtaposition of a variety of structurally intact oncogenes to enhancer elements of the immunoglobulin (Ig)- or the T-cell receptor *(TCR)* loci.

A paradigmatic example of this category is the Burkitt lymphoma-associated t(8;14) translocation, which brings the *MYC* oncogene into the vicinity of regulatory elements of the *IgH* locus, resulting in deregulated expression of *MYC (15,16).* In this subset of malignancies the mechanism behind the chromosomal breakpoint in one of the partner genes seems apparent as double strand breaks arise normally in the *Ig* and *TCR* loci during V(D)J recombination. However, little is known about the underlying mechanisms of chromosomal breaks in *MYC.*

The breakpoints occur at sites with no homology to V(D)J or switch recombinase recognition sequences, suggesting that they are unlikely to be dependent upon these activities. Interesting progress has been made recently in elucidating the mechanism behind the t(14;18), which is characteristic of follicular lymphoma. Through this translocation, the *BCL2* gene is relocated into the *IgH* locus.

The breakpoints in *BCL2* occur within a confined 150 basepair region (Mbr) and it has been shown that the RAG complex, which is the normal enzyme for DNA cleavage at V(D)J segments, is capable of introducing nicks in the Mbr. Importantly, the *BCL2* Mbr is not a V(D)J recombination signal. Instead, the RAG complex nicks the Mbr because it adopts a nonB DNA structure, providing a basis for the fragility of the *BCL2* locus.

Whether similar structural features may explain the juxtaposition of other oncogenes to the *Ig* or *TCR* loci remains to be investigated. The second and by far the most common outcome of balanced translocations in neoplasia is the creation of fusion genes, resulting in the expression of chimeric fusion proteins. In such fusion genes, double-strand breaks typically (but not exclusively) occur in introns within so called "breakpoint clusters regions" that can vary in size from a few basepairs to hundreds of kilobases.

So far, no consistent consensus sequences or specific repetitive elements have been identified within the many breakpoint clusters identified. Many recombinogeneic motifs and repetitive elements have been reported to coincide with translocation breakpoints in neoplasia, *e.g., Alu* sequences, topoisomerase II cleavage sites, and chi-like elements, but given their frequent genome-wide occurrence it is difficult, with a few exceptions, to currently ascribe them a direct role in generating reciprocal translocations.

Only a few systematic studies on the importance of the nucleotide composition and the occurrence of recombination-associated motifs in neoplasia-associated translocation breakpoints are currently available. A recent survey examined up to 125 by on either side of translocation and gross deletion breakpoints in neoplasia and constitutional disorders.

The majority of the breakpoints were derived from translocations in hematologic malignancies and polypyrimidine tracts and specific recombination-associated motifs (e.g., DNA polymerise pause sites/frameshift hotspots, IgHclass swith sites, heptamer/nonamer V [D]J recombination signal sequences, translin binding sites, and chi elements) were found to be over-represented in the translocation breakpoints.

Specific DNA breakage has been implicated in the genesis of some translocations, mainly those involving the *MLL* gene, known to fuse to a wide variety of partner genes in acute myeloid and lymphocytic leukemias. Rearrangement of *MLL* is also frequent in therapy-related leukemias, occurring in patients previously treated with topoisomerase II inhibitors.

The genomic breakpoints in *MLL* in therapy-related cases often coincide with topoisomerase II cleavage sites, DNase I-hypersensitive sites, and scaffold attachment regions, indicating that these chromatin structural elements may influence the location of these translocation breakpoints. Data are also at hand suggesting that sublethal apoptotic signals, through the activation of endonucleases, may result in site-specific cleavage of *MLL* and other genes forming fusion genes.

The great majority of all translocations that have been studied to date, mainly in leukemias and mesenchymal tumors, show no extensive regions of homology at the breakpoints, and are, hence, believed to arise through (error-prone) nonhomologous end joining. A prerequisite for end joining to take place is, of course, a spatial proximity of the broken chromosome ends.

Available data on several genes involved in frequent and specific translocations, e.g., *IgH/MYC* and *BCR/ABLJ,* show that, indeed, such genes are preferentially positioned in close proximity relative to each other in the nucleus, suggesting that the formation of translocations also is determined in part by higher order of spatial organization of the genome.

When trying to understand the features underlying the observed confinement of chromosomal breakpoints into clusters, it is important to consider that they represent the result of selection on the protein level. In fact, transcription factors and tyrosine kinase-encoding proteins are overrepresented among proteins involved in fusion genes, and critical functional domains of these have to remain intact to provide a selective advantage.

Hence, numerous chromosomal breaks, with scattered distribution, may occur in somatic cells, either because of endogenous or exogenous genotoxic factors, but are never brought to our attention because of their failure to provide a selective edge.

It is likely that other genome structural features of importance for generating neoplasiaassociated genomic rearrangements will be identified in the near future. An entirely novel mechanism that may prove important will be discussed in detail in the following sections.

ISOCHROMOSOME IN MAN

Isochromosome 17q [i(17q)] constitutes the most common isochromosome in human neoplasia and has been described both as a primary and as a secondary abnormality, suggesting an important pathogenetic role in tumor development as well as in progression. The i(17q) is frequent

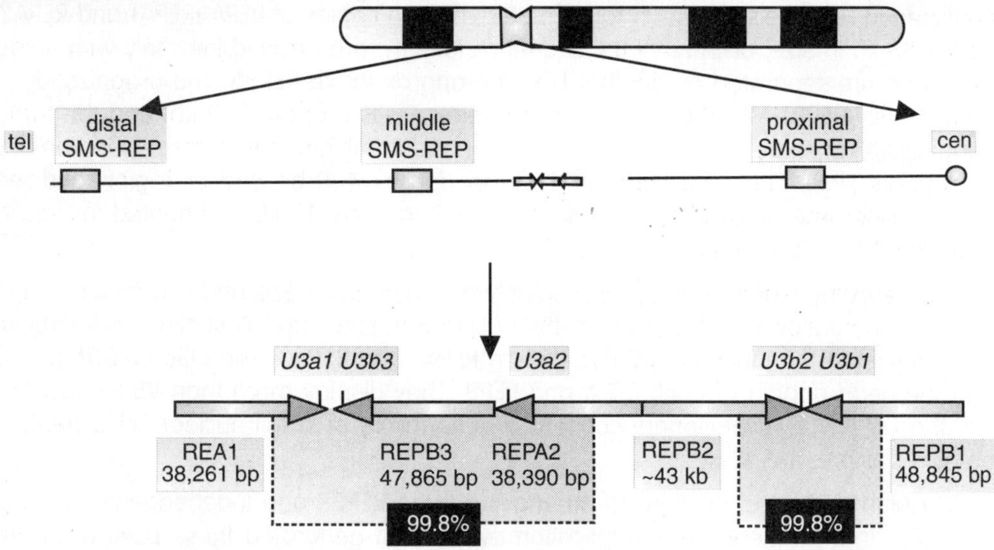

Figure 22.1: Schematic representation of the Isochromosome 17q [i(17q)] breakpoint region.

in hematologic malignancies, in particular as a secondary abnormality in addition to the characteristic t(9;22)(q34;q11) in CML, but is also the most common chromosomal aberration in primitive neuroectodermal tumors and medulloblastomas.

Fioretos et al. characterized the i(17q) in more detail, using fluorescence *in situ* hybridization (FISH) with several yeast artificial chomosome (YAC) clones on a series of leukemias, and showed that the i(17q), in most cases, was a dicentric isochromosome (isodicentric), and not monocentric as suggested by conventional G-banding. Consistent with this observation, the breakpoints were shown to cluster within 17p11 implying that the i(17q) formally should be designated idic(17)(p11).

Because of its less cumbersome and more descriptive designation, and also forhistorical reasons, "i(17q)" prevails in the literature. Interestingly, the majority of the 17p11 breakpoints were shown to occur within a 900 kb YAC clone (828b9) located in the Smith-Magenis syndrome (SMS) common deletion region, suggesting that this germ line genetically unstable region also could be of importance in generating the i(17q) somatic event.

Given the genetically unstable nature of the SMS common deletion region, further efforts were undertaken to delineate the i(17q) breakpoint region in 17p11. FISH mapping using a large set of bacterial artificial chromosome (BAC), P1 artificial chromosome, and fosmid clones covering the breakpoint, together with extensive sequence comparison and the analysis of restriction maps of individual clones, enabled the establishment of a physical map of an approx 240-kb genomic interval including the i(17q) breakpoint cluster region.

This region, localized between the middle and proximal SMS-REPs, was shown to contain several interesting features. First, two types of distinct LCRs, designated REPA and REPB, present in two and three copies, respectively, were identified within this region. Secondly, their location and orientation with respect to each other strongly suggested that they play an important role in the formation of the i(17q) rearrangement.

As outlined in Figure elsewhere in this chapter, the two copies of REPA (REPA1 and REPA2) are both approx 38 kb in size, display 99.8% sequence identity, are arrayed inversely with respect to each other, and are separated by another LCR, the approx 48 kb REPB3. The organization of the three REPB copies (REPB1, -2, and -3) is more complex. REPB 1 (approx 49 kb) and the truncated REPB2 (approx 43 kb) share 99.8% sequence identity and are inverted with respect to one another (inverted repeats [IRs]). They are separated by an approx 400-bp spacer region and, hence, represent a palindrome or cruciform structure. REPB 3 (approx 48 kb) is oriented inversely with respect to REPB2 (99.8% identity).

Another interesting feature is the presence of two *U3b* genes *(U3b2* and *U3b1)* that are located in an inverted orientation in the center of this palindromic structure. A similar DNA structure of 8731 by is present in the junction region between REPA1 and REPB3. The *U3a* or *U3b* genes are located at the head-portion of each REPA and REPB. They display more than 98.5% identity and belong to the *U3 (RNU3)* evolutionary conserved gene family of small nuclear RNAs required for the processing of pre- 18SrRNA.

The very high degree of similarities among the three REPBs and independently among two REPAs indicates that the segmental duplication events that generated these LCRs are of recent origin, as suggested by evolutionary studies of other highly homologous LCRs, or alternatively, undergo frequent gene conversion. A final remarkable feature of this region is the high content (approx 30%) of *Alu* repeats. Greater than 80% of the 4 kb flanking the center of the cruciform are repetitive sequences, with 50% consisting of *Alu* sequences.

Furthermore, each of the five REPs is flanked by *Alu* sequences. Thus, as recently suggested, the high *Alu* repeat content could be important in the genesis of the large segmental duplications present in this region. The large approx 86-kb long cruciform generated by REPB1 and REPB2 is a genomic architectural feature that potentially could trigger a recombination event resulting in the i (l7q)-formation.

Indeed, IRs or cruciforms are known to generate hairpins, facilitated by intrastrand pairing of complementary single-strand DNA sequences assembled in the lagging strand during DNA replication. A double-strand break (DSB) introduced in such a hairpin can result in deletions, or either inter- or intrachromosomal recombination events, mediated by the homologous recombination machinery.

Critical factors implicated in the rate of such rearrangements are the extent of sequence identity between the inverted repeats, their size, and the length of the spacer (sequences present between the cruciform sequences). The rate of genomic rearrangements rises as the length of the repeat increases and the spacer sequence length decreases. The cruciform identified in the i(17q) breakpoint cluster region is characterized by two inverted repeats, approx 48 kb and approx 43 kb in length, separated by a 400-bp spacer.

If broken, DNA ends generated in the hairpin/cruciform structures of the i(17q) breakpoint can trigger the DSB-repair (DSBR) machinery. A subsequent nonallelic homologous recombination (NAHR) event between the repeats with opposite orientations located in the two sister chromatids (i.e., sister chromatid exchange) can result in the formation of an isodicentric chromosome 17 and of an acentric fragment.

Similar to i(17q), breakpoint analyses of constitutional isochromosomes of the long arm of the

X chromosome, "i(Xq)," and of the most common constitutional isodicentric chromosomes, i.e., "inv dup(15)" and "inv dup(22)," suggest that these isodicentric abnormalities also may result from NAHR between inverted LCRs.

As to the role of IRs or cruciform in constitutional genomic rearrangements, such genome architecture features, albeit of smaller size, have been identified at the breakpoints of the most common non-Robertsonian translocation, t(11;22)(q23;q 11) and of the translocation t(1 7;22)(q 11 ;q 11) described in two patients with neurofibromatosis type 1.

Moreover, genomic characterization of the *AZFc* region on chromosome Y, which is deleted in close to 20% of patients with oligo- or azoospermia, revealed a 3-Mb sequence that is deleted via homologous recombination events mediated by two direct repeats, present at the edges of this structure.

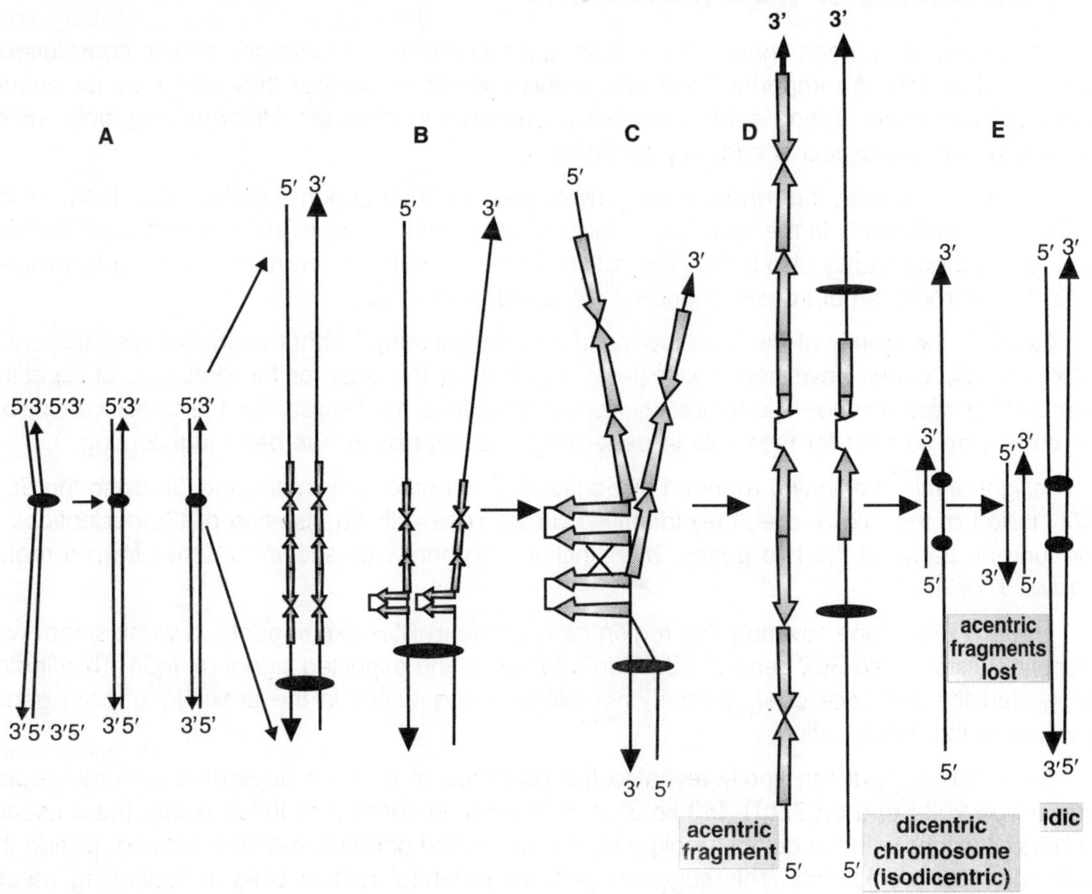

Figure 22.2: Molecular mechanism for i(1 7q)-formation. (A) The division of a metaphase chromosome is shown. The double strand DNA of each chromatid is depicted in red and blue and the orientation with 3' and 5'. (B) The formation of a REPB I and REPB2 palindrome with (C) subsequent breakage and (D) reunion between palindromes on sister chromatids results in the origin of both dicentric and acentric structures. (E) The dicentric and acentric structures after replication.

The breakpoint region of i(17q) resides within the SMS commonly deleted region and is flanked by the middle and proximal SMS-REPs. Furthermore, several other repeat elements have been reported telomeric and centromeric to the i(17q) breakpoint region.

It is unclear why the breakpoints in i(17q) are clustered in the delineated 240-kb region and not in the other closely located repeats, but it could well be that only the complex nature of the .+5breakpoint region allows this chromosomal abnormality to be formed, or, alternatively, that only such rearrangements provide a selective advantage to the neoplastic cells.

The i(17q) breakpoint region could, of course, also facilitate erroneous meiotic recombination, but the resulting gross genomic imbalances (i.e., whole arm deletion of 17p and duplications of 17q), are unlikely to be compatible with normal embryonic development.

MEDIATORS OF REARRANGEMENTS

LCRs have, thus, been shown to be important mediators of rearrangements in constitutional genomic disorders. An important question in this context is whether they also play an equally important role in generating neoplasia-associated genomic changes. Unfortunately, only limited knowledge on this aspect is currently available.

It is likely, however, that an increasing awareness of such genome architectural features will yield important insights in the near future. Indeed, apart from a role in i(17q)-formation, LCRs have been implicated recently also in the generation of one of the most common neoplasia-associated reciprocal chromosomal translocation, the t(9;22)(q34;q11) in CML.

Not only the vicinity of the breakpoints of this chromosomal abnormality, but also the entire *BCR* and *ABLI* genes have been extensively searched in the past for the presence of repetitive elements or recombinogeneic motifs, that could provide a mechanism for the translocation, but no convincing support for their role in generating this aberration has been forthcoming.

In light of this, the finding reported by Saglio et al. is particularly interesting. Studying the *BCR/ABL1* fusion at the mRNA level, they identified a rare case with an insertion of 126 nucleotides at the junction between the two genes. This junction fragment was shown to derive from a region 1.4 Mb 5' *of ABL1.*

Using a BAC clone covering this region as a probe in FISH experiments, a weak signal was identified distal to the *BCR* gene at 22q 11 in addition to the expected signal at 9q34. This finding suggested the existence of significantly homologous sequences in the proximity of both genes involved in the translocation.

Sequence comparison finally revealed the presence of a 76-kb duplicated genomic region present on 9q34 and on 22q11, 150 kb 3' of *BCR*. Thus, in contrast to i(17q), where the presence of LCRs coincides with the actual breakpoint, the duplicated genomic region is located outside the *BCR* and *ABLI* breakpoints. This suggests another potential role of LCRs in facilitating mitotic chromosomal exchange: by bringing two genes into proximity of each other, the likelihood for recombination between them would be dramatically increased.

Subsequent random breakage and end-joining of the two genes, possibly guided by repetitive elements or sequence motifs in the vicinity of the breakpoints, followed by selection for cells producing inframe *BCR/ABL1* fusion products, would result in a clonal expansion and clinically manifest leukemia. To date, no systematic study investigating the presence of LCRs in the proximity of genes reported to form fusion genes has been reported, but in light of the t(9;22) example, such analyses may well prove fruitful.

Could LCRs also play a role in mediating other chromosomal abnormalities characterizing human neoplasia? Although far from proven, it has been suggested that the amplification of candidate oncogenes in high-grade ostesarcoma could be LCR-mediated, because the amplification profiles in these tumors coincide with the genomic localization of LCRs in the Charcot-MarieTooth disease type 1A and SMS regions.

Another question that arises is whether genomic regions known to be deleted, or less frequently duplicated, in *constitutional* genomic disorders could be rearranged also in neoplastic disorders. Genomic regions known to be *meiotically* unstable and resulting in the well-known constitutional microdeletion syndromes may be unstable also in cancer cells, not least because of the dysfunctional DNA repair and decreased apoptotic activity characterizing such cells in general.

It is important to stress, however, that fundamentally different selection mechanisms are likely to be active in neoplastic cells and congenital microdeletion syndromes. Whereas neoplastic cells will become selected because of growth advantage at the cellular level allowing gross genomic alterations, imbalances in congenital disorders have to be compatible with embryonic development.

It cannot be excluded, however, that genes contained within regions lost in microdeletion syndromes, in the haploinsufficient state may confer selective advantage to tumor cells. Finally, an interesting, but entirely speculative, aspect to address in the future will be if certain LCRs, yet to be detected or already implicated in neoplasia, display copy number and/or structural polymorphisms that could render certain individuals more prone to acquire structural somatic chromosomal changes of importance for tumor development or progression.

Most likely, such polymorphisms or "risk alleles" would be of low penetrance, but could nevertheless provide important additional information in a clinical setting.

CONCLUDING REMARKS

It seems clear that LCRs, like in constitutional disorders, may play an important, previously unrecognized, role in mediating chromosomal rearrangements also in neoplasia. However, it is still too early to postulate a more general role for their involvement in cancers. Such conclusions will have to await more systematic studies comparing their genomic localization and organization with that of chromosomal breakpoints in neoplasia.

In principle, this should be possible already at this stage for balanced chromosomal translocations, which, to date, have been shown to involve more than 250 genes. High-resolution array-based comparative genomic hybridization methods and improvement of other technologies, allowing genome-wide detection of subtle genomic imbalances in the cancer genome, are likely to yield further insights into commonly lost or gained chromosomal regions.

With improved annotations of the human genome, this will facilitate the identification of genome structural features that may be critical mediators of tumor typespecific genomic rearrangements or of genetic changes occurring in several tumor entities.

Having identified such features, their isolation and introduction into various experimental systems will be important further steps to undertake in order to validate their suggested involvement in generating neoplasia-associated genetic alterations.

Index

A

B

C

D

E

F

G

H

I

M

N

O

P

Q

R

S

T

U

V

W

Y

Z